Algebraic and Topological Aspects of Representation Theory

CONTEMPORARY
MATHEMATICS

791

Algebraic and Topological Aspects of Representation Theory

Virtual AMS Special Session
Geometric and Algebraic Aspects
of Quantum Groups and Related Topics
November 20–21, 2021

Mee Seong Im
Bach Nguyen
Arik Wilbert
Editors

AMS AMERICAN MATHEMATICAL SOCIETY

2020 *Mathematics Subject Classification.* Primary 14R15, 16E40, 16W20, 17B37, 35Q53, 18M05, 57K45, 18M30; Secondary 57K18, 57K16.

Library of Congress Cataloging-in-Publication Data

Names: AMS Special Session on Geometric and Algebraic Aspects of Quantum Groups and Related Topics (2021), author. | Im, Mee Seong, 1980– editor. | Nguyen, Bach, 1989– editor. | Wilbert, Arik, 1988– editor.

Title: Algebraic and topological aspects of representation theory : virtual AMS Special Session on Geometric and Algebraic Aspects of Quantum Groups and Related Topics, November 20–21, 2021 / Mee Seong Im, Bach Nguyen, Arik Wilbert, editors.

Description: Providence, Rhode Island : American Mathematical Society, [2024] | Series: Contemporary mathematics, 0271-4132 ; volume 791 | Includes bibliographical references.

Identifiers: LCCN 2023048362 | ISBN 9781470470340 (paperback) | ISBN 9781470476045 (ebook)

Subjects: LCSH: Representations of groups–Congresses. | AMS: Algebraic geometry – Affine geometry – Jacobian problem. | Associative rings and algebras – Homological methods – (Co)homology of rings and algebras (e.g. Hochschild, cyclic, dihedral, etc.). | Associative rings and algebras – Rings and algebras with additional structure – Automorphisms and endomorphisms. | Nonassociative rings and algebras – Lie algebras and Lie superalgebras – Quantum groups (quantized enveloping algebras) and related deformations. | Partial differential equations – Equations of mathematical physics and other areas of application – KdV-like equations (Korteweg-de Vries).

Classification: LCC QA176 .A475 2024 | DDC 515/.7223–dc23/eng20231103

LC record available at https://lccn.loc.gov/2023048362

∞ The paper used in this book is acid-free and falls within the guidelines
established to ensure permanence and durability.
Visit the AMS home page at `https://www.ams.org/`

10 9 8 7 6 5 4 3 2 1 29 28 27 26 25 24

Contents

Preface

Noncommutative algebras and noncommutative algebraic geometry have been an active field of research for the past several decades, with many important applications in mathematical physics, representation theory, number theory, combinatorics, geometry, low-dimensional topology, and category theory. Our purpose for this special session is to bring together junior and established researchers on (noncommutative) algebras, especially quantum algebras, (noncommutative) algebraic geometry, low-dimensional topology, and related areas to celebrate and share their contributions in the area. The aim of this event is to initiate collaborations among the junior mathematicians and create connections to further advance their careers.

Many of our speakers were graduate students, postdocs, and junior faculty members from campuses in the southeastern region, such as University of Georgia, Georgia Southern University, Georgia Tech, University of South Alabama, University of Alabama, Louisiana State University, Tulane University, and University of Louisiana at Lafayette. To help accomplish this broader approach to integrating young researchers into the field, more than 70% of the invited speakers were graduate students and junior faculty members. Each of the sessions consisted of 4 to 5 early career speakers and one more established speaker. The goal has been to pair junior speakers with senior faculty speakers who have direct research overlap to help facilitate this collaboration.

Due to COVID-19, our special session was moved online, with the AMS moving quickly to build a platform to accommodate all the special sessions. When this change was announced in going from an in-person meeting to an online meeting, our speakers promptly, and without hesitation, agreed to give virtual talks. This was the beginning of an emergence of online conferences, colloquiums and seminars from 2020 to 2022.

All manuscripts in this Contemporary Mathematics volume contain original research, written by speakers and their collaborators in our special session. Many papers in this volume also discuss new concepts with detailed examples and current trends with novel and important results, all of which are invaluable contributions to the mathematics community.

Mee Seong Im
Bach Nguyen
Arik Wilbert

List of Participants

Mikhail Khovanov (Plenary Speaker)
Columbia University
New York, NY

Lev Rozansky (Plenary Speaker)
University of North Carolina
Chapel Hill, NC

Alisina Azhang
Louisiana State University
Baton Rouge, LA

Tamanna Chatterjee
Louisiana State University
Baton Rouge, LA
University of Georgia
Athens, GA

Jason Gaddis
Miami University
Oxford, OH

David Galban
University of Georgia
Athens, GA

Shengnan Huang
Northeastern University
Boston, MA

Andy Jenkins
University of Georgia
Athens, GA

Garrett Johnson
North Carolina Central University
Durham, NC

Elizabeth Jurisich
College of Charleston
Charleston, SC

Scott Larson
University of Georgia
Athens, GA

Kyu-Hwan Lee
University of Connecticut
Mansfield, CT

Kyungyong Lee
University of Alabama
Tuscaloosa, AL

Yiqiang Li
University at Buffalo
Buffalo, NY

Tolulope Oke
Texas A&M University
College Station, TX

Shifra Reif
Institute for Advanced Study
Princeton, NJ
Bar-Ilan University
Ramat Gan, Israel

Radmila Sazdanovic
North Carolina State University
Raleigh, NC

Joshua Sussan
Medgar Evers College
City University of New York
Brooklyn, NY

Kurt Trampel
University of Notre Dame
Notre Dame, IN

Anton Zeitlin
Louisiana State University
Baton Rouge, LA

Contemporary Mathematics
Volume **791**, 2024
https://doi.org/10.1090/conm/791/15869

On the two-dimensional Jacobian conjecture: Magnus' formula revisited, III

Jacob Glidewell, William E. Hurst, Kyungyong Lee, and Li Li

Abstract. This article is part of an ongoing investigation of the two-dimensional Jacobian conjecture. In the second paper of this series, we introduced the remainder vanishing conjecture, which implies the two-dimensional Jacobian conjecture. In this paper we give several examples to illustrate how to prove the remainder vanishing conjecture.

1. Introduction

The Jacobian conjecture, raised by Keller [**26**], has been studied by many mathematicians: a partial list of related results includes [**1–15, 18, 21, 22, 24, 25, 27–29, 31–46**]. A survey is given in [**16, 17**]. In this series of papers we exclusively deal with the plane case. Hence whenever we write the Jacobian conjecture, we mean the two-dimensional Jacobian conjecture.

Jacobian conjecture. *Let $f, g \in \mathbb{C}[x, y]$. Consider the polynomial map*

$$\pi : \mathbb{C}[x, y] \longrightarrow \mathbb{C}[x, y]$$

given by $\pi(x) = f$ *and* $\pi(y) = g$. *If*

$$(\partial f/\partial x)(\partial g/\partial y) - (\partial g/\partial x)(\partial f/\partial y) \in \mathbb{C} \setminus \{0\},$$

then π is bijective.

In the first paper [**23**] of the series, we generalized Magnus' formula [**30**]. In the second paper [**19**], we introduced the remainder vanishing conjecture (which is recalled in Conjecture 2.4) and proved the following theorem as an application of the generalized Magnus' formula.

Theorem 1.1 ([**19**]). *The remainder vanishing conjecture implies the Jacobian conjecture.*

In the current (third) paper we give several examples to illustrate how to show the remainder vanishing conjecture, after recalling its statement. In the forthcoming paper(s) including [**20**], we plan to prove the remainder vanishing conjecture.

2020 *Mathematics Subject Classification.* Primary 14R15; Secondary 13F20.

This paper grew out of an undergraduate research project for JG and WEH at the University of Alabama. KL was supported by the University of Alabama, Korea Institute for Advanced Study, and the NSF grant DMS 2042786.

2. The remainder vanishing conjecture

In this section, we recall the remainder vanishing conjecture. The statement may look complicated, but every object appearing here is directly computable.

Let $\mathcal{R} = \mathbb{C}[x, y]$ and let $W = \{(0, 1), (1, 1)\} \subseteq \mathbb{Z}^2$. An element $w = (u, 1) \in W$ is called a *direction*. To each such a direction we consider its w-grading on $\mathcal{R}$. So

$$\mathcal{R} = \bigoplus_{n\in\mathbb{Z}} \mathcal{R}_n^w,$$

where $\mathcal{R}_n^w = \mathcal{R}_n^{(u,1)}$ is the $\mathbb{C}$-vector space generated by the monomials $x^i y^j$ with $ui + j = n$. A non-zero element f of $\mathcal{R}_n^w$ is called a w-homogeneous element of $\mathcal{R}$ and n is called its w-degree, denoted $w\text{-}\deg(f)$. The element of highest w-degree in the homogeneous decomposition of a non-zero polynomial f is called its w-leading form and is denoted by f_+. The w-degree of f is by definition $w\text{-}\deg(f_+)$. For any $f \in \mathcal{R}$ and any $w = (u, 1) \in W$, we write the w-homogeneous decomposition

$$f = \sum_i f_i^w,$$

where $f_i^w := f_i^{(u,1)} \in \mathcal{R}_i^w$.

2.1. Set-up. For any integers r_1 and r_2, the set $\{x \in \mathbb{Z} : r_1 \le x \le r_2\}$ will be denoted by the usual notation $[r_1, r_2]$. Define

$$\mathcal{Q} = \{(a, b, m, n) \in \mathbb{Z}_{>0}^4 : m < n,\ a|m,\ a|n,\ \gcd(a, b) = 1 \text{ and } 2 \le a < b\}.$$

Fix $(a, b, m, n) \in \mathcal{Q}$. Define Δ to be the set of positive common divisors of $\frac{m}{a}$ and $\frac{n}{a}$, and fix $\delta \in \Delta$. Let

$$\mathcal{I} = \left\{ i \in [0, m(n-m)] : \frac{i}{m} \in \mathbb{Z} \text{ or } \frac{i+1}{n-m} \in \mathbb{Z} \right\}$$

and fix $i \in \mathcal{I}$.

Define $u, L, \mathrm{u}_E, \mathrm{u}_F$ in two cases as below. Note that the two cases cannot hold simultaneously because of [**19**, Lemma 2.16].

$$\text{(2.1)} \quad \begin{cases} u = 0, L = x + 1, \mathrm{u}_E = \frac{n-m}{\delta a}, \text{ and } \mathrm{u}_F = \lfloor i(n-m)\varepsilon \rfloor, & \text{if } \frac{i}{m} \in \mathbb{Z}; \\ u = 1, L = x + y, \mathrm{u}_E = \frac{m}{\delta a}, \text{ and } \mathrm{u}_F = \lfloor (i+1)m\varepsilon \rfloor, & \text{if } \frac{i+1}{n-m} \in \mathbb{Z}, \end{cases}$$

where $\varepsilon = \frac{1}{m(n-m)}$. Let $\widetilde{\Gamma}_0, \widetilde{\Gamma}_1, \ldots, \widetilde{x}_0, \widetilde{x}_1, \ldots, \widetilde{c}_1, \widetilde{c}_2, \ldots$ be (infinitely many) algebraically independent variables and define

$$R_1 = \mathbb{C}[\widetilde{\Gamma}_0, \widetilde{\Gamma}_1, \ldots, \widetilde{x}_0, \widetilde{x}_1, \ldots, \widetilde{c}_1, \widetilde{c}_2, \ldots] \quad \text{and} \quad R_2 = \begin{cases} \mathbb{C}[x^{\pm 1}, y^{\pm \frac{1}{m}}] & \text{if } u = 0; \\ \mathbb{C}[x^{\pm \frac{1}{n}}, y^{\pm 1}] & \text{if } u = 1. \end{cases}$$

For simplicity, let

$$a_i = \lfloor (i+1)m\varepsilon \rfloor \text{ and } b_i = \lfloor (i+1)m\varepsilon \rfloor + \lfloor i(n-m)\varepsilon \rfloor.$$

Define
(2.2)

$$\begin{cases} w=(u,1)\in W; \\ d=um+n; \\ e=bd/a; \\ \mathfrak{m}=d+e-u-2; \\ \mathrm{v}_E=d/(\delta a); \\ \mathrm{v}_F=\begin{cases} \mathrm{u}_F+\frac{(un-um+m)(a-1)}{a}-1, \text{ if } i>\frac{m(n-m)(a-1)}{a}; \\ ua_i+b_i-1, \text{ if } i\le\frac{m(n-m)(a-1)}{a} \text{ and } \frac{a\delta\lfloor i(n-m)\varepsilon\rfloor}{n-m}=\frac{a\delta\lfloor (i+1)m\varepsilon\rfloor}{m}\in\mathbb{Z}; \\ ua_i+b_i, \text{ otherwise}; \end{cases} \\ \widetilde{F}^\circ=\sum_{j=\mathrm{u}_F}^{\mathrm{v}_F}\tau^{j-\mathrm{u}_F}\widetilde{x}_j+\sum_{j=0}^{\mathrm{u}_F-1}\widetilde{x}_j\in R_1[\tau]; \\ \widetilde{E}^\circ=\tau^{\mathrm{v}_E-\mathrm{u}_E}+\sum_{j=\mathrm{u}_E}^{\mathrm{v}_E-1}\tau^{j-\mathrm{u}_E}\widetilde{\Gamma}_j+\sum_{j=0}^{\mathrm{u}_E-1}\widetilde{\Gamma}_j\in R_1[\tau]. \end{cases}$$

EXAMPLE 2.1. Fix $(a,b,m,n)=(2,3,4,8)$, $\delta=1$, and $i=15$. Then $\frac{i+1}{n-m}\in\mathbb{Z}$. So

$$u=1,\quad d=12,\quad e=18,\quad L=x+y,\quad \mathrm{u}_E=2,\quad \mathrm{v}_E=6,\quad \mathrm{u}_F=4,\quad \mathrm{v}_F=7.$$

Let $\mathrm{Hom}(R_1[\tau],R_2)$ be the set of ring homomorphisms from $R_1[\tau]$ to R_2. For $w\in W$, we define the subset $\mathcal{S}^w$ of $\mathrm{Hom}(R_1[\tau],R_2)$ by

$$\mathcal{S}^{(0,1)}=\{S\in\mathrm{Hom}(R_1[\tau],R_2)\,:\,S(\tau)=(x+1)y^{n/m}\};$$
$$\mathcal{S}^{(1,1)}=\{S\in\mathrm{Hom}(R_1[\tau],R_2)\,:\,S(\tau)=x^{m/n}(x+y)\}.$$

For $r\in\mathbb{Z}$, let $[r]_+=\max(0,r)$. This notation is commonly used for objects in the theory of cluster algebras.

DEFINITION 2.2. Let $\mathbb{T}$ be the set of Tschirnhausen polynomials, that is,

$$\mathbb{T}=\big\{\alpha(z)=z^k+e_{k-1}z^{k-1}+\ldots+e_0z^0\in\mathbb{C}[z]\,:\,k\in\mathbb{Z}_{>0},\ e_{k-1}=0, \text{ and } e_{k-2},\ldots,e_0\in\mathbb{C}\big\}.$$

For a polynomial $f\in\mathbb{C}[x,y]$, let

$$\mathcal{E}(f)=\{E\in\mathbb{C}[x,y]\,:\,f=\alpha(E)\text{ for some }\alpha(z)\in\mathbb{T}\}.$$

Note that $f\in\mathcal{E}(f)$. An element $f\in\mathbb{C}[x,y]$ with $\mathcal{E}(f)=\{f\}$ is called a *principal* polynomial.

Fix $E^\circ,F^\circ\in\mathbb{C}[x,y]$, $S\in\mathcal{S}^w$, and $\alpha^\circ\in\mathbb{T}$ such that
(2.3)

$$\begin{cases} S(\widetilde{c}_\beta)\in\mathbb{C}\text{ for all }\beta\in[1,\mathfrak{m}]; \\ F^\circ=S(\widetilde{F}^\circ), w\text{-deg}(F^\circ)\le\mathrm{v}_F, \text{ and } (F^\circ)^w_j=S(\tau^{[j-\mathrm{u}_F]_+}\widetilde{x}_j)\text{ for } j\in[0,\mathrm{v}_F]; \\ E^\circ=S(\widetilde{E}^\circ), w\text{-deg}(E^\circ)\le\mathrm{v}_E, \text{ and } (E^\circ)^w_j=S(\tau^{[j-\mathrm{u}_E]_+}\widetilde{\Gamma}_j)\text{ for } j\in[0,\mathrm{v}_E-1]; \\ \deg(\alpha^\circ)=\delta a. \end{cases}$$

Give a grading on $R_1[\tau]$ as follows:

$$\deg(\tau)=\begin{cases} n/m, & \text{if } w=(0,1); \\ m/n+1, & \text{if } w=(1,1), \end{cases}$$

$\deg(\widetilde{\Gamma}_j) = j - [j - \mathrm{u}_E]_+ \deg(\tau)$, and $\deg(\widetilde{x}_j) = j - [j - \mathrm{u}_F]_+ \deg(\tau)$. Note that

$$\deg(\tau^{[j-\mathrm{u}_E]_+}\widetilde{\Gamma}_j) = \deg(\tau^{[j-\mathrm{u}_F]_+}\widetilde{x}_j) = j.$$

For any nonzero element $f \in R_1[\tau]$ which is $\mathbb{Z}$-graded (i.e. f is of the form $f = \sum_{j\in\mathbb{Z}} f_j$), define $\deg(f) = \max\{j : f_j \neq 0\}$, and define the homogenization map

$$h(f) = \sum_{j \le \deg(f)} f_j t^{\deg(f)-j} \in R_1[\tau, t].$$

Note that all terms in $\alpha^\circ(\widetilde{E}^\circ) + \widetilde{F}^\circ$ are of integer degrees, so we can apply the homogenization map and get

$$H = h\big(\alpha^\circ(\widetilde{E}^\circ) + \widetilde{F}^\circ\big).$$

For any $y_0, \ldots, y_n \in R_1[\tau^{\pm 1}]$ and for any $A = r/s$ with $r \in \mathbb{Z}$ and $s \in \mathbb{Z}_{>0}$, we have the following identity in the ring $R_1[\tau^{\pm1}, y_0^{\pm 1/s}][[t]]$ (where we fix a choice of $y_0^{1/s}$):
(2.4)

$$\begin{aligned}&(y_0 + y_1 t + \ldots + y_n t^n)^A\\ &= \sum_{v_1,\ldots,v_n \in \mathbb{Z}_{\ge 0}} \frac{A(A-1)\cdots(A - \sum_{j=1}^n v_j + 1)}{\prod_{j=1}^n v_j!} y_0^{A - \sum_{j=1}^n v_j} y_1^{v_1} \cdots y_n^{v_n} t^{v_1 + \cdots + n v_n}.\end{aligned}$$

For any $\widetilde{F} \in R_1[\tau^{\pm1}][[t]]$, denote

$$[\widetilde{F}]_{t^j} = \text{ the coefficient of } t^j \text{ in } \widetilde{F},$$

which is an element in $R_1[\tau^{\pm 1}]$.

For any $\widetilde{F} \in R_1[\tau^{\pm 1}]$, denote

$$[\widetilde{F}]_{\tau^j} = \text{ the coefficient of } \tau^j \text{ in } \widetilde{F},$$

which is an element in R_1.

Let $r = \gcd(m, n)$. For each $\mu \in [0, \mathfrak{m}]$ and each $B \subseteq \{\beta \in [1, \mathfrak{m}] : r(e-\beta)/d \in \mathbb{Z}\}$, let

$$\widetilde{G}_{B,e-\mu} = [H^{\frac{e}{d}}]_{t^\mu} + \sum_{\beta \in B \cap [1,\mu]} \widetilde{c}_\beta [H^{\frac{e-\beta}{d}}]_{t^{\mu-\beta}} \in R_1[\tau^{\pm 1}], \tag{2.5}$$

and let $\mathrm{j}(\mu) = \min\{\mathrm{j} \in \mathbb{Z}_{\ge 0} : [\widetilde{G}_{B,e-\mu}]_{\tau^{-j}} = 0 \text{ for all } j > \mathrm{j}\}$. For $k \in [1, \mathrm{j}(\mu)]$, define $\widetilde{G}_{B,e-\mu,k}$ by

$$\widetilde{G}_{B,e-\mu,k} = \sum_{j=\mathrm{j}(\mu)+1-k}^{\mathrm{j}(\mu)} \tau^{\mathrm{j}(\mu)-j}[\widetilde{G}_{B,e-\mu}]_{\tau^{-j}} \in R_1[\tau].$$

If $B = \{\beta \in [1, \mathfrak{m}] : r(e-\beta)/d \in \mathbb{Z}\}$, then we simply denote $\widetilde{G}_{B,e-\mu}$ by $\widetilde{G}_{e-\mu}$, and $\widetilde{G}_{B,e-\mu,k}$ by $\widetilde{G}_{e-\mu,k}$.

2.2. The statement of the remainder vanishing conjecture. For $k \in \mathbb{Z}_{\geq 0}$, define the following natural projection as a ring homomorphism:

$$P_k : R_2 \longrightarrow R_2/(L^k).$$

Note that the codomain of P_1 is an integral domain, in particular,
- if $P_1(fg) = 0$, then $P_1(f) = 0$ or $P_1(g) = 0$;
- if $P_1(f^j) = 0$ for some integer $j > 0$, then $P_1(f) = 0$.

Indeed, this follows immediately from the fact that $(x+1)$ is irreducible in the coordinate ring $\mathbb{C}[x^{\pm 1}, y^{\pm \frac{1}{m}}]$, and $x+y$ is irreducible in $\mathbb{C}[x^{\pm \frac{1}{n}}, y^{\pm 1}]$.

DEFINITION 2.3. Fix $(a,b,m,n) \in \mathcal{Q}$, $\delta \in \Delta$, and $i \in \mathcal{I}$, which will determine the objects appearing in (2.1) and (2.2). Fix $S, E^\circ, F^\circ, \alpha^\circ$ so that (2.3) is satisfied. We say that a subset B of $\{\beta \in [1, \mathfrak{m}] : r(e-\beta)/d \in \mathbb{Z}\}$ *is supported with respect to* $(S, E^\circ, F^\circ, \alpha^\circ)$ if

$$\left\{ \begin{array}{ll} P_k(S(\widetilde{G}_{B,e-\mu,k})) = 0 & \text{for } \mu \in [0, \mathfrak{m}] \text{ and } k \in [1, \mathfrak{j}(\mu)]; \\ S(\widetilde{G}_{B,e-\mu}) = 0 & \text{for } \mu \in [e+1, \mathfrak{m}]. \end{array} \right. \tag{2.6}$$

Note that if $B' \subseteq B$ and B' is supported with respect to $(S, E^\circ, F^\circ, \alpha^\circ)$, then there exists $S' \in \mathcal{S}^w$ such that $S'(\widetilde{c}_\beta) = 0$ for every $\beta \in B \setminus B'$, and that each of B' and B is supported with respect to $(S', E^\circ, F^\circ, \alpha^\circ)$.

Denote $x_j = S(\widetilde{x}_j)$ for $0 \leq j \leq \mathrm{v}_F$, and $\Gamma_j = S(\widetilde{\Gamma}_j)$ for $0 \leq j \leq \mathrm{v}_E - 1$.

Now we recall the remainder vanishing conjecture proposed in [**19**].

CONJECTURE 2.4 (The Remainder Vanishing Conjecture). *Assume the hypotheses as in Definition* 2.3. *Fix* $\ell \in [\mathrm{u}_F, \mathrm{v}_F]$. *Suppose that*

(2.7)
B is supported with respect to $(S, E^\circ, F^\circ, \alpha^\circ)$ and $P_1(x_j) = 0$ for all $j \in [\ell+1, \mathrm{v}_F]$.

If E° is a principal polynomial, then either $P_1(x_\ell) = 0$ or there exists a proper subset of B which is supported with respect to $(S, E^\circ, F^\circ, \alpha^\circ)$.

Note that when $\ell = \mathrm{v}_F$, the condition "$P_1(x_j) = 0$ for all $j \in [\ell+1, \mathrm{v}_F]$" is vacuous. Also note that Conjecture 2.4 has the following consequence by applying it iteratively:

Consequence of Conjecture 2.4. *Assume the same hypotheses. Fix $\ell \in [\mathrm{u}_F, \mathrm{v}_F]$. Suppose that E° is a principal polynomial, B is a subset of $\{\beta \in [1, \mathfrak{m}] : r(e-\beta)/d \in \mathbb{Z}\}$ supported with respect to $(S, E^\circ, F^\circ, \alpha^\circ)$, and $P_1(x_j) = 0$ for all $j \in [\ell+1, \mathrm{v}_F]$. Then $P_1(x_\ell) = 0$.*

There is a systematic way to solve (2.7). In the next section, we give some examples to illustrate how to solve it. In the forthcoming paper(s) including [**20**], we plan to prove the remainder vanishing conjecture.

3. Examples for the remainder vanishing conjecture

In this section, we give a few examples to illustrate a proof of the Remainder Vanishing Conjecture (Conjecture 2.4) for some special cases.

EXAMPLE 3.1. Fix $(a,b,m,n) = (2,3,2,4)$. Then $\Delta = \{1\}$, so $\delta = 1$. Then $\varepsilon = \frac{1}{4}$. The range of i is $\mathcal{I} = \{i \in [0, m(n-m)] : \frac{i}{m} \in \mathbb{Z} \text{ or } \frac{i+1}{n-m} \in \mathbb{Z}\} = \{4,3,2,1,0\}$. Since $\delta = \max \Delta$, the polynomial E° is principal for each $i \in \mathcal{I}$.

- For $i = 4$: we have $u = 0$, $w = (0,1)$, $L = x+1$, $d = 4$, $e = 6$, $\mathfrak{m} = 8$, $\mathrm{u}_E = 1$, $\mathrm{v}_E = 2$, $\mathrm{u}_F = \mathrm{v}_F = 2$,

$\widetilde{F}^\circ = \widetilde{x}_2 + \widetilde{x}_1 + \widetilde{x}_0$, $F^\circ = x_2 + x_1 + x_0$,

$\widetilde{E}^\circ = \tau^1 + \widetilde{\Gamma}_1 + \widetilde{\Gamma}_0$, $E^\circ = (x+1)y^2 + \Gamma_1 + \Gamma_0$, and $\alpha^\circ(z) = z^2 + c$ where $c \in \mathbb{C}$ is a constant, so $\alpha^\circ(\widetilde{E}^\circ) = (\tau + \widetilde{\Gamma}_1 + \widetilde{\Gamma}_0)^2 + c$. Then

$$\begin{aligned} H &= h((\tau + \widetilde{\Gamma}_1 + \widetilde{\Gamma}_0)^2 + c + \widetilde{x}_2 + \widetilde{x}_1 + \widetilde{x}_0) \\ &= (\tau^2) + (2\tau\widetilde{\Gamma}_1)t + (2\tau\widetilde{\Gamma}_0 + \widetilde{\Gamma}_1^2 + \widetilde{x}_2)t^2 + (2\widetilde{\Gamma}_1\widetilde{\Gamma}_0 + \widetilde{x}_1)t^3 + (\widetilde{\Gamma}_0^2 + c + \widetilde{x}_0)t^4. \end{aligned}$$

There is only one value of ℓ to be considered: $\ell = 2 (= \mathrm{v}_F = \mathrm{u}_F)$. We want to show that $P_1(x_2) = 0$. A simple computation shows:

$\widetilde{G}_{e-0} = \tau^3$,

$\widetilde{G}_{e-1} = 3\widetilde{\Gamma}_1\tau^2$,

$\widetilde{G}_{e-2} = (3\widetilde{\Gamma}_1^2 + \frac{3}{2}\widetilde{x}_2)\tau + (\widetilde{c}_2 + 3\widetilde{\Gamma}_0)\tau^2$,

$\widetilde{G}_{e-3} = (\widetilde{\Gamma}_1^3 + \frac{3}{2}\widetilde{\Gamma}_1\widetilde{x}_2) + (2\widetilde{c}_2\widetilde{\Gamma}_1 + 6\widetilde{\Gamma}_0\widetilde{\Gamma}_1 + \frac{3}{2}\widetilde{x}_1)\tau$.

The above four expressions will not be used, since they do not contain negative powers of τ. In contrast,

$\widetilde{G}_{e-4} = \frac{3}{8}\widetilde{x}_2^2\tau^{-1} + (\widetilde{c}_2\widetilde{\Gamma}_1^2 + 3\widetilde{\Gamma}_0\widetilde{\Gamma}_1^2 + \frac{3}{2}\widetilde{\Gamma}_1 + \widetilde{c}_2\widetilde{x}_2 + \frac{3}{2}\widetilde{\Gamma}_0\widetilde{x}_2) + (2\widetilde{c}_2\widetilde{\Gamma}_0 + 3\widetilde{\Gamma}_0^2 + \widetilde{c}_4 + \frac{3}{2}\widetilde{x}_0)\tau$.

So $\widetilde{G}_{e-4,1} = \frac{3}{8}\widetilde{x}_2^2$, hence $P_1(S(\widetilde{G}_{e-4,1})) = P_1(\frac{3}{8}x_2^2) = 0$, which implies $P_1(x_2) = 0$ as desired.

• For $i = 3$: we have $u = 1$, $w = (1,1)$, $L = x + y$, $d = 6$, $e = 9$, $\mathfrak{m} = 12$, $\mathrm{u}_E = 1$, $\mathrm{v}_E = 3$, $\mathrm{u}_F = 2$, $\mathrm{v}_F = 3$,

$\widetilde{F}^\circ = \tau\widetilde{x}_3 + \widetilde{x}_2 + \widetilde{x}_1 + \widetilde{x}_0$, $F^\circ = x^{1/2}(x+y)x_3 + x_2 + x_1 + x_0$,

$\widetilde{E}^\circ = \tau^2 + \tau\widetilde{\Gamma}_2 + \widetilde{\Gamma}_1 + \widetilde{\Gamma}_0$, and $E^\circ = x(x+y)^2 + x^{1/2}(x+y)\Gamma_2 + \Gamma_1 + \Gamma_0$.

Let I be the R_1-ideal generated by $\{\widetilde{G}_{e-\mu,1} \mid \mu = 7, 8, 9\}$, where

$$\begin{aligned} \widetilde{G}_{e-7,1} &= -\frac{3}{8}\widetilde{\Gamma}_2\widetilde{x}_3^2 + \frac{3}{4}\widetilde{x}_2\widetilde{x}_3, \\ \widetilde{G}_{e-8,1} &= \frac{3}{8}\widetilde{\Gamma}_2^2\widetilde{x}_3^2 - \frac{3}{4}\widetilde{\Gamma}_2\widetilde{x}_2\widetilde{x}_3 - \frac{3}{8}\widetilde{\Gamma}_1\widetilde{x}_3^2 + \frac{3}{8}\widetilde{x}_2^2, \\ \widetilde{G}_{e-9,1} &= -\frac{3}{8}\widetilde{\Gamma}_2^3\widetilde{x}_3^2 + \frac{3}{4}\widetilde{\Gamma}_2^2\widetilde{x}_2\widetilde{x}_3 + \frac{3}{4}\widetilde{\Gamma}_1\widetilde{\Gamma}_2\widetilde{x}_3^2 - \frac{3}{8}\widetilde{\Gamma}_2\widetilde{x}_2^2 - \frac{3}{4}\widetilde{\Gamma}_1\widetilde{x}_2\widetilde{x}_3 - \frac{1}{16}\widetilde{x}_3^3. \end{aligned}$$

Let $\ell = 3$. It is easy to verify that

$$-16\widetilde{\Gamma}_1\widetilde{G}_{e-7,1} - 16\widetilde{\Gamma}_2\widetilde{G}_{e-8,1} - 16\widetilde{G}_{e-9,1} = \widetilde{x}_3^3,$$

so $\widetilde{x}_3^3$ is in I. Using the assumption that $P_1(S(\widetilde{G}_{e-\mu,1})) = 0$ for $\mu = 7, 8, 9$ (see (2.6)), we conclude that $P_1(x_3) = 0$.

Let $\ell = 2$. Then $P_1(x_3) = 0$ implies that $0 = P_1(S(\widetilde{G}_{e-8,1})) = P_1(\frac{3}{8}\widetilde{x}_2^2)$, which gives $P_1(x_2) = 0$. So we are done with the case of $i = 3$.

• For $i = 2$: we have $u = 0$, $w = (0,1)$, $L = x + 1$, $d = 4$, $e = 6$, $\mathfrak{m} = 8$, $\mathrm{u}_E = 1$, $\mathrm{v}_E = 2$, $\mathrm{u}_F = 1$, $\mathrm{v}_F = 1$,

$\widetilde{F}^\circ = \widetilde{x}_1 + \widetilde{x}_0$, $F^\circ = x_1 + x_0$,

$\widetilde{E}^\circ = \tau + \widetilde{\Gamma}_1 + \widetilde{\Gamma}_0$, and $E^\circ = (x+1)y^2 + \Gamma_1 + \Gamma_0$.

We want to show that $P_1(x_1) = 0$. But this can be proved in the same way as in the last paragraph of the case $i = 4$.

• For $i = 1$: we have $u = 1$, $w = (1,1)$, $L = x + y$, $d = 6$, $e = 9$, $\mathfrak{m} = 12$, $\mathrm{u}_E = 1$, $\mathrm{v}_E = 3$, $\mathrm{u}_F = 1$, $\mathrm{v}_F = 2$,

$\widetilde{F}^\circ = \tau\widetilde{x}_2 + \widetilde{x}_1 + \widetilde{x}_0$, $F^\circ = x^{1/2}(x+y)x_2 + x_1 + x_0$,

$\widetilde{E}^\circ = \tau^2 + \tau\widetilde{\Gamma}_2 + \widetilde{\Gamma}_1 + \widetilde{\Gamma}_0$, and $E^\circ = x(x+y)^2 + x^{1/2}(x+y)\Gamma_2 + \Gamma_1 + \Gamma_0$.

We want to show that $P_1(x_2) = P_1(x_1) = 0$. Indeed,

$$\begin{aligned}
\widetilde{G}_{e-9,1} &= -\frac{3}{8}\widetilde{\Gamma}_2\widetilde{x}_2^2 + \frac{3}{4}\widetilde{x}_1\widetilde{x}_2,\\
\widetilde{G}_{e-10,1} &= \frac{3}{8}\widetilde{\Gamma}_2^2\widetilde{x}_2^2 - \frac{3}{4}\widetilde{\Gamma}_2\widetilde{x}_1\widetilde{x}_2 - \frac{3}{8}\widetilde{\Gamma}_1\widetilde{x}_2^2 + \frac{3}{8}\widetilde{x}_1^2,\\
\widetilde{G}_{e-10,2} &= \frac{3}{8}\widetilde{\Gamma}_2^2\widetilde{x}_2^2 - \frac{3}{4}\widetilde{\Gamma}_2\widetilde{x}_1\widetilde{x}_2 - \frac{3}{8}\widetilde{\Gamma}_1\widetilde{x}_2^2 + \frac{3}{8}\widetilde{x}_1^2 + (\frac{1}{2}\widetilde{c}_6\widetilde{x}_2 + \frac{3}{4}\widetilde{x}_0\widetilde{x}_2)\tau,\\
\widetilde{G}_{e-11,2} &= -\frac{3}{8}\widetilde{\Gamma}_2^3\widetilde{x}_2^2 + \frac{3}{4}\widetilde{\Gamma}_2^2\widetilde{x}_1\widetilde{x}_2 + \frac{3}{4}\widetilde{\Gamma}_1\widetilde{\Gamma}_2\widetilde{x}_2^2 - \frac{3}{8}\widetilde{\Gamma}_2\widetilde{x}_1^2 - \frac{3}{4}\widetilde{\Gamma}_1\widetilde{x}_1\widetilde{x}_2\\
&\quad + \left(\frac{1}{2}\widetilde{c}_6\widetilde{\Gamma}_2\widetilde{x}_2 - \frac{3}{4}\widetilde{\Gamma}_2\widetilde{x}_0\widetilde{x}_2 - \frac{3}{8}\widetilde{\Gamma}_0\widetilde{x}_2^2 + \frac{1}{2}\widetilde{c}_6\widetilde{x}_1 + \frac{3}{4}\widetilde{x}_0\widetilde{x}_1\right)\tau,\\
\widetilde{G}_{e-12,2} &= \frac{3}{8}\widetilde{\Gamma}_2^4\widetilde{x}_2^2 - \frac{3}{4}\widetilde{\Gamma}_2^3\widetilde{x}_1\widetilde{x}_2 - \frac{9}{8}\widetilde{\Gamma}_1\widetilde{\Gamma}_2^2\widetilde{x}_2^2 + \frac{3}{8}\widetilde{\Gamma}_2^2\widetilde{x}_1^2 + \frac{3}{2}\widetilde{\Gamma}_1\widetilde{\Gamma}_2\widetilde{x}_1\widetilde{x}_2 + \frac{3}{8}\widetilde{\Gamma}_1^2\widetilde{x}_2^2\\
&\quad - \frac{3}{8}\widetilde{\Gamma}_1\widetilde{x}_1^2 + \left(\frac{1}{2}\widetilde{c}_6\widetilde{\Gamma}_2^2\widetilde{x}_2 + \frac{3}{4}\widetilde{\Gamma}_2^2\widetilde{x}_0\widetilde{x}_2 + \frac{3}{4}\widetilde{\Gamma}_0\widetilde{\Gamma}_2\widetilde{x}_2^2 - \frac{1}{2}\widetilde{c}_6\widetilde{\Gamma}_2\widetilde{x}_1 - \frac{3}{4}\widetilde{\Gamma}_2\widetilde{x}_0\widetilde{x}_1\right.\\
&\quad \left. -\frac{1}{2}\widetilde{c}_6\widetilde{\Gamma}_1\widetilde{x}_2 - \frac{3}{4}\widetilde{\Gamma}_1\widetilde{x}_0\widetilde{x}_2 - \frac{3}{4}\widetilde{\Gamma}_0\widetilde{x}_1\widetilde{x}_2 - \frac{1}{16}\widetilde{x}_2^3\right)\tau.
\end{aligned}$$

It is easy to verify that

$$-\widetilde{\Gamma}_1\widetilde{G}_{e-10,2} - \widetilde{\Gamma}_2\widetilde{G}_{e-11,2} - \widetilde{G}_{e-12,2} = \widetilde{R}\tau, \text{ where } \widetilde{R} = -\frac{3}{8}\widetilde{\Gamma}_0\widetilde{\Gamma}_2\widetilde{x}_2^2 + \frac{3}{4}\widetilde{\Gamma}_0\widetilde{x}_1\widetilde{x}_2 + \frac{1}{16}\widetilde{x}_2^3.$$

Since we assume (2.7) holds, we must have $P_2(S(\widetilde{R}\tau)) = 0$. Using the fact that $P_2(S(\widetilde{R}\tau)) = 0 \Rightarrow L^2|S(\widetilde{R})S(\tau) \Rightarrow L|S(\widetilde{R}) \Rightarrow P_1(S(\widetilde{R})) = 0$, we get $P_1(S(\widetilde{R})) = 0$. Since

$$-16\widetilde{\Gamma}_0\widetilde{G}_{e-9,1} + 16\widetilde{R} = \widetilde{x}_2^3$$

and

$$\left(-8\widetilde{\Gamma}_0\widetilde{\Gamma}_1\widetilde{\Gamma}_2 + \frac{8}{3}\widetilde{\Gamma}_2\widetilde{x}_1 + \frac{4}{3}\widetilde{\Gamma}_1\widetilde{x}_2\right)\widetilde{G}_{e-9,1} + \frac{8}{3}\widetilde{x}_1\widetilde{G}_{e-10,1} + 8\widetilde{\Gamma}_1\widetilde{\Gamma}_2\widetilde{R} = \widetilde{x}_1^3,$$

we conclude that $P_1(S(\widetilde{x}_2^3)) = P_1(S(\widetilde{x}_1^3)) = 0$, thus $P_1(x_2) = P_1(x_1) = 0$. So we are done with the case of $i = 1$.

- For $i = 0$: we have $u = 0$, $w = (0, 1)$, $L = x + y$, $d = 4$, $e = 6$, $\mathfrak{m} = 8$, $\mathrm{u}_E = 1$, $\mathrm{v}_E = 3$, $\mathrm{u}_F = 0$, and $\mathrm{v}_F = -1$. Since $\mathrm{u}_F > \mathrm{v}_F$, there is nothing to prove.

This completes the verification of Conjecture 2.4 for $(a, b, m, n) = (2, 3, 2, 4)$.

For the rest, we give two more complicated examples.

EXAMPLE 3.2. We continue from Example 2.1. If $\mu \in [0, 12]$, then $\mathrm{j}(\mu) = 0$. If $\mu \in [13, 19]$, then $\mathrm{j}(\mu) = \mu - 12$. We verify Conjecture 2.4 for $\ell = 7 (= \mathrm{v}_F)$, that is, we want to show $P_1(x_7) = 0$.

By way of contradiction, suppose that $P_1(x_7) \neq 0$.

For a monomial $M = \prod_j \Gamma_j^{g_j} x_j^{e_j}$ in Γ_j's and x_j's, we define **x**-degree of M by **x**-deg$(M) = \sum_j e_j$. If every term of a polynomial in Γ_j's and x_j's has the same **x**-degree, then we say that it is **x**-homogeneous. Consider the order between $x_4, \ldots, x_7$, which is given by $x_4 > x_5 > x_6 > x_7$.

When $\mu = 13$, we get

$$S(\widetilde{G}_{e-\mu,1}) = -\frac{3}{8}V_{4,7},$$

where

$$V_{4,7} = -2x_4x_7 + \Gamma_5^3x_7^2 - 2\Gamma_5^2x_6x_7 - 2\Gamma_4\Gamma_5x_7^2 + \Gamma_5x_6^2 \\ + 2\Gamma_5x_5x_7 + 2\Gamma_4x_6x_7 + \Gamma_3x_7^2 - 2x_5x_6.$$

Note that $-2x_4x_7$ is the leading term of $V_{4,7}$, and that $V_{4,7}$ is $\mathbf{x}$-homogeneous.

When $\mu = 14$, we get

$$S(\widetilde{G}_{e-\mu,1}) = \frac{3}{8}p_{14},$$

where $p_{14} = \Gamma_5^4x_7^2 - 2\Gamma_5^3x_6x_7 - 3\Gamma_4\Gamma_5^2x_7^2 + \Gamma_5^2x_6^2 + 2\Gamma_5^2x_5x_7 + 4\Gamma_4\Gamma_5x_6x_7 + \Gamma_4^2x_7^2 + 2\Gamma_3\Gamma_5x_7^2 - 2\Gamma_5x_5x_6 - \Gamma_4x_6^2 - 2\Gamma_5x_4x_7 - 2\Gamma_4x_5x_7 - 2\Gamma_3x_6x_7 - \Gamma_2x_7^2 + x_5^2 + 2x_4x_6$.

To cancel the term that has x_4x_7 as a factor, let $V_{4,6} = p_{14} - \Gamma_5V_{4,7}$. Then $V_{4,6} = 2x_4x_6 - \Gamma_4\Gamma_5^2x_7^2 + 2\Gamma_4\Gamma_5x_6x_7 + \Gamma_4^2x_7^2 + \Gamma_3\Gamma_5x_7^2 - \Gamma_4x_6^2 - 2\Gamma_4x_5x_7 - 2\Gamma_3x_6x_7 - \Gamma_2x_7^2 + x_5^2$. Note that $2x_4x_6$ is the leading term of $V_{4,6}$, and that $V_{4,6}$ is $\mathbf{x}$-homogeneous.

Since $P_1(S(\widetilde{G}_{e-\mu,1})) = 0$ for $\mu \in [13, 14]$, we get $P_1(V_{4,7}) = P_1(p_{14}) = 0$, which implies $P_1(V_{4,6}) = 0$.

When $\mu = 15$, let

$$p_{15} = -16S(\widetilde{G}_{e-\mu,1}).$$

This is more complicated than p_{14} so we will not write it here. But all we need is a simpler expression which is obtained by canceling any term that has x_4x_7 or x_4x_6 as a factor. Let

$$V_{4,5} = p_{15} + 6(\Gamma_4 - \Gamma_5^2)V_{4,7} - 6\Gamma_5V_{4,6},$$

so that $V_{4,5} = -12x_4x_5 + 6\Gamma_3\Gamma_5^2x_7^2 - 12\Gamma_3\Gamma_5x_6x_7 - 6\Gamma_3\Gamma_4x_7^2 - 6\Gamma_2\Gamma_5x_7^2 + 6\Gamma_3x_6^2 + 12\Gamma_3x_5x_7 + 12\Gamma_2x_6x_7 + x_7^3$. We have $P_1(V_{4,5}) = 0$, but $V_{4,5}$ is not $\mathbf{x}$-homogeneous, which will help us prove what we want. Let $(V_{4,5})_2 = -12x_4x_5 + 6\Gamma_3\Gamma_5^2x_7^2 - 12\Gamma_3\Gamma_5x_6x_7 - 6\Gamma_3\Gamma_4x_7^2 - 6\Gamma_2\Gamma_5x_7^2 + 6\Gamma_3x_6^2 + 12\Gamma_3x_5x_7 + 12\Gamma_2x_6x_7$, so that $(V_{4,5})_2$ is $\mathbf{x}$-homogeneous.

When $\mu = 16$, let

$$p_{16} = 16S(\widetilde{G}_{e-\mu,1}).$$

As before, we cancel any term that has x_4x_7, x_4x_6, or x_4x_5 as a factor. Let $V_{4,4}$ be the resulting expression, that is,

$$V_{4,4} = 6x_4^2 - 6\Gamma_2\Gamma_5^2x_7^2 + 12\Gamma_2\Gamma_5x_6x_7 + 6\Gamma_2\Gamma_4x_7^2 + 2\Gamma_5x_7^3 - 6\Gamma_2x_6^2 - 12\Gamma_2x_5x_7 - 3x_6x_7^2.$$

We have $P_1(V_{4,4}) = 0$

When $\mu = 17$, let

$$p_{17} = -16S(\widetilde{G}_{e-\mu,1}).$$

As before, we cancel any term that has x_4x_7, x_4x_6, x_4x_5, or x_4^2 as a factor. Let $W_{5,7}$ be the resulting expression, that is,

$$W_{5,7} = x_7(3\Gamma_5^2x_7^2 - 6\Gamma_5x_6x_7 - 2\Gamma_4x_7^2 + 3x_6^2 + 3x_5x_7).$$

It is remarkable that every term in $W_{5,7}$ has a common factor x_7, and that $W_{5,7}$ is $\mathbf{x}$-homogeneous. Let $V_{5,7} = W_{5,7}/x_7$, that is,

$$V_{5,7} = 3x_5x_7 + 3\Gamma_5^2x_7^2 - 6\Gamma_5x_6x_7 - 2\Gamma_4x_7^2 + 3x_6^2.$$

Since $P_1(W_{5,7}) = 0 \neq P_1(x_7)$, we get $P_1(V_{5,7}) = 0$.

Continuing in this fashion, the expression obtained from $S(\widetilde{G}_{e-18,1})$ becomes

$$V_{5,6} = -12x_5x_6 - 3\Gamma_5^3x_7^2 + 6\Gamma_5^2x_6x_7 - 6\Gamma_4\Gamma_5x_7^2 - 3\Gamma_5x_6^2 + 14\Gamma_4x_6x_7 + 3\Gamma_3x_7^2,$$

and the expression obtained from $S(\widetilde{G}_{e-19,1})$ becomes

$$\begin{aligned}V_{5,5} =&12x_5^2 - 111\Gamma_5^4x_7^2 + 222\Gamma_5^3x_6x_7 + 86\Gamma_4\Gamma_5^2x_7^2 - 111\Gamma_5^2x_6^2 - 108\Gamma_5^2x_5x_7\\ &- 26\Gamma_4\Gamma_5x_6x_7 - 4\Gamma_4^2x_7^2 + 7\Gamma_3\Gamma_5x_7^2 + 12\Gamma_4x_6^2 - 12\Gamma_3x_6x_7 - 4\Gamma_2x_7^2.\end{aligned}$$

We have $P_1(V_{5,6}) = P_1(V_{5,5}) = 0$.

The key observation is that $V_{4,7}$, $V_{4,6}$, $(V_{4,5})_2$, $V_{5,7}$, $V_{5,6}$, and $V_{5,5}$ are all **x**-homogeneous of degree 2. Consider the following relation between them:

$$\begin{aligned}-x_7(V_{4,5})_2 =& \left(-24\Gamma_5^2x_7 + 33\Gamma_5x_6 + 16\Gamma_4x_7 - 30x_5\right)V_{4,7}\\ &+ (-15\Gamma_5x_7 + 24x_6)\,V_{4,6}\\ &+ \left(95\Gamma_5^3x_7 - 127\Gamma_5^2x_6 - 22\Gamma_4\Gamma_5x_7 + 16\Gamma_5x_5 + 20\Gamma_4x_6 + 2\Gamma_3x_7 - 16x_4\right)V_{5,7}\\ &+ \left(\frac{15}{4}\Gamma_5^2x_7 - 5\Gamma_5x_6 - 2\Gamma_4x_7 + 4x_5\right)V_{5,6}\\ &+ \left(\frac{9}{4}\Gamma_5x_7 - 3x_6\right)V_{5,5}.\end{aligned}$$

Since $P_1(V_{4,7}) = P_1(V_{4,6}) = P_1(V_{5,7}) = P_1(V_{5,6}) = P_1(V_{5,5}) = 0 \neq P_1(x_7)$, we get $P_1((V_{4,5})_2) = 0$. Since $P_1(V_{4,5}) = 0$ and $x_7^3 = V_{4,5} - (V_{4,5})_2$, we get $P_1(x_7^3) = 0$, which gives $P_1(x_7) = 0$, but this contradicts the assumption that $P_1(x_7) \neq 0$. Therefore $P_1(x_7)$ must be equal to 0. This completes the example.

In Example 3.2, we did not use the assumption that E° is a principal polynomial. In the next example, we need to use this condition.

EXAMPLE 3.3. Fix $(a, b, m, n) = (2, 3, 4, 8)$, $\delta = 1$, and $i = 4$. Then $i/m \in \mathbb{Z}$. So

$$w = (0, 1),\quad L = x + 1,\quad \mathrm{u}_E = 2,\quad \mathrm{v}_E = 4,\quad \mathrm{u}_F = 1,\quad \mathrm{v}_F = 2,\quad d = 8,\quad e = 12.$$

Assume that E° is principal. Then there are three possibilities: [1]

Case 1: $4\Gamma_2 - \Gamma_3^2 \neq 8(x+1)y^4$.

Case 2: Γ_3 does not divide Γ_1.

Case 3: $\Gamma_0 - (\Gamma_1/\Gamma_3)^2$ is not a constant.

Here we will discuss Case 1 only. In fact, for the discussion below, we will assume a weaker condition that:

$$4\Gamma_2 - \Gamma_3^2 \text{ is not divisible by } L.$$

We verify Conjecture 2.4 for $\ell = 2 (= \mathrm{v}_F)$, that is, we want to show $P_1(x_2) = 0$.

[1] Assume that E° is not principal, then there exist a constant $c \in \mathbb{C}$ and a polynomial $P = (x+1)y^2 + P_1y + P_0 \in \mathcal{E}(E^\circ)$ such that $E^\circ = P^2 + c$, where P_1 and P_0 are polynomials in x. Expanding, we get $(x+1)^2y^4 + (x+1)y^2\Gamma_3 + \Gamma_2 + \Gamma_1 + \Gamma_0 = (x+1)^2y^4 + 2(x+1)y^3P_1 + (2(x+1) + P_1^2)y^2 + 2P_1P_0y + (P_0^2 + c)$, thus $\Gamma_3 = 2yP_1$, $\Gamma_2 = (2(x+1)y^2 + P_1^2)y^2$, $\Gamma_1 = 2P_1P_0y$, and $\Gamma_0 = P_0^2 + c$. Note that $4\Gamma_2 - \Gamma_3^2 = 8(x+1)y^4 + 4P_1^2y^2 - 4y^2P_1^2 = 8(x+1)y^4$ which is divisible by $L = x+1$. So E° is not principal if and only if

$$\begin{cases} 4\Gamma_2 - \Gamma_3^2 = 8(x+1)y^4, \text{ and}\\ \Gamma_3 | \Gamma_1, \text{ and}\\ \Gamma_0 - (\Gamma_1/\Gamma_3)^2 \text{ is a constant.}\end{cases}$$

(if these conditions are satisfied, we can solve P by $P_1 = \Gamma_3/(2y)$ and $P_0 = \Gamma_1/(2yP_1)$).

Suppose that $B = \{0, 2, 4, 6, 8, \ldots, 18\}$ is supported. If $\mu \in [0, 7]$, then $\mathfrak{j}(\mu) = 0$. If $\mu = 8$, then $\mathfrak{j}(\mu) = 1$. When $\mu = 8$, we get

$$S(\widetilde{G}_{e-\mu,1}) = \frac{5}{1024} c_2(4\Gamma_2 - \Gamma_3^2)^3,$$

which implies $c_2 = 0$. Hence $B = \{0, 4, 6, 8, 10, 12, 14, 16, 18\}$ is supported.

Now suppose that $B = \{0, 4, 6, 8, 10, 12, 14, 16, 18\}$ is supported. Then $\mathfrak{j}(\mu) = 0$ for $\mu < 10$ and $\mathfrak{j}(\mu) = 1$ for $\mu = 10$. When $\mu = 10$, we get

$$S(\widetilde{G}_{e-\mu,1}) = \frac{3}{128} c_6(4\Gamma_2 - \Gamma_3^2)^2,$$

which implies $c_6 = 0$. Hence $B = \{0, 4, 8, 10, 12, 14, 16, 18\}$ is supported.

Next suppose that $B = \{0, 4, 8, 10, 12, 14, 16, 18\}$ is supported. Then $\mathfrak{j}(\mu) = 0$ for $\mu < 12$ and $\mathfrak{j}(\mu) = 1$ for $\mu = 12$. When $\mu = 12$, we get

$$S(\widetilde{G}_{e-\mu,1}) = \frac{1}{8} c_{10}(4\Gamma_2 - \Gamma_3^2),$$

which implies $c_{10} = 0$. Hence $B = \{0, 4, 8, 12, 14, 16, 18\}$ is supported.

This time suppose that $B = \{0, 4, 8, 12, 14, 16, 18\}$ is supported. Then $\mathfrak{j}(\mu) = 0$ for $\mu < 13$ and $\mathfrak{j}(\mu) = \mu - 12$ for $\mu \in [13, 14]$. When $\mu \in [13, 14]$, we get

$$S(\widetilde{G}_{e-13,1}) = \frac{3}{8} x_2(2x_1 - \Gamma_3 x_2)$$

and

$$S(\widetilde{G}_{e-14,1}) = \frac{3}{8}(\Gamma_3^2 x_2^2 - 2\Gamma_3 x_1 x_2 - \Gamma_2 x_2^2 + x_1^2).$$

Let $g_1 = x_2(2x_1 - \Gamma_3 x_2)$ and $g_2 = \Gamma_3^2 x_2^2 - 2\Gamma_3 x_1 x_2 - \Gamma_2 x_2^2 + x_1^2$, and it is easy to verify that $(4\Gamma_2 - \Gamma_3^2)x_2^3 = (-3\Gamma_3 x_2 + 2x_1)g_1 + (-4x_2)g_2$.

So $P_1((4\Gamma_2 - \Gamma_3^2)x_2^3) = 0$. Since $P_1(4\Gamma_2 - \Gamma_3^2) \neq 0$, we must have $P_1(x_2) = 0$.

Acknowledgements

We would like to thank David Wright for valuable discussion, and Christian Valqui for numerous helpful suggestions. We also thank Rob Lazarsfeld, Lenny Makar-Limanov, and Avinash Sathaye for their correspondences, and thank the referee for the valuable suggestions.

References

[1] S. S. Abhyankar, *Lectures on expansion techniques in algebraic geometry*, Tata Institute of Fundamental Research Lectures on Mathematics and Physics, vol. 57, Tata Institute of Fundamental Research, Bombay, 1977. MR542446

[2] S. S. Abhyankar, *Some remarks on the Jacobian question*, Proc. Indian Acad. Sci. Math. Sci. **104** (1994), no. 3, 515–542, DOI 10.1007/BF02867118. With notes by Marius van der Put and William Heinzer; Updated by Avinash Sathaye. MR1314394

[3] S. S. Abhyankar and T. T. Moh, *Embeddings of the line in the plane*, J. Reine Angew. Math. **276** (1975), 148–166. MR379502

[4] K. Adjamagbo and A. van den Essen, *Eulerian operators and the Jacobian conjecture. III*, J. Pure Appl. Algebra **81** (1992), no. 2, 111–116, DOI 10.1016/0022-4049(92)90001-V. MR1176017

[5] H. Appelgate and H. Onishi, *The Jacobian conjecture in two variables*, J. Pure Appl. Algebra **37** (1985), no. 3, 215–227, DOI 10.1016/0022-4049(85)90099-4. MR797863

[6] H. Bass, E. H. Connell, and D. Wright, *The Jacobian conjecture: reduction of degree and formal expansion of the inverse*, Bull. Amer. Math. Soc. (N.S.) **7** (1982), no. 2, 287–330, DOI 10.1090/S0273-0979-1982-15032-7. MR663785

[7] A. Belov-Kanel and M. Kontsevich, *The Jacobian conjecture is stably equivalent to the Dixmier conjecture* (English, with English and Russian summaries), Mosc. Math. J. **7** (2007), no. 2, 209–218, 349, DOI 10.17323/1609-4514-2007-7-2-209-218. MR2337879

[8] M. de Bondt and D. Yan, *Irreducibility properties of Keller maps*, Algebra Colloq. **23** (2016), no. 4, 663–680, DOI 10.1142/S1005386716000560. MR3563558

[9] P. Cassou-Noguès, *Newton trees at infinity of algebraic curves*, Affine algebraic geometry, CRM Proc. Lecture Notes, vol. 54, Amer. Math. Soc., Providence, RI, 2011, pp. 1–19, DOI 10.1090/crmp/054/01. MR2768630

[10] Nguyen Van Chau, *Plane Jacobian conjecture for simple polynomials*, Ann. Polon. Math. **93** (2008), no. 3, 247–251, DOI 10.4064/ap93-3-5. MR2403658

[11] C. C.-a. Cheng, T. Sakkalis, and S. S. S. Wang, *A case of the Jacobian conjecture*, J. Pure Appl. Algebra **96** (1994), no. 1, 15–18, DOI 10.1016/0022-4049(94)90082-5. MR1297436

[12] C. Cheng and S. Wang, *Radial similarity of Newton polygons*, Automorphisms of affine spaces (Curaçao, 1994), Kluwer Acad. Publ., Dordrecht, 1995, pp. 157–167. MR1352697

[13] E. Connell and J. Zweibel, *Subgroups of polynomial automorphisms*, Bull. Amer. Math. Soc. (N.S.) **23** (1990), no. 2, 401–406, DOI 10.1090/S0273-0979-1990-15944-0. MR1039423

[14] L. M. Drużkowski, *An effective approach to Keller's Jacobian conjecture*, Math. Ann. **264** (1983), no. 3, 303–313, DOI 10.1007/BF01459126. MR714105

[15] A. van den Essen and H. Tutaj, *A remark on the two-dimensional Jacobian conjecture*, J. Pure Appl. Algebra **96** (1994), no. 1, 19–22, DOI 10.1016/0022-4049(94)90083-3. MR1297437

[16] A. van den Essen, *Polynomial automorphisms and the Jacobian conjecture* (English, with English and French summaries), Algèbre non commutative, groupes quantiques et invariants (Reims, 1995), Sémin. Congr., vol. 2, Soc. Math. France, Paris, 1997, pp. 55–81. MR1601194

[17] A. van den Essen, *Polynomial automorphisms and the Jacobian conjecture*, Progress in Mathematics, vol. 190, Birkhäuser Verlag, Basel, 2000, DOI 10.1007/978-3-0348-8440-2. MR1790619

[18] A. van den Essen, D. Wright, and W. Zhao, *On the image conjecture*, J. Algebra **340** (2011), 211–224, DOI 10.1016/j.jalgebra.2011.04.036. MR2813570

[19] J. Glidewell, W. E. Hurst, K. Lee and L. Li, *On the two-dimensional Jacobian conjecture: Magnus' formula revisited, II*, arXiv:2205.12792.

[20] J. Glidewell, W. E. Hurst, K. Lee and L. Li, *On the two-dimensional Jacobian conjecture: Magnus' formula revisited, IV*, in preparation.

[21] J. Gwoździewicz, *Injectivity on one line*, Bull. Soc. Sci. Łódź 7 (1993), 59–60, Série: Recherches sur les déformationes XV.

[22] R. C. Heitmann, *On the Jacobian conjecture*, J. Pure Appl. Algebra **64** (1990), no. 1, 35–72, DOI 10.1016/0022-4049(90)90005-3. MR1055020

[23] W. E. Hurst, K. Lee, L. Li and G. D. Nasr, *On the two-dimensional Jacobian conjecture: Magnus' formula revisited, I*, arXiv:2201.06613.

[24] E.-M.G.M. Hubbers, *The Jacobian Conjecture: Cubic Homogeneous Maps in Dimension Four*, Master's thesis, University of Nijmegen, 1994, directed by A.R.P. van den Essen.

[25] P. Jędrzejewicz and J. Zieliński, *An approach to the Jacobian conjecture in terms of irreducibility and square-freeness*, Eur. J. Math. **3** (2017), no. 2, 199–207, DOI 10.1007/s40879-017-0145-5. MR3652577

[26] O.-H. Keller, *Ganze Cremona-Transformationen* (German), Monatsh. Math. Phys. **47** (1939), no. 1, 299–306, DOI 10.1007/BF01695502. MR1550818

[27] M. Kirezci, *The Jacobian conjecture. I, II* (English, with Turkish summary), İstanbul Tek. Üniv. Bül. **43** (1990), no. 3, 421–436, 451–457. MR1092112

[28] J. Lang, *Jacobian pairs. II*, J. Pure Appl. Algebra **74** (1991), no. 1, 61–71, DOI 10.1016/0022-4049(91)90049-8. MR1129130

[29] J. Lang and S. Maslamani, *Some results on the Jacobian conjecture in higher dimension*, J. Pure Appl. Algebra **94** (1994), no. 3, 327–330, DOI 10.1016/0022-4049(94)90013-2. MR1285548

[30] A. Magnus, *Volume-preserving transformations in several complex variables*, Proc. Amer. Math. Soc. **5** (1954), 256–266, DOI 10.2307/2032232. MR65655

[31] A. Magnus, *On polynomial solutions of a differential equation*, Math. Scand. **3** (1955), 255–260 (1956), DOI 10.7146/math.scand.a-10443. MR78453

[32] L. Makar-Limanov, *On the Newton polygon of a Jacobian mate*, Automorphisms in birational and affine geometry, Springer Proc. Math. Stat., vol. 79, Springer, Cham, 2014, pp. 469–476, DOI 10.1007/978-3-319-05681-4_25. MR3229366

[33] L. G. Makar-Limanov, *On the Newton polyhedron of a Jacobian pair* (Russian, with Russian summary), Izv. Ross. Akad. Nauk Ser. Mat. **85** (2021), no. 3, 127–137, DOI 10.4213/im9067; English transl., Izv. Math. **85** (2021), no. 3, 457–467. MR4265370

[34] L. Makar-Limanov and U. Umirbaev, *The Freiheitssatz for Poisson algebras*, J. Algebra **328** (2011), 495–503, DOI 10.1016/j.jalgebra.2010.08.015. MR2745580

[35] J. H. McKay and S. S. S. Wang, *A note on the Jacobian condition and two points at infinity*, Proc. Amer. Math. Soc. **111** (1991), no. 1, 35–43, DOI 10.2307/2047856. MR1034887

[36] G. H. Meisters and C. Olech, *Power-exact, nilpotent, homogeneous matrices*, Linear and Multilinear Algebra **35** (1993), no. 3-4, 225–236, DOI 10.1080/03081089308818260. MR1308692

[37] M. Nagata, *Two-dimensional Jacobian conjecture*, Algebra and topology 1988 (Taejŏn, 1988), Korea Inst. Tech., Taejŏn, 1988, pp. 77–98. MR1022249

[38] M. Nagata, *Some remarks on the two-dimensional Jacobian conjecture*, Chinese J. Math. **17** (1989), no. 1, 1–7. MR1007872

[39] Y. Nakai and K. Baba, *A generalization of Magnus' theorem*, Osaka Math. J. **14** (1977), no. 2, 403–409. MR485879

[40] A. Nowicki and Y. Nakai, *On Appelgate-Onishi's lemmas*, J. Pure Appl. Algebra **51** (1988), no. 3, 305–310, DOI 10.1016/0022-4049(88)90069-2. MR946581

[41] A. Nowicki and Y. Nakai, *Correction to: "On Appelgate-Onishi's lemmas" [J. Pure Appl. Algebra* **51** *(1988), no. 3, 305–310; MR0946581 (89h:13007)]*, J. Pure Appl. Algebra **58** (1989), no. 1, 101, DOI 10.1016/0022-4049(89)90055-8. MR996177

[42] M. Oka, *On the boundary obstructions to the Jacobian problem*, Kodai Math. J. **6** (1983), no. 3, 419–433, DOI 10.2996/kmj/1138036806. MR717330

[43] C. Valqui, J. A. Guccione, and J. J. Guccione, *On the shape of possible counterexamples to the Jacobian conjecture*, J. Algebra **471** (2017), 13–74, DOI 10.1016/j.jalgebra.2016.08.039. MR3569178

[44] S. S. S. Wang, *A Jacobian criterion for separability*, J. Algebra **65** (1980), no. 2, 453–494, DOI 10.1016/0021-8693(80)90233-1. MR585736

[45] A.V. Yagzhev, *On Keller's problem*, Siberian Math. J. **21** (1980), 747–754.

[46] J. T. Yu, *On the Jacobian conjecture: reduction of coefficients*, J. Algebra **171** (1995), no. 2, 515–523, DOI 10.1006/jabr.1995.1024. MR1315909

Department of Mathematics, University of Alabama, Tuscaloosa, Alabama 35487
Email address: `jbglidewell@crimson.ua.edu`

Department of Mathematics, University of Alabama, Tuscaloosa, Alabama 35487
Email address: `wehurst@crimson.ua.edu`

Department of Mathematics, University of Alabama, Tuscaloosa, Alabama 35487; and Korea Institute for Advanced Study, Seoul 02455, Republic of Korea
Email address: `kyungyong.lee@ua.edu; klee1@kias.re.kr`

Department of Mathematics and Statistics, Oakland University, Rochester, Michigan 48309
Email address: `li2345@oakland.edu`

Contemporary Mathematics
Volume **791**, 2024
https://doi.org/10.1090/conm/791/15870

Homotopy lifting maps on Hochschild cohomology and connections to deformation of algebras using reduction systems

Tolulope Oke

ABSTRACT. We describe the Gerstenhaber bracket structure on Hochschild cohomology of Koszul quiver algebras in terms of homotopy lifting maps. There is a projective bimodule resolution of Koszul quiver algebras that admits a comultiplicative structure. Introducing new scalars, we describe homotopy lifting maps associated to Hochschild cocycles using the comultiplicative structure. We show that the scalars can be described by some recurrence relations and we give several examples where these scalars appear in the literature. In particular, for a member of a family of quiver algebras, we describe Hochschild 2-cocycles and their associated homotopy lifting maps and determine the Maurer-Cartan elements of the quiver algebra in two ways: (i) by the use of homotopy lifting maps and (ii) by the use of a combinatorial star product that arises from the deformation of algebras using reduction systems.

1. Introduction

The Hochschild cohomology $\mathrm{HH}^*(\Lambda)$ of an associative algebra Λ possesses a multiplicative map called the cup product making it into a graded commutative ring. The ring structure of Hochschild cohomology of certain path algebras was determined using quiver techniques. For instance, if a path algebra is Koszul, its resolution possesses a comultiplicative structure and the cup product structure on its Hochschild cohomology can be presented using this comultiplicative structure. This cup product was described in [**6**].

In addition to the cup product on Hochschild cohomology ring is the Gerstenhaber bracket making $\mathrm{HH}^*(\Lambda)$ into a graded Lie/Gerstenhaber algebra. The bracket plays an important role in the theory of deformation of algebras. The theory of deformation of algebras employs techniques in algebraic and noncommutative geometry to describe variations of the associative multiplicative structure on any algebra. In a recent article, S. Barmeier and Z. Wang [**1**] have introduced a technique for finding families of deformation of quiver algebras with relations using *reduction* systems. Reduction systems were introduced by Bergman [**2**] in the late seventies. In particular, for an algebra $\Lambda = kQ/I$ where Q is a finite quiver, there is

2020 *Mathematics Subject Classification.* Primary 16E40; 16S37; 16S80, 16S15 .
The author thanks Severin Barmeier for useful discussions and for assisting with the calculations in Section 6.

associated a reduction system R useful in determining a projective bimodule resolution of Λ reminiscent of the Bardzell resolution [**4**]. It was shown in [**1**] that there is an equivalence of formal deformations between (i) deformations of the associative algebra Λ, (ii) deformations of the reduction system R and (iii) deformations of the relations in I.

The Gerstenhaber bracket can be difficult to compute in general settings. Several works have been carried out to interpret the bracket as well as make computations of the bracket accessible for a large class of algebras for instance in [**7, 9**]. In [**16**], Y. Volkov introduced a method in which the bracket is defined in terms of a homotopy lifting map. This method works for any arbitrary projective bimodule resolution of the algebra. In earlier works [**13**], we present a general formula for homotopy lifting maps associated to cocycles on Hochschild cohomology of Koszul path algebras.

The resolution introduced in [**6**] had scalars $c_{p,j}(n,i,r)$ appearing in the definition of the differentials on the resolution. These scalars made it possible to give a closed formula for the cup product structure on Hochschild cohomology. In Section 3, we present new scalars $b_{m,r}(m-n+1,s)$ associated to homotopy lifting maps on Hochschild cocycles using the scalars $c_{p,j}(n,i,r)$ of the comultiplicative relations. We show that the scalars $b_{m,r}(m-n+1,s)$ can be described using some recurrence relations and present the Gerstenhaber bracket structure using these scalars. We give several examples where these scalars appear in the literature in Section 4. In Section 5, we introduce a family of quiver algebras that has been extensively studied in [**12, 13**]. For the algebra A_1 from the family, we find Hochschild 2-cocycles and their associated homotopy lifting maps. We show that $\mathrm{HH}^2(A_1)$ is generated as a vector space by five Maurer-Cartan elements.

Relevant results about Hochschild cohomology, Gerstenhaber bracket, quiver algebras and deformation of algebras using reduction systems were recalled in the preliminaries. The deformation of an algebra involves altering the associative multiplicative structure on the algebra. The candidates for determining how the multiplicative structure is altered can be described using a combinatorial star product ($\star$) [**1**] and this product can be used to describe Maurer-Cartan elements. In Section 6, we describe the Maurer-Cartan elements of the algebra A_1 using ($\star$), showing that $\mathrm{HH}^2(A_1)$ is a five dimensional k-vector space.

2. Preliminaries

The Hochschild cohomology of an associative k-algebra Λ was originally defined using the following projective resolution known as the bar resolution.

$$\text{(2.1)}\quad \mathbb{B}_\bullet: \qquad \cdots \to \Lambda^{\otimes(n+2)} \xrightarrow{\delta_n} \Lambda^{\otimes(n+1)} \xrightarrow{\delta_{n-1}} \cdots \xrightarrow{\delta_2} \Lambda^{\otimes 3} \xrightarrow{\delta_1} \Lambda^{\otimes 2} \; (\xrightarrow{\mu} \Lambda)$$

where μ is multiplication and the differentials δ_n are given by

$$\text{(2.2)}\qquad \delta_n(a_0 \otimes a_1 \otimes \cdots \otimes a_{n+1}) = \sum_{i=0}^{n} (-1)^i a_0 \otimes \cdots \otimes a_i a_{i+1} \otimes \cdots \otimes a_{n+1}$$

for all $a_0, a_1, \ldots, a_{n+1} \in \Lambda$. This resolution consists of Λ-bimodules or left modules over the enveloping algebra $\Lambda^e = \Lambda \otimes \Lambda^{op}$, where Λ^{op} is the opposite algebra. The resolution is sometimes written $\mathbb{B}_\bullet \xrightarrow{\mu} \Lambda$ with μ referred to as the augmentation map. Let M be a finitely generated left Λ^e-module, the Hochschild cohomology of Λ with coefficients in M denoted $\mathrm{HH}^*(\Lambda, M)$ is obtained by applying the functor

$\mathrm{Hom}_{\Lambda^e}(-, M)$ to the complex $\mathbb{B}_\bullet$, and then taking the cohomology of the resulting cochain complex. That is

$$\mathrm{HH}^*(\Lambda, M) := \bigoplus_{n\geq 0} \mathrm{HH}^n(\Lambda, M) = \bigoplus_{n\geq 0} \mathrm{H}^n(\mathrm{Hom}_{\Lambda^e}(\mathbb{B}_n, M)).$$

If we let $M = \Lambda$, we then define $\mathrm{HH}^*(\Lambda) := \mathrm{HH}^*(\Lambda, \Lambda)$ to be the Hochschild cohomology of Λ. An element $\chi \in \mathrm{Hom}_{\Lambda^e}(\mathbb{B}_m, \Lambda)$ is a cocycle if $(\delta^*(\chi))(\cdot) := \chi\delta(\cdot) = 0$. There is an isomorphism of the abelian groups $\mathrm{Hom}_{\Lambda^e}(\mathbb{B}_m, \Lambda) \cong \mathrm{Hom}_k(\Lambda^{\otimes m}, \Lambda)$, so we can also view χ as an element of $\mathrm{Hom}_k(\Lambda^{\otimes m}, \Lambda)$.

The Gerstenhaber bracket of two cocycles $\chi \in \mathrm{Hom}_k(\Lambda^{\otimes m}, \Lambda)$ and $\theta \in \mathrm{Hom}_k(\Lambda^{\otimes n}, \Lambda)$ at the chain level is given by

$$[\chi, \theta] = \chi \circ \theta - (-1)^{(m-1)(n-1)}\theta \circ \chi \tag{2.3}$$

where $\chi \circ \theta = \sum_{j=1}^{m} (-1)^{(n-1)(j-1)} \chi \circ_j \theta$ with

$$\begin{aligned}(\chi \circ_j \theta)&(a_1 \otimes \cdots \otimes a_{m+n-1}) \\ &= \chi(a_1 \otimes \cdots \otimes a_{j-1} \otimes \theta(a_j \otimes \cdots \otimes a_{j+n-1}) \otimes a_{j+n} \otimes \cdots \otimes a_{m+n-1}).\end{aligned}$$

This induces a well defined map $[\cdot\ ,\cdot] : \mathrm{HH}^m(\Lambda) \times \mathrm{HH}^n(\Lambda) \to \mathrm{HH}^{m+n-1}(\Lambda)$ on cohomology.

Gerstenhaber bracket using homotopy lifting: We present an equivalent definition of the Gerstenhaber bracket presented by Y. Volkov in [**16**] and reformulated with a sign change by S. Witherspoon in Theorem (2.4). We assume that A is an algebra over the field k and take $\mathbb{P} \xrightarrow{\mu_{\mathbb{P}}} A$ to be a projective resolution of A as an A^e-module with differential $d^{\mathbb{P}}$ and augmentation map $\mu_{\mathbb{P}}$. We take $\mathbf{d}$ to be the differential on the Hom complex $\mathrm{Hom}_{\Lambda^e}(\mathbb{P}, \mathbb{P})$ defined for any degree n map $g : \mathbb{P} \to \mathbb{P}[-n]$ as

$$\mathbf{d}(g) := d^{\mathbb{P}} g - (-1)^n g d^{\mathbb{P}}$$

where $\mathbb{P}[-n]$ is a shift in homological dimension with $(\mathbb{P}[-n])_m = \mathbb{P}_{m-n}$. In the following definition, the notation $\sim$ is used for two cocycles that are cohomologous, that is, they differ by a coboundary.

DEFINITION 2.1. Let $\Delta_{\mathbb{P}} : \mathbb{P} \to \mathbb{P} \otimes_A \mathbb{P}$ be a chain map lifting the identity map on $A \cong A \otimes_A A$ and suppose that $\eta \in \mathrm{Hom}_{A^e}(\mathbb{P}_n, A)$ is a cocycle. A module homomorphism $\psi_\eta : \mathbb{P} \to \mathbb{P}[1-n]$ is called a **homotopy lifting** map of η with respect to $\Delta_{\mathbb{P}}$ if

$$\begin{aligned}\mathbf{d}(\psi_\eta) &= (\eta \otimes 1_{\mathbb{P}} - 1_{\mathbb{P}} \otimes \eta)\Delta_{\mathbb{P}} \qquad \text{and} \\ \mu_{\mathbb{P}}\psi_\eta &\sim (-1)^{n-1}\eta\psi\end{aligned} \tag{2.4}$$

for some $\psi : \mathbb{P} \to \mathbb{P}[1]$ for which $\mathbf{d}(\psi) = (\mu_{\mathbb{P}} \otimes 1_{\mathbb{P}} - 1_{\mathbb{P}} \otimes \mu_{\mathbb{P}})\Delta_{\mathbb{P}}$.

EXAMPLE 2.2. Let us consider a homotopy lifting formula for a cocycle β using the bar resolution $\mathbb{B}$. Suppose that $\beta \in \mathrm{Hom}_{\Lambda^e}(\mathbb{B}_n, A) \cong \mathrm{Hom}_k(A^{\otimes n}, A)$, then one way to define a homotopy lifting map $\psi_\beta : \mathbb{B} \to \mathbb{B}[1-n]$ for the cocycle β is the following:

$$\begin{aligned}&\psi_\beta(1 \otimes a_1 \otimes \cdots \otimes a_{m+n-1} \otimes 1) = \\ &\sum_{i=1}^{m} (-1)^{(m-1)(i-1)} 1 \otimes a_1 \otimes \cdots \otimes a_{i-1} \otimes g(a_i \otimes \cdots \otimes a_{i+n-1}) \otimes a_{i+n} \otimes \cdots \otimes a_{m+n-1} \otimes 1.\end{aligned}$$

We compute an example in which $\beta \in \mathrm{Hom}_k(\Lambda^{\otimes 2}, \Lambda)$. In degree 3, $\psi_\beta : \mathbb{B}_3 \to \mathbb{B}_2$. Using the differentials on the bar resolution given in Equation (2.2) and the diagonal map $\Delta_{\mathbb{B}}$ later given in Equation (2.8), we have

$$\begin{aligned}\delta\psi_\beta(1\otimes a_1\otimes a_2\otimes a_3\otimes 1) &= \beta(a_1\otimes a_2)\otimes a_3\otimes 1 - 1\otimes\beta(a_1\otimes a_2)a_3\otimes 1\\ &\quad + 1\otimes\beta(a_1\otimes a_2)\otimes a_3 - a_1\otimes\beta(a_2\otimes a_3)\otimes 1\\ &\quad + 1\otimes a_1\beta(a_2\otimes a_3)\otimes 1 - 1\otimes a_1\otimes\beta(a_2\otimes a_3)\end{aligned}$$

and

$$\begin{aligned}\psi_\beta\delta(1\otimes a_1\otimes a_2\otimes a_3\otimes 1) &= a_1\otimes\beta(a_2\otimes a_3)\otimes 1 - 1\otimes\beta(a_1a_2\otimes a_3)\otimes 1\\ &\quad + 1\otimes\beta(a_1\otimes a_2a_3)\otimes 1 - 1\otimes\beta(a_1\otimes a_2)\otimes a_3.\end{aligned}$$

Therefore

$$\begin{aligned}&(\delta\psi_\beta + \psi_\beta\delta)(1\otimes a_1\otimes a_2\otimes a_3\otimes 1)\\ &\qquad = \beta(a_1\otimes a_2)\otimes a_3\otimes 1 - 1\otimes a_1\otimes\beta(a_2\otimes a_3).\end{aligned}$$

On the other hand,

$$\begin{aligned}&(\beta\otimes 1 - 1\otimes\beta)\Delta_{\mathbb{B}}(1\otimes a_1\otimes a_2\otimes a_3\otimes 1)\\ &\quad = (\beta\otimes 1 - 1\otimes\beta)\Big((1\otimes 1)\otimes_\Lambda(1\otimes a_1\otimes a_2\otimes a_3\otimes 1)\\ &\qquad + (1\otimes a_1\otimes 1)\otimes_\Lambda(1\otimes a_2\otimes a_3\otimes 1) + (1\otimes a_1\otimes a_2\otimes 1)\otimes_\Lambda(1\otimes a_3\otimes 1)\\ &\qquad + (1\otimes a_1\otimes a_2\otimes a_3\otimes 1)\otimes_\Lambda(1\otimes 1)\Big)\\ &\quad = \beta(a_1\otimes a_2)\otimes a_3\otimes 1 - 1\otimes a_1\otimes\beta(a_2\otimes a_3).\end{aligned}$$

So we see that Equation (2.4) holds in degree 3 i.e.

$$\delta\psi_\beta - (-1)^{2-1}\psi_\beta\delta = (\beta\otimes 1 - 1\otimes\beta)\Delta_{\mathbb{B}}.$$

REMARK 2.3. Suppose that $\mathbb{K}$ is the Koszul resolution, then it is a differential graded coalgebra i.e. $(\Delta_{\mathbb{K}}\otimes 1_{\mathbb{K}})\Delta_{\mathbb{K}} = (1_{\mathbb{K}}\otimes\Delta_{\mathbb{K}})\Delta_{\mathbb{K}}$ and $(d\otimes 1 + 1\otimes d)\Delta_{\mathbb{K}} = \Delta_{\mathbb{K}}d$. Furthermore, the augmentation map $\mu : \mathbb{K}\to\Lambda$ makes $(\mu\otimes 1_{\mathbb{K}})\Delta_{\mathbb{K}} - (1_{\mathbb{K}}\otimes\mu)\Delta_{\mathbb{K}} = 0$. We can therefore set $\psi = 0$ in the second part of Equation (2.4), so that we have $\mu\psi_\eta \sim 0$. Next, we set $\psi_\eta(\mathbb{K}_{n-1}) = 0$ and the second relations of Equation (2.4) is satisfied. To check if a map is a homotopy lifting map, it is sufficient to verify the first equation in (2.4) if the resolution is Koszul.

The following is a theorem of Y. Volkov which is equivalent to the definition of the bracket presented earlier in Equation (2.3).

THEOREM 2.4. [**16**, Theorem 4] *Let $(\mathbb{P}, \mu_{\mathbb{P}})$ be a A^e-projective resolution of the algebra A, and let $\Delta_{\mathbb{P}} : \mathbb{P}\to\mathbb{P}\otimes_A\mathbb{P}$ be a diagonal map. Let $\eta : \mathbb{P}_n\to A$ and $\theta : \mathbb{P}_m\to A$ be cocycles representing two classes. Suppose that ψ_η and ψ_θ are homotopy liftings for η and θ respectively. Then the Gerstenhaber bracket of the classes of η and θ can be represented by the class of the element*

$$[\eta,\theta]_{\Delta_{\mathbb{P}}} = \eta\psi_\theta - (-1)^{(m-1)(n-1)}\theta\psi_\eta.$$

Quiver algebras: A quiver is a directed graph with the allowance of loops and multiple arrows. A quiver Q is sometimes denoted as a quadruple (Q_0, Q_1, o, t) where Q_0 is the set of vertices in Q, Q_1 is the set of arrows in Q, and $o, t : Q_1 \longrightarrow Q_0$ are maps which assign to each arrow $a\in Q_1$, its origin vertex $o(a)$ and terminal vertex $t(a)$ in Q_0. A path in Q is a sequence of arrows $a = a_1a_2 \cdots a_{n-1}a_n$ such that the terminal vertex of a_i is the same as the origin vertex of a_{i+1}, using the convention of concatenating paths from left to right. The quiver algebra or path algebra kQ is defined as a vector space having all paths in Q as a basis. Vertices are regarded as paths of length 0, an arrow is a path of length 1, and so on. We take multiplication on kQ as concatenation of paths. Two paths a and b satisfy $ab = 0$ if $t(a) \neq o(b)$. This multiplication defines an associative algebra over k. By taking kQ_i to be the k-vector subspace of kQ with paths of length i as basis, $kQ = \bigoplus_{i\geq 0} kQ_i$ can be viewed as an $\mathbb{N}$-graded vector space. Two paths are parallel if they have the same origin and terminal vertex. A relation on a quiver Q is a linear combination of parallel paths in Q. A quiver together with a set of relations is called a quiver with relations. Letting I be an ideal of the path algebra kQ, we denote by (Q, I) the quiver Q with relations I. The quotient $\Lambda = kQ/I$ is called the quiver algebra associated with (Q, I). Suppose that Λ is graded by positive integers and is Koszul, the degree 0 component Λ_0 is isomorphic to k or copies of k and Λ_0 has a linear graded projective resolution $\mathbb{L}$ as a right Λ-module [**8, 15**].

An algorithmic approach to finding such a minimal projective resolution $\mathbb{L}$ of Λ_0 was given in [**5**]. The modules $\mathbb{L}_n$ are right Λ-modules for each n. There is a "comultiplicative structure" on $\mathbb{L}$ and this structure was used to find a minimal projective resolution $\mathbb{K} \to \Lambda$ of modules over the enveloping algebra of Λ in [**6**]. A non-zero element $x \in kQ$ is called uniform if it is a linear combination of paths each having the same origin vertex and the same terminal vertex: In other words, $x = \sum_j c_j w_j$ with scalars $c_j \neq 0$ for all j and each path w_j are of equal length having the same origin vertex and the same terminal vertex. For $R = kQ$, it was shown in [**5**] that there are integers t_n and uniform elements $\{f_i^n\}_{i=0}^{t_n}$ such that the right projective resolution $\mathbb{L} \to \Lambda_0$ is obtained from a filtration of R. This filtration is given by the following nested family of right ideals:

$$\cdots \subseteq \bigoplus_{i=0}^{t_n} f_i^n R \subseteq \bigoplus_{i=0}^{t_{n-1}} f_i^{n-1} R \subseteq \cdots \subseteq \bigoplus_{i=0}^{t_1} f_i^1 R \subseteq \bigoplus_{i=0}^{t_0} f_i^0 R = R$$

where for each n, $\mathbb{L}_n = \bigoplus_{i=0}^{t_n} f_i^n R / \bigoplus_{i=0}^{t_n} f_i^n I$ and the differentials on $\mathbb{L}$ are induced by the inclusions $\bigoplus_{i=0}^{t_n} f_i^n R \subseteq \bigoplus_{i=0}^{t_{n-1}} f_i^{n-1} R$. Furthermore, it was shown in [**5**] that with some choice of scalars, the $\{f_i^n\}_{i=0}^{t_n}$ satisfying the comultiplicative equation of (2.5) make $\mathbb{L}$ minimal. In other words, for $0 \leq i \leq t_n$, there are scalars $c_{pq}(n, i, r)$ such that

$$f_i^n = \sum_{p=0}^{t_r} \sum_{q=0}^{t_{n-r}} c_{pq}(n, i, r) f_p^r f_q^{n-r} \tag{2.5}$$

holds and $\mathbb{L}$ is a minimal resolution. To construct the above multiplicative equation for example, we can take $\{f_i^0\}_{i=0}^{t_0}$ to be the set of vertices, $\{f_i^1\}_{i=0}^{t_1}$ to be the set of arrows, $\{f_i^2\}_{i=0}^{t_2}$ to be the set of uniform relations generating the ideal I, and define $\{f_i^n\}_{i=0}^{t_n} (n \geq 3)$ recursively, that is in terms of f_i^{n-1} and f_j^1. We presented the comultiplicative structure of a family of quiver algebras in [**12**] and use the homotopy lifting technique to show that for some members of the family, the Hochschild cohomology ring modulo the weak Gerstenhaber ideal generated by homogeneous nilpotent elements is not finitely generated.

The resolution $\mathbb{L}$ and the comultiplicative structure (2.5) were used to construct a minimal projective resolution $\mathbb{K} \to \Lambda$ of modules over the enveloping algebra $\Lambda^e = \Lambda \otimes \Lambda^{op}$ on which we now define Hochschild cohomology. This minimal projective resolution $\mathbb{K}$ of Λ^e-modules associated to Λ was given in [**6**] and now restated with slight notational changes below.

THEOREM 2.5. [**6**, Theorem 2.1] *Let $\Lambda = kQ/I$ be a Koszul algebra, and let $\{f_i^n\}_{i=0}^{t_n}$ define a minimal resolution of Λ_0 as a right Λ-module. A minimal projective resolution $(\mathbb{K}, d)$ of Λ over Λ^e is given by*

$$\mathbb{K}_n = \bigoplus_{i=0}^{t_n} \Lambda o(f_i^n) \otimes_k t(f_i^n)\Lambda$$

for $n \geq 0$, where the differential $d_n : \mathbb{K}_n \to \mathbb{K}_{n-1}$ applied the basis element $\varepsilon_i^n = (0, \ldots, 0, o(f_i^n) \otimes_k t(f_i^n), 0, \ldots, 0)$, $0 \leq i \leq t_n$ with $o(f_i^n) \otimes_k t(f_i^n)$ in the i-th position, is given by

$$(2.6) \quad d_n(\varepsilon_i^n) = \sum_{j=0}^{t_{n-1}} \Big(\sum_{p=0}^{t_1} c_{p,j}(n,i,1) f_p^1 \varepsilon_j^{n-1} + (-1)^n \sum_{q=0}^{t_1} c_{j,q}(n,i,n-1) \varepsilon_j^{n-1} f_q^1 \Big)$$

and $d_0 : \mathbb{K}_0 \to \Lambda$ is the multiplication map. In particular, Λ is a linear module over Λ^e.

We note that for each n and i, $\{\varepsilon_i^n\}_{i=0}^{t_n}$ is a basis of $\mathbb{K}_n$ as a Λ^e-module. The scalars $c_{p,j}(n,i,r)$ are those appearing in (2.5) and $f_*^1 := \overline{f_*^1}$ is the residue class of f_*^1 in $\bigoplus_{i=0}^{t_1} f_i^1 R / \bigoplus_{i=0}^{t_n} f_i^1 I$. Using the comultiplicative structure of Equation (2.5), a cup product formula on Hochschild cohomology of Koszul quiver algebra was presented in [**3**] using the resolution $\mathbb{K}$.

We recall the definition of the reduced bar resolution of algebras defined by quivers and relations. If Λ_0 is isomorphic to m copies of k, take $\{e_1, e_2, \ldots, e_m\}$ to be a complete set of primitive orthogonal central idempotents of Λ. In this case Λ is not necessarily an algebra over Λ_0. If Λ_0 is isomorphic to k, then Λ is an algebra over Λ_0. For convenience, we use the same notation $\mathbb{B}$ for both the bar resolution and the reduced bar resolution. The reduced bar resolution $(\mathbb{B}, \delta)$, where $\mathbb{B}_n := \Lambda^{\otimes_{\Lambda_0}(n+2)}$ is the $(n+2)$-fold tensor product of Λ over Λ_0 and uses the same differential as the usual bar resolution presented in Equation (2.2). The resolution $\mathbb{K}$ can be embedded naturally into the reduced bar resolution $\mathbb{B}$. There is a map $\iota : \mathbb{K} \to \mathbb{B}$ defined by $\iota(\varepsilon_r^n) = 1 \otimes \widetilde{f_r^n} \otimes 1$ such that $\delta\iota = \iota d$, where

$$(2.7) \quad \widetilde{f_j^n} = \sum c_{j_1 j_2 \cdots j_n} f_{j_1}^1 \otimes f_{j_2}^1 \otimes \cdots \otimes f_{j_n}^1 \quad \text{if} \quad f_j^n = \sum c_{j_1 j_2 \cdots j_n} f_{j_1}^1 f_{j_2}^1 \cdots f_{j_n}^1$$

for some scalar $c_{j_1 j_2 \cdots j_n}$. It was shown in [**3**, Proposition 2.1] that ι is indeed an embedding. By taking $\Delta_{\mathbb{B}} : \mathbb{B} \to \mathbb{B} \otimes_\Lambda \mathbb{B}$ to be the following comultiplicative map (or diagonal map) on the bar resolution,

$$(2.8) \quad \Delta_{\mathbb{B}}(a_0 \otimes \cdots \otimes a_{n+1}) = \sum_{i=0}^{n} (a_0 \otimes \cdots \otimes a_i \otimes 1) \otimes_\Lambda (1 \otimes a_{i+1} \otimes \cdots \otimes a_{n+1}),$$

it was also shown in [**3**, Proposition 2.2] that the diagonal map $\Delta_{\mathbb{K}} : \mathbb{K} \to \mathbb{K} \otimes_\Lambda \mathbb{K}$ on the complex $\mathbb{K}$ has the following form.

$$(2.9) \qquad \Delta_{\mathbb{K}}(\varepsilon_r^n) = \sum_{v=0}^{n} \sum_{p=0}^{t_v} \sum_{q=0}^{t_{n-v}} c_{p,q}(n,r,v) \varepsilon_p^v \otimes_\Lambda \varepsilon_q^{n-v}.$$

The compatibility of $\Delta_{\mathbb{K}}, \Delta_{\mathbb{B}}$ and ι means that $(\iota\otimes\iota)\Delta_{\mathbb{K}} = \Delta_{\mathbb{B}}\iota$ where $(\iota\otimes\iota)(\mathbb{K}\otimes_\Lambda \mathbb{K}) = \iota(\mathbb{K})\otimes_\Lambda \iota(\mathbb{K}) \subseteq \mathbb{B}\otimes_\Lambda \mathbb{B}$.

Deformation of algebras using reduction system: The theory of deformation of algebras has a much more wider scope than the contents of this article. There has been survey articles covering several aspects of algebraic deformation theory - from deformations arising from noncommutative geometry to formal, infinitesimal and graded deformations. References were made to such articles in [**15**, chapter 5]. Although many results are known for deformation of commutative algebras, little is known of deformation of quiver and path algebras. Let $\Lambda = kQ/I$ and Q a finite quiver. There is associated to Λ, a reduction system $R = \{(s,\varphi_s)\}$ given by Definition 2.6 and the Gerstenhaber bracket equips Hochschild cohomology $\mathrm{HH}^*(\Lambda)$ with a DG Lie algebra structure which controls the theory of deformation of R. Using a combinatorial star product, it was shown in [**1**] that the deformations of Λ is equivalent to the deformations of the reduction system R which is also equivalent to the deformation of the relations in I. There is a map $\varphi : kQ \to kIrr_S$ defined by $\varphi(s) = \varphi_s$ sending a path in the quiver algebra to an irreducible path. Let $\pi : kQ \to \Lambda$ be the projection map. The combinatorial star product on Λ defined on irreducible paths $u, v \in kQ$ is the image of $\pi(u) \star \pi(v)$ in Λ after performing right-most *reductions* on the path $uv \in kQ$.

Suppose that (Λ_τ, μ_τ) is a formal deformation (Definition 2.10) of the associative multiplication on Λ, one way to describe the deformed multiplication μ_τ is by obtaining a necessary and sufficient condition for the associativity of the combinatorial star product. This condition is given by Equation (2.13). There is a projective bimodule resolution $\mathfrak{p}(Q,R)$ arising from a reduction system R and the combinatorial star product can be used to describe Maurer-Cartan elements in $\mathfrak{p}(Q,R)\otimes\mathfrak{m}$, where $(N,\mathfrak{m})$ is a complete local Noetherian k-algebra. In Section 6, we use the combinatorial star product to determine Maurer-Cartan elements of $\mathrm{HH}^2(A_1)$, thus determining a family of deformations of the algebra A_1. In the future, it will be interesting to find meaningful ways to describe Maurer-Cartan elements that were obtained using the star product (as in Example 6.1) in terms of those obtained by the homotopy lifting technique (as in Example 5.2) and vice versa. We now give a result from [**2**] on reduction systems.

DEFINITION 2.6. Let $\Lambda = kQ/I$ be a path algebra with finite quiver Q. A reduction system R for kQ is a set of pairs

$$R = \{(s,\varphi_s) \mid s\in S, \varphi_s \in kQ\},$$

where

- S is a subset of paths of length greater than or equal to 2 such that s is not a subpath of s' when $s \neq s' \in S$,
- s and φ_s are parallel paths,
- for each $(s,\varphi_s)\in R$, φ_s is irreducible or a linear combination of irreducible paths.

We say a path is *irreducible* if it does not contain elements in S as a subpath.

DEFINITION 2.7. Given a two-sided ideal I of kQ, we say that a reduction system R satisfies the diamond condition $(\diamond)$ for I if

- I is equal to the two-sided ideal generated by the set $\{s-\varphi_s\}_{(s,\varphi_s)\in R}$ and
- every path is *reduction unique*.

We call a reduction system R finite if R is a finite set.

DEFINITION 2.8. Let R be a reduction system for kQ and p, q, r be paths of length at least 1. A path pqr of length at least 3 is an *overlap ambiguity* of R if $pq, qr \in S$. The set of all paths having one overlap is also the set of all 1-ambiguity and is denoted S_3.

We now state the diamond lemma as provided in [**2**].

THEOREM 2.9. [**2**, Thm 1.2] *Let $R = \{(s, \varphi_s)\}$ be a reduction system for kQ and let $S = \{s \mid (s, \varphi_s) \in R\}$. Denote by $I = \langle s - \varphi_s \rangle_{s \in S}$ the corresponding two-sided ideal of $\Lambda = kQ/I$. If R is reduction finite, the following are equivalent:*

- *all overlap ambiguities of R are resolvable*
- *R is reduction unique, that is R satisfies $(\diamond)$ for I*
- *the image of the set of irreducible paths under the projection $\pi : kQ \to \Lambda$ forms a k-basis for Λ.*

It is known that for any two-sided ideal I of the quiver algebra kQ/I, there is always a choice of a reduction system R satisfying the diamond condition $(\diamond)$ of Definition 2.7. This is a result of S. Chouhy and A. Solotar [**4**, Prop 2.7, Thm 4.1, Thm 4.2]. Furthermore, there is a projective bimodule resolution $\mathfrak{p}(Q, R)$ associated to the algebra $\Lambda = kQ/I$ useful in extracting information about Hochschild cohomology.

Let B be a k-algebra and let τ be an indeterminate. The ring $B[[\tau]]$ is the ring of formal power series in τ with coefficients in B. The ring $B[[\tau]]$ is a $k[[\tau]]$-module if we identify k with the subalgebra $k \cdot 1$ of B. The multiplication in B is usually denoted by concatenation while the multiplication in $B[[\tau]]$ is given as

$$\left(\sum_{i\geq 0} a_i\tau^i\right)\left(\sum_{j\geq 0} b_j\tau^j\right) = \sum_{m\geq 0}\left(\sum_{i+j=m} a_i b_j \tau^m\right)$$

We are interested in a new associative structure on B. All such associative structure provides a deformation B_τ of the algebra B.

DEFINITION 2.10. A *formal* deformation (B_τ, μ_τ) of B (also called a deformation of A over $k[[\tau]]$) is an associative bilinear multiplication $\mu_\tau : B[[\tau]] \otimes B[[\tau]] \to B[[\tau]]$ such that in the quotient by the ideal (τ), the multiplication $\mu_\tau(b_1, b_2)$ coincides with the multiplication in B for all $b_1, b_2 \in B[[\tau]]$.

The multiplication μ_τ above is determined by products of pairs of elements of B, so that for every $a, b \in B$

$$\mu_\tau(a, b) = ab + \mu_1(a, b)\tau + \mu_2(a, b)\tau^2 + \mu_3(a, b)\tau^3 + \cdots \tag{2.10}$$

where ab is the usual multiplication in B and $\mu_i : B \otimes_k B \to B$ are bilinear maps. If we denote the usual multiplication in B by μ, we may denote a deformation of (B, μ) by (B_τ, μ_τ) where

$$\mu_\tau = \mu + \mu_1\tau + \mu_2\tau^2 + \mu_3\tau^3 + \cdots \tag{2.11}$$

REMARK 2.11. The *first order term* i.e. the bilinear map μ_1 is called an *infinitesimal deformation* if it is a Hochschild 2-cocycle. Furthermore, if μ_1 is an infinitesimal deformation, it defines a deformation of B over $k[\tau]/(\tau^2)$ and satisfies

$$\mu_1(ab, c) + \mu_1(a, b)c = \mu_1(a, bc) + a\mu_1(b, c).$$

This equation is derived from the associative property that the bilinear map μ_τ.

While formal deformations are deformations over $k[[\tau]]$, algebraic deformations are deformations over $k[\tau]$. The idea of making a formal deformation into an algebraic deformation is called the *algebraization of formal deformations* and were examined in details with respect to reduction systems by S. Barmeier and Z. Wang in [**1**]. One of their main results is the following theorem.

THEOREM 2.12. [**1**, Thm 7.1] *Given any finite quiver Q and any two-sided ideal of relations I, let $\Lambda = kQ/I$ be the quotient algebra and let R be any reduction system satisfying the diamond condition ($\diamond$) for I. There is an equivalence of formal deformation problems between*

(i) *deformations of the associative algebra structure on Λ,*
(ii) *deformations of the reduction system R,*
(iii) *deformations of the relations I.*

This equivalence can be made explicit by considering combinatorial star products which produce a deformation of Λ from a deformation of the reduction system R. In [**1**, Section 5], it was established that there are comparison morphisms $F_\bullet, G_\bullet$ between the bar resolution $\mathbb{B}$ and the resolution $\mathfrak{p}(Q, R)$,

$$\mathfrak{p}_\bullet \xrightarrow{F_\bullet} \mathbb{B}_\bullet \xrightarrow{G_\bullet} \mathfrak{p}_\bullet$$

coming from reduction systems and the combinatorial star product can be defined in terms of these morphisms.

Given a reduction system $R = \{(s, \varphi_s)\}$ for the algebra $B = kQ/I$ determined by S as in Definition 2.6, we view $\varphi \in \mathrm{Hom}(kS, kIrr_S)$ with $\varphi(s) = \varphi_s$. There is an isomorphism $B \cong kIrr_S$ so that $\mathrm{Hom}(kS, kIrr_S) \cong \mathrm{Hom}(kS, B)$ and a Hochschild 2-cochain in $\mathrm{Hom}(B \otimes_k B, B)$ may be viewed as an element $\varphi \in \mathrm{Hom}(kS, B)$. Taking $\mathfrak{m} = (\tau)$ as the maximal ideal of $k[[\tau]]$, it was shown that the map $\tilde{\varphi} \in \mathrm{Hom}(kS, B) \otimes \mathfrak{m}$ given by

$$\widetilde{\varphi} = \widetilde{\varphi}_1 t + \widetilde{\varphi}_2 t^2 + \widetilde{\varphi}_3 t^3 + \cdots \tag{2.12}$$

are candidates for deformations of R determined by φ. Moreover, the deformation of R given by $\varphi + \widetilde{\varphi}$ is a deformation of the algebra B if and only if $\widetilde{\varphi}$ is a Maurer-Cartan element of the L_∞ algebra structure on the resolution $\mathfrak{p}(Q, R) \otimes \mathfrak{m}$. More precisely, if $uvw \in S_3$ are overlaps such that $uv, vw \in S$, then $\widetilde{\varphi}$ satisfies the Maurer-Cartan equation if and only if

$$(\pi(u) \star \pi(v)) \star \pi(w) = \pi(u) \star (\pi(v) \star \pi(w)). \tag{2.13}$$

We recall that the product $\star$ defined on irreducible paths u, v is the image of $\pi(u) \star \pi(v)$ in Λ under the map $\pi : kQ \to \Lambda$ after performing right most reductions on the path $uv \in kQ$ using the reduction system. Indeed, the combinatorial star product $\star$ defines an associative structure on the algebra (B_τ, μ_τ) and we can also write $a \star b = ab + \mu_1(a, b)$.

In Sections 5, we explicitly find Maurer-Cartan elements of $\mathrm{HH}^2(A_1)$ by first solving for Hochschild 2-cocycles μ_1 of Equation (2.11) and then showing that they satisfy the equation $d^*\mu + \frac{1}{2}[\mu, \mu] = 0$ using homotopy lifting maps to define the bracket. The combinatorial star product solves the Maurer-Cartan equation by construction. We check in Section 6 that for the same algebra A_1 and a choice reduction system R, the maps $\tilde{\varphi}$ of Equation (2.12) describing the combinatorial star product solves the Maurer-Cartan equation given previously in terms of the homotopy lifting maps in Section 5.

3. Main Results

In what follows, we consider a finite quiver Q and a quiver algebra $\Lambda = kQ/I$ that is Koszul i.e. I is an admissible ideal generated by paths of length 2. We assume the quiver Q has arrows labelled $f_1^1, f_2^1, \dots, f_{t_1}^1$ for some integer t_1. We suppose further that for each n, there are uniform elements $f_1^n, f_2^n, \dots, f_{t_n}^n$, for some integer t_n defining a minimal projective resolution $\mathbb{K}$ of Λ as given by Theorem (2.5). Let $\eta : \mathbb{K}_n \to \Lambda$ be a Hochschild cocycle such that for some index i, $\eta(\varepsilon_i^n) = f_w^1$ (resp. $\eta(\varepsilon_i^n) = f_w^1 f_{w'}^1$) with $0 \leq w, w' \leq t_1$ and $\eta(\varepsilon_j^n) = 0$ for all $i \neq j$. We also write $\eta = \begin{pmatrix} 0 & \cdots & 0 & (f_w^1)^{(i)} & 0 & \cdots & 0 \end{pmatrix}$ for this type of cocycle (resp. $\eta = \begin{pmatrix} 0 & \cdots & 0 & (f_w^1 f_{w'}^1)^{(i)} & 0 & \cdots & 0 \end{pmatrix}$). Let $\Delta_{\mathbb{K}} : \mathbb{K} \to \mathbb{K} \otimes_\Lambda \mathbb{K}$ be the diagonal map. Results from **[15, 16]** establish that there exist maps $\psi_\eta : \mathbb{K} \to \mathbb{K}[1-n]$ such that

$$d\psi_\eta - (-1)^{1-n}\psi_\eta d = (\eta \otimes 1 - 1 \otimes \eta)\Delta_{\mathbb{K}} \tag{3.1}$$

for Koszul algebras. These maps are called *homotopy lifting* maps for η. How would we define such a map explicitly in terms of the basis elements ε_r^n? Can we give a closed formula or expression of the Gerstenhaber bracket using an explicitly described versions of these maps? These are among the questions we address in this section.

In order to distinguish the index n which is the degree of the cocycle η, we will index the resolution $\mathbb{K}$ by m so that each $\mathbb{K}_m$ is free and generated by $\{\varepsilon_r^m\}_{r=0}^{t_m}$. For an n-cocycle, the map $\psi_\eta : \mathbb{K}_\bullet \to \mathbb{K}_\bullet$ associated to η is of degree $1-n$ so that for a fixed m, $\psi_\eta : \mathbb{K}_m \to \mathbb{K}_{m-n+1}$. Suppose that $\mathbb{K}_{m-n+1}$ is generated by $\{\varepsilon_{r'}^{m-n+1}\}_{r'=0}^{t_{m-n+1}}$, fundamental results from linear algebra means such a map is a $t_{m-n+1} \times t_m$ matrix when the modules are considered as left Λ^e-modules. In particular, for $\eta = \begin{pmatrix} 0 & \cdots & 0 & (f_w^1)^{(i)} & 0 & \cdots & 0 \end{pmatrix}$, such a map is defined on ε_r^m by

$$\psi_\eta(\varepsilon_r^m) = \sum_{j=0}^{t_{m-n+1}} \lambda_j(m,r)\varepsilon_j^{m-n+1} \tag{3.2}$$

and for $\eta = \begin{pmatrix} 0 & \cdots & 0 & (f_w^1 f_{w'}^1)^{(i)} & 0 & \cdots & 0 \end{pmatrix}$, such a map is defined on ε_r^m by

$$\psi_\eta(\varepsilon_r^m) = \sum_{j=0}^{t_{m-n+1}} \lambda_j(m,r) f_w^1 \varepsilon_j^{m-n+1} + \lambda'_j(m,r)\varepsilon_j^{m-n+1} f_{w'}^1 \tag{3.3}$$

where $\lambda_j(m,r), \lambda'_j(m,r) \in \Lambda^e$ in general. For details about these maps, see **[13]**. We now restrict Equation (3.2) to the special case where for some $j = r'$, $\lambda_j(m,r) = b_{m,r}(m-n+1, r')$ is a scalar and $\lambda_j(m,r) = 0$ for all $j \neq r'$, that is

$$\psi_\eta(\varepsilon_r^m) = b_{m,r}(m-n+1, r')\varepsilon_{r'}^{m-n+1} \tag{3.4}$$

and restrict Equation (3.3) to the special case where all $\lambda_j(m,r), \lambda'_j(m,r)$ are zero except for some indices s and s' with $\lambda_j(m,r) = \lambda_{m,r}(m-n+1, s) \neq 0$ and $\lambda'_j(m,r) = \lambda_{m,r}(m-n+1, s') \neq 0$. That is,

$$\psi_\eta(\varepsilon_r^m) = \lambda_{m,r}(m-n+1, s) f_w^1 \varepsilon_s^{m-n+1} + \lambda_{m,r}(m-n+1, s')\varepsilon_{s'}^{m-n+1} f_{w'}^1. \tag{3.5}$$

It was shown in [**13**] that Equation (3.1) holds under certain conditions on the scalars $b_{m,r}(m-n+1,r')$, and therefore the special maps given by Equations (3.4) are indeed homotopy lifting maps for the associated cocycles. A similar argument holds for the map given by Equation (3.5).

Our motivation for defining the maps the way they were defined comes from several examples that were computed. We note in particular that the scalars $b_{m,r}(m-n+1,r')$ satisfy some recurrence relations given by Equation (3.6). We observe that if $\psi_\eta(\varepsilon_{\bar{r}}^{m-1}) = b_{m-1,\bar{r}}(m-n,r'')\varepsilon_{r''}^{m-n}$, we can obtain the scalars $b_{m,r}(m-n+1,r')$ in terms of the scalars $b_{m-1,\bar{r}}(m-n,r'')$ and the scalars $c_{pq}(n,i,r)$ coming from the comultiplicative structure on $\mathbb{K}$. The following diagram is not commutative but gives a pictures of this idea:

$$\begin{array}{ccccccccc}
\mathbb{K} := \cdots & \longrightarrow & \mathbb{K}_{m+1} & \longrightarrow & \mathbb{K}_m & \longrightarrow & \mathbb{K}_{m-1} & \longrightarrow & \cdots \\
\downarrow{\scriptstyle \psi_{\bar\eta}} & & \downarrow{\scriptstyle \psi_{\eta}} & & \downarrow{\scriptstyle \psi_{\bar\eta}} & & \downarrow{\scriptstyle \psi_{\bar\eta}} & & \\
\mathbb{K}[1-n] := \cdots & \longrightarrow & \mathbb{K}_{m-n+2} & \longrightarrow & \mathbb{K}_{m-n+1} & \longrightarrow & \mathbb{K}_{m-n} & \longrightarrow & \cdots
\end{array}$$

• We can obtain $\psi_{\bar\eta}|_{\mathbb{K}_{m+1}}$ from $\psi_{\bar\eta}|_{\mathbb{K}_m}$ for every m using the scalars $b_{m,r}(m-n+1,r^*)$ and $c_{pq}(n,i,r)$. From Remark 2.3, the scalars $b_{n-1,r^*}(0,r^{**}) = 0$ for all r^*, r^{**} since $\psi_{\bar\eta}|_{\mathbb{K}_{n-1}}$ is the zero map.

For a finite quiver Q, let $\Lambda = kQ/I$ be a quiver algebra that is Koszul and let $\mathbb{K}_m$ be the projective bimodule resolution of Λ with basis $\{\varepsilon_r^m\}_{r=0}^{t_m}$ as given by Theorem 2.5. For the specific modules $\mathbb{K}_{m-n+1}$, $\mathbb{K}_{m-n}$, and $\mathbb{K}_{m-1}$, let the basis elements be $\{\varepsilon_{r'}^{m-n+1}\}_{r'=0}^{t_{m-n+1}}$, $\{\varepsilon_{r''}^{m-n}\}_{r''=0}^{t_{m-n}}$, and $\{\varepsilon_{\bar r}^{m-1}\}_{\bar r=0}^{t_{m-1}}$ respectively. For the indices $r,\bar r, r'', r'$, let the scalars $b_{m,r}(*,**)$ be such that the following recurrence relations hold;

$$\begin{aligned}
(3.6)\qquad & (i)\ b_{m,r}(m-n+1,r')c_{rr''}(m-n+1,r',1)\\
&= (-1)^{m-1}b_{m-1,\bar r}(m-n,r'')c_{r\bar r}(m,r,1) + c_{ir''}(m,r,n),\\
& b_{m,r}(m-n+1,r')c_{r''r}(m-n+1,r',m-n)\\
&= (-1)^{m-1}b_{m-1,\bar r}(m-n,r'')c_{\bar r r}(m,r,m-1)\\
(3.7)\qquad & -(-1)^{n(m-n)}c_{r''i}(m,r,m-n),
\end{aligned}$$

and for every pair of indices $(p,q)\neq(r,r'),(p,q)\neq(r'',r)$,

$$\begin{aligned}
& (ii)\ b_{m,r}(m-n+1,r')c_{pq}(m-n+1,r',*)\\
&= (-1)^{m-1}b_{m-1,\bar *}(m-n,*'')c_{pq}(m,r,*).
\end{aligned}$$

We start with the following Lemma.

LEMMA 3.1. *Let Q be a finite quiver and $\Lambda = kQ/I$ a quiver algebra that is Koszul. Suppose that $\eta : \mathbb{K}_n \to \Lambda$ is a cocycle such that*

$$\eta = \begin{pmatrix} 0 & \cdots & 0 & (f_w^1)^{(i)} & 0 & \cdots & 0 \end{pmatrix}, \qquad 0 \le w \le t_1.$$

If $\psi_\eta : \mathbb{K} \to \mathbb{K}[1-n]$ is defined as $\psi_\eta(\varepsilon_r^m) = b_{m,r}(m-n+1,r')\varepsilon_{r'}^{m-n+1}$ and the recursion equation (3.6) hold, then $d\psi_\eta - (-1)^{1-n}\psi_\eta d = (\eta\otimes 1 - 1\otimes\eta)\Delta_{\mathbb{K}}$.

PROOF. We prove this result in the following way; for the free modules $\mathbb{K}_{m-1}, \mathbb{K}_m, \mathbb{K}_{m-n}, \mathbb{K}_{m-n+1}$, we define the special case of the map ψ_η given by Equation (3.4). We then use the left and right hand side of (3.1) to derive the recurrence relations. This is equivalent to saying that under these conditions, equation (3.1) holds provided the recurrence relations of (3.6) hold.

Let us suppose we have a quiver Q generated by two arrows $\{f_r^1, f_s^1\}$ and each $\mathbb{K}_n$ is free of rank 2. For each m let $\{\varepsilon_{\bar{r}}^{m-1}, \varepsilon_{\bar{s}}^{m-1}\}$, $\{\varepsilon_r^m, \varepsilon_s^m\}$, $\{\varepsilon_{r'}^{m-n+1}, \varepsilon_{s'}^{m-n+1}\}$, and $\{\varepsilon_{r''}^{m-n}, \varepsilon_{s''}^{m-n}\}$ be a basis for $\mathbb{K}_{m-1}$, $\mathbb{K}_m$, $\mathbb{K}_{m-n+1}$, and $\mathbb{K}_{m-n}$ respectively. A possible example of this scenario is given in Example (4.3). The differential given by (2.6) on ε_r^m for this special case for instance, is given by

$$\begin{aligned}
d(\varepsilon_r^m) &= c_{r\bar{r}}(m,r,1)f_r^1\varepsilon_{\bar{r}}^{m-1} + c_{\bar{r}r}(m,r,m-1)\varepsilon_{\bar{r}}^{m-1}f_r^1\\
&+ c_{r\bar{s}}(m,r,1)f_r^1\varepsilon_{\bar{s}}^{m-1} + c_{\bar{r}s}(m,r,m-1)\varepsilon_{\bar{r}}^{m-1}f_s^1\\
&+ c_{s\bar{r}}(m,r,1)f_s^1\varepsilon_{\bar{r}}^{m-1} + c_{\bar{s}r}(m,r,m-1)\varepsilon_{\bar{s}}^{m-1}f_r^1\\
&+ c_{s\bar{s}}(m,r,1)f_s^1\varepsilon_{\bar{s}}^{m-1} + c_{\bar{s}s}(m,r,m-1)\varepsilon_{\bar{s}}^{m-1}f_s^1
\end{aligned}$$

and a similar formula can be written for $d(\varepsilon_s^m)$.

Let us recall that $\eta = \begin{pmatrix}0 & \cdots & 0 & (f_w^1)^{(i)} & 0 & \cdots & 0\end{pmatrix}$ means that $\eta(\varepsilon_i^n) = f_w^1$ with $w = r$ or $w = s$ and $\eta(\varepsilon_j^n) = 0$ for all $j \neq i$. From the hypothesis, we define $\psi_\eta : \mathbb{K}_m \to \mathbb{K}_{m-n+1}$ by $\psi_\eta(\varepsilon_r^m) = b_{m,r}(m-n+1,r')\varepsilon_{r'}^{m-n+1}$, and $\psi_\eta(\varepsilon_s^m) = b_{m,s}(m-n+1,s')\varepsilon_{s'}^{m-n+1}$, and $\psi_\eta : \mathbb{K}_{m-1} \to \mathbb{K}_{m-n}$ is defined by $\psi_\eta(\varepsilon_{\bar{r}}^{m-1}) = b_{m-1,\bar{r}}(m-n,r'')\varepsilon_{r''}^{m-n}$, and $\psi_\eta(\varepsilon_{\bar{s}}^{m-1}) = b_{m-1,\bar{s}}(m-n,s'')\varepsilon_{s''}^{m-n}$.

Using Equation(3.1), the expression $(d\psi_\eta - (-1)^{m-1}\psi_\eta d)(\varepsilon_r^m)$ becomes

$$d(b_{m,r}(m-n+1,r')\varepsilon_{r'}^{m-n+1}) - (-1)^{m-1}\psi_\eta d(\varepsilon_r^m)$$

and is equal to

$$\begin{aligned}
&b_{m,r}(m-n+1,r')\Big(c_{rr''}(m-n+1,r',1)f_r^1\varepsilon_{r''}^{m-n}\\
&\quad+ c_{r''r}(m-n+1,r',m-n)\varepsilon_{r''}^{m-n}f_r^1 + c_{rs''}(m-n+1,r',1)f_r^1\varepsilon_{s''}^{m-n}\\
&\quad+ c_{r''s}(m-n+1,r',m-n)\varepsilon_{r''}^{m-n}f_s^1 + c_{sr''}(m-n+1,r',1)f_s^1\varepsilon_{r''}^{m-n}\\
&\quad+ c_{s''r}(m-n+1,r',m-n)\varepsilon_{s''}^{m-n}f_r^1 + c_{ss''}(m-n+1,r',1)f_s^1\varepsilon_{s''}^{m-n}\\
&\quad+ c_{s''s}(m-n+1,r',m-n)\varepsilon_{s''}^{m-n}f_s^1\Big)\\
&\quad- (-1)^{m-1}\Big(b_{m-1,\bar{r}}(m-n,r'')c_{r\bar{r}}(m,r,1)f_r^1\varepsilon_{r''}^{m-n}\\
&\quad+ b_{m-1,\bar{r}}(m-n,r'')c_{\bar{r}r}(m,r,m-1)\varepsilon_{r''}^{m-n}f_r^1\\
&\quad+ b_{m-1,\bar{s}}(m-n,s'')c_{r\bar{s}}(m,r,1)f_r^1\varepsilon_{s''}^{m-n}\\
&\quad+ b_{m-1,\bar{r}}(m-n,r'')c_{\bar{r}s}(m,r,m-1)\varepsilon_{r''}^{m-n}f_s^1\\
&\quad+ b_{m-1,\bar{r}}(m-n,r'')c_{s\bar{r}}(m,r,1)f_s^1\varepsilon_{r''}^{m-n}\\
&\quad+ b_{m-1,\bar{s}}(m-n,s'')c_{\bar{s}r}(m,r,m-1)\varepsilon_{s''}^{m-n}f_r^1\\
&\quad+ b_{m-1,\bar{s}}(m-n,s'')c_{s\bar{s}}(m,r,1)f_s^1\varepsilon_{s''}^{m-n}\\
&\quad+ b_{m-1,\bar{s}}(m-n,s'')c_{\bar{s}s}(m,r,m-1)\varepsilon_{s''}^{m-n}f_s^1\Big).
\end{aligned}$$

On the other hand, the diagonal map is given by
$\Delta_{\mathbb{K}}(\varepsilon_r^m) = \sum_{x+y=r}\sum_{u+v=m} c_{x,y}(m,r,u)\varepsilon_x^u \otimes_\Lambda \varepsilon_y^v$. We obtain a non-zero in the expansion of $(\eta\otimes 1 - 1\otimes\eta)\Delta_{\mathbb{K}}(\varepsilon_r^m)$ whenever $x = i$ and $y = i$. This means that

$$\begin{aligned}
&(\eta\otimes 1 - 1\otimes\eta)\sum_{x+y=r}\sum_{u+v=m} c_{x,y}(m,r,u)\varepsilon_x^u \otimes_\Lambda \varepsilon_y^v\\
&= (\eta\otimes 1)(c_{i,y}(m,r,n)\varepsilon_i^n \otimes_\Lambda \varepsilon_y^{m-n}) - (1\otimes\eta)(c_{x,i}(m,r,m-n)\varepsilon_x^{m-n}\otimes_\Lambda \varepsilon_i^n)\\
&= c_{i,y}(m,r,n)f_w^1\varepsilon_y^{m-n} - (-1)^{n(m-n)}c_{xi}(m,r,m-n)\varepsilon_x^{m-n}f_w^1,
\end{aligned}$$

for $\{x,y\} = \{r'',s''\}$ with $i+y=r, x+i=r$ and some arrow f_w^1. After collecting common terms, the expression $(d\psi_\eta - (-1)^{m-1}\psi_\eta d)(\varepsilon_r^m)$ which is the left hand side of Equation (3.1) becomes

$$\begin{aligned}
&\Big(b_{m,r}(m-n+1,r')c_{rr''}(m-n+1,r',1)\\
&\quad - (-1)^{m-1}b_{m-1,\bar r}(m-n,r'')c_{r\bar r}(m,r,1)\Big)f_r^1\varepsilon_{r''}^{m-n}\\
&\quad + \Big(b_{m,r}(m-n+1,r')c_{r''r}(m-n+1,r',m-n)\\
&\quad - (-1)^{m-1}b_{m-1,\bar r}(m-n,r'')c_{\bar r r}(m,r,m-1)\Big)\varepsilon_{r''}^{m-n}f_r^1\\
&\quad + \Big(b_{m,r}(m-n+1,r')c_{rs''}(m-n+1,r',1)\\
&\quad - (-1)^{m-1}b_{m-1,\bar s}(m-n,s'')c_{r\bar s}(m,r,1)\Big)f_r^1\varepsilon_{s''}^{m-n}\\
&\quad + \Big(b_{m,r}(m-n+1,r')c_{r''s}(m-n+1,r',m-n)\\
&\quad - (-1)^{m-1}b_{m-1,\bar r}(m-n,r'')c_{\bar r s}(m,r,m-1)\Big)\varepsilon_{r''}^{m-n}f_s^1\\
&\quad + \Big(b_{m,r}(m-n+1,r')c_{sr''}(m-n+1,r',1)\\
&\quad - (-1)^{m-1}b_{m-1,\bar r}(m-n,r'')c_{s\bar r}(m,r,1)\Big)f_s^1\varepsilon_{r''}^{m-n}\\
&\quad + \Big(b_{m,r}(m-n+1,r')c_{s''r}(m-n+1,r',m-n)\\
&\quad - (-1)^{m-1}b_{m-1,\bar s}(m-n,s'')c_{\bar s r}(m,r,m-1)\Big)\varepsilon_{s''}^{m-n}f_r^1\\
&\quad + \Big(b_{m,r}(m-n+1,r')c_{ss''}(m-n+1,r',1)\\
&\quad - (-1)^{m-1}b_{m-1,\bar s}(m-n,s'')c_{s\bar s}(m,r,1)\Big)f_s^1\varepsilon_{s''}^{m-n}\\
&\quad + \Big(b_{m,r}(m-n+1,r')c_{s''s}(m-n+1,r',m-n)\\
&\quad - (-1)^{m-1}b_{m-1,\bar s}(m-n,s'')c_{\bar s s}(m,r,m-1)\Big)\varepsilon_{s''}^{m-n}f_s^1.
\end{aligned}$$

The expression $(\eta\otimes 1 - 1\otimes\eta)\Delta_{\mathbb{K}}(\varepsilon_r^m)$, which is the right hand side of Equation (3.1), still remains as

$$c_{i,y}(m,r,n)f_w^1\varepsilon_y^{m-n} - (-1)^{n(m-n)}c_{xi}(m,r,m-n)\varepsilon_x^{m-n}f_w^1.$$

We observe the following about indices w, x, y. The index w is either r or s, the index y is either r'' or s'' and the index x is either r'' or s''. We notice that the

additional constraint that $i+y=r$ and $i+x=r$ implies that whenever $y=r''$ we must have $x=r''$ and whenever $y=s''$, we must have $x=s''$. We therefore have the following four cases:
Case I: Whenever $w=r$, $y=x=r''$, $i+r''=r$, we have the following set of *recurrence* relations on the scalars $b_{m,r}(m-n+1,r')$,

$$\begin{aligned}
&b_{m,r}(m-n+1,r')c_{rr''}(m-n+1,r',1)\\
&\quad-(-1)^{m-1}b_{m-1,\bar{r}}(m-n,r'')c_{r\bar{r}}(m,r,1)=c_{ir''}(m,r,n)\\
&b_{m,r}(m-n+1,r')c_{r''r}(m-n+1,r',m-n)\\
&\quad-(-1)^{m-1}b_{m-1,\bar{r}}(m-n,r'')c_{\bar{r}r}(m,r,m-1)\\
&\qquad=-(-1)^{n(m-n)}c_{r''i}(m,r,m-n)\\
&b_{m,r}(m-n+1,r')c_{rs''}(m-n+1,r',1)\\
&\quad-(-1)^{m-1}b_{m-1,\bar{s}}(m-n,s'')c_{r\bar{s}}(m,r,1)=0
\end{aligned}$$

$$\begin{aligned}
&b_{m,r}(m-n+1,r')c_{r''s}(m-n+1,r',m-n)\\
&\quad-(-1)^{m-1}b_{m-1,\bar{r}}(m-n,r'')c_{\bar{r}s}(m,r,m-1)=0\\
&b_{m,r}(m-n+1,r')c_{sr''}(m-n+1,r',1)\\
&\quad-(-1)^{m-1}b_{m-1,\bar{r}}(m-n,r'')c_{s\bar{r}}(m,r,1)=0\\
&b_{m,r}(m-n+1,r')c_{s''r}(m-n+1,r',m-n)\\
&\quad-(-1)^{m-1}b_{m-1,\bar{s}}(m-n,s'')c_{\bar{s}r}(m,r,m-1)=0\\
&b_{m,r}(m-n+1,r')c_{ss''}(m-n+1,r',1)\\
&\quad-(-1)^{m-1}b_{m-1,\bar{s}}(m-n,s'')c_{s\bar{s}}(m,r,1)=0\\
&b_{m,r}(m-n+1,r')c_{s''s}(m-n+1,r',m-n)\\
&\quad-(-1)^{m-1}b_{m-1,\bar{s}}(m-n,s'')c_{\bar{s}s}(m,r,m-1)=0.
\end{aligned}$$

We note that all the equations of Case (I) above can be expressed more succinctly to mean that whenever $w=r$, $i+r''=r$ and for all $s\neq r$

$$\begin{aligned}
&b_{m,r}(m-n+1,r')c_{rr''}(m-n+1,r',1)\\
&\quad=(-1)^{m-1}b_{m-1,\bar{r}}(m-n,r'')c_{r\bar{r}}(m,r,1)+c_{ir''}(m,r,n),\\
&b_{m,r}(m-n+1,r')c_{r''r}(m-n+1,r',m-n)\\
&\quad=(-1)^{m-1}b_{m-1,\bar{r}}(m-n,r'')c_{\bar{r}r}(m,r,m-1)\\
&\qquad-(-1)^{n(m-n)}c_{r''i}(m,r,m-n),
\end{aligned}$$

and for every indices such that $(p,q)\neq(r,r'')$ and $(p,q)\neq(r'',r)$, $b_{m,r}(m-n+1,r')c_{pq}(m-n+1,r',*)=(-1)^{m-1}b_{m-1,\bar{*}}(m-n,*'')c_{pq}(m,r,*)$.
Case II: Whenever $w=s$, $y=x=r''$, $i+r''=r$, we have the following set of

recurrence relations on the scalars $b_{m,r}(m-n+1,r')$,

$$\begin{aligned}
&b_{m,r}(m-n+1,r')c_{sr''}(m-n+1,r',1)\\
&\quad=(-1)^{m-1}b_{m-1,\bar{r}}(m-n,r'')c_{r\bar{r}}(m,r,1)+c_{ir''}(m,r,n),\\
&b_{m,r}(m-n+1,r')c_{r''s}(m-n+1,r',m-n)\\
&\quad=(-1)^{m-1}b_{m-1,\bar{r}}(m-n,r'')c_{\bar{r}r}(m,r,m-1)\\
&\qquad-(-1)^{n(m-n)}c_{r''i}(m,r,m-n),
\end{aligned}$$

and for every indices $(p,q)\neq(s,r''),(p,q)\neq(r'',s)$,
$b_{m,r}(m-n+1,r')c_{pq}(m-n+1,r',*)=(-1)^{m-1}b_{m-1,\bar{*}}(m-n,*'')c_{pq}(m,r,*)$.
Case III: Whenever $w=r$, $y=x=s''$, $i+s''=r$, we have the following set of *recurrence* relations on the scalars $b_{m,r}(m-n+1,r')$,

$$\begin{aligned}
&b_{m,r}(m-n+1,r')c_{rs''}(m-n+1,r',1)\\
&\quad=(-1)^{m-1}b_{m-1,\bar{s}}(m-n,s'')c_{r\bar{s}}(m,r,1)+c_{is''}(m,r,n),\\
&b_{m,r}(m-n+1,r')c_{s''r}(m-n+1,r',m-n)\\
&\quad=(-1)^{m-1}b_{m-1,\bar{s}}(m-n,s'')c_{\bar{s}r}(m,r,m-1)\\
&\qquad-(-1)^{n(m-n)}c_{s''i}(m,r,m-n),
\end{aligned}$$

and for every indices $(p,q)\neq(r,s''),(p,q)\neq(s'',r)$
$b_{m,r}(m-n+1,r')c_{pq}(m-n+1,r',*)=(-1)^{m-1}b_{m-1,\bar{*}}(m-n,*'')c_{pq}(m,r,*)$.

Case IV: Whenever $w=s$, $y=x=s''$, $i+s''=r$, we have the following set of *recurrence* relations on the scalars $b_{m,r}(m-n+1,r')$,

$$\begin{aligned}
&b_{m,r}(m-n+1,r')c_{ss''}(m-n+1,r',1)\\
&\quad=(-1)^{m-1}b_{m-1,\bar{s}}(m-n,s'')c_{s\bar{s}}(m,r,1)+c_{is''}(m,r,n),\\
&b_{m,r}(m-n+1,r')c_{s''s}(m-n+1,r',m-n)\\
&\quad=(-1)^{m-1}b_{m-1,\bar{s}}(m-n,s'')c_{\bar{s}s}(m,r,m-1)\\
&\qquad-(-1)^{n(m-n)}c_{s''i}(m,r,m-n),
\end{aligned}$$

and for every indices $(p,q)\neq(s,s''),(p,q)\neq(s'',s)$
$b_{m,r}(m-n+1,r')c_{pq}(m-n+1,r',*)=(-1)^{m-1}b_{m-1,\bar{*}}(m-n,*'')c_{pq}(m,r,*)$.

More generally, if $\mathbb{K}_m$ is generated by $\{\varepsilon_r^m\}_{r=1}^{t_m}$, $\mathbb{K}_{m-1}$ generated by $\{\varepsilon_{\bar{r}}^{m-1}\}_{\bar{r}=1}^{t_{m-1}}$, $\mathbb{K}_{m-n}$ by $\{\varepsilon_{r''}^{m-n}\}_{r''=1}^{t_{m-n}}$, and $\mathbb{K}_{m-n+1}$ by $\{\varepsilon_{r'}^{m-n+1}\}_{r'=1}^{t_{m-n+1}}$, the following relations hold for all r,r',r'' and $\bar{r}$

$$\begin{aligned}
(i)\ &b_{m,r}(m-n+1,r')c_{rr''}(m-n+1,r',1)\\
&\quad=(-1)^{m-1}b_{m-1,\bar{r}}(m-n,r'')c_{r\bar{r}}(m,r,1)+c_{ir''}(m,r,n),\\
&b_{m,r}(m-n+1,r')c_{r''r}(m-n+1,r',m-n)\\
&\quad=(-1)^{m-1}b_{m-1,\bar{r}}(m-n,r'')c_{\bar{r}r}(m,r,m-1)\\
&\qquad-(-1)^{n(m-n)}c_{r''i}(m,r,m-n),
\end{aligned}$$

and for every pair of indices $(p,q) \neq (r,r'')$, $(p,q) \neq (r'',r)$,

$$(ii)\ b_{m,r}(m-n+1,r')c_{pq}(m-n+1,r',*) \\ = (-1)^{m-1}b_{m-1,\bar{*}}(m-n,*'')c_{pq}(m,r,*).$$

This concludes the proof. □

THEOREM 3.2. *Let Q be a finite quiver and $\Lambda = kQ/I$ a quiver algebra that is Koszul. Suppose that $\eta : \mathbb{K}_n \to \Lambda$ is a cocycle such that $\eta = \begin{pmatrix} 0 & \cdots & 0 & (f_w^1)^{(i)} & 0 & \cdots & 0 \end{pmatrix}$, $0 \leq w \leq t_1$ and f_w^1 is path of length 1. There are scalars $\lambda_{m,r}(m-n+1,r')$ such that the map $\psi_\eta : \mathbb{K} \to \mathbb{K}[1-n]$ associated to η and defined in degree m by*

$$\psi_\eta(\varepsilon_r^m) = \lambda_{m,r}(m-n+1,r')\varepsilon_{r'}^{m-n+1}$$

is a homotopy lifting map for η.

PROOF. Let $\lambda_{m,r}(m-n+1,r') = b_{m,r}(m-n+1,r')$ be the scalars satisfying the recurrence relations of (3.6). By Lemma 3.1, the equation $d\psi_\eta - (-1)^{1-n}\psi_\eta d = (\eta \otimes 1 - 1 \otimes \eta)\Delta_{\mathbb{K}}$ holds, so ψ_η is a homotopy lifting map. □

THEOREM 3.3. *Let Q be a finite quiver and let $\Lambda = kQ/I$ be a quiver algebra that is Koszul. Suppose that $\eta : \mathbb{K}_n \to \Lambda$ is a cocycle such that $\eta = \begin{pmatrix} 0 & \cdots & 0 & (f_w^1 f_{w'}^1)^{(i)} & 0 & \cdots & 0 \end{pmatrix}$ for some $0 \leq w, w' \leq t_1$ where f_w^1 and $f_{w'}^1$ are paths of length 1. Then there exist scalars $\lambda_{m,r}(m-n+1,s)$ and $\lambda_{m,r}(m-n+1,s')$ such that $\psi_\eta : \mathbb{K}_m \to \mathbb{K}_{m-n+1}$ defined by*

$$\psi_\eta(\varepsilon_r^m) = \lambda_{m,r}(m-n+1,s)f_w^1\varepsilon_s^{m-n+1} + \lambda_{m,r}(m-n+1,s')\varepsilon_{s'}^{m-n+1}f_{w'}^1$$

for all ε_r^m is a homotopy lifting map for η.

PROOF. Similar to Lemma 3.1, we can write a recurrence relations on the scalars given in Equation (3.5) so that

$$d\psi_\eta - (-1)^{1-n}\psi_\eta d = (\eta \otimes 1 - 1 \otimes \eta)\Delta_{\mathbb{K}}$$

holds true. See [**14**, Lemma 5.20] and [**14**, Theorem 5.23] for details about this. □

3.1. A special case of Theorem 3.2. We consider a special case in which each Λ^e-module $\mathbb{K}_n$ is generated by one element. This case arises for example, from a quiver with one arrow (a loop) on a vertex e_1. We also give a concrete example in Example (4.1). Let $I = (x^n)$ be an ideal of the path algebra kQ. The quiver algebra of interest here is not only Morita equivalent to the truncated polynomial ring $A = k[x]/(x^n)$, they are isomorphic. This is the case where $f_r^n = x^n$ where $r = 0$ for all n and $\varepsilon_r^n = 1 \otimes \widetilde{f_r^n} \otimes 1$. From the Preliminaries (2), there are scalars $c_{p,q}(m,r,u)$ for which the diagonal map is given by

$$\Delta_{\mathbb{K}}(\varepsilon_r^m) = \sum_{u+v=m} c_{i,j}(m,r,u)\varepsilon_i^u \otimes_\Lambda \varepsilon_j^v, \tag{3.8}$$

with $i = j = r = 0$. Also with $p = r = 0$, the differential takes the form

$$d(\varepsilon_r^m) = c_{p,r}(m,r,1)f_p^1\varepsilon_r^{m-1} + (-1)^m c_{r,p}(m,r,m-1)\varepsilon_r^{m-1}f_p^1.$$

Let $\chi : \mathbb{K}_n \to A$ be an n-cocycle defined by $\chi(\varepsilon_r^n) = f_r^1$. According to Theorem (3.2), a homotopy lifting map for χ can be given by

$$\psi_{\chi_m}(\varepsilon_r^m) = b_{m,r}(m-n+1,r)\varepsilon_r^{m-n+1}, \qquad r = 0.$$

We can determine $b_{m,r}(m-n+1,r)$ from the previous scalar $b_{m-1,r}(m-n,r)$. In order words, the conditions (i) and (ii) of Equation (3.6) is a *recurrence relation.* We know from Defintion (2.1) that homotopy lifting maps satisfy

$$\begin{aligned}&(d\psi_\chi-(-1)^{m-1}\psi_\chi d)(\varepsilon_r^m)=(\chi\otimes 1-1\otimes\chi)\Delta_{\mathbb{K}}(\varepsilon_r^m), \qquad \text{so then}\\ &d(b_{m,r}(m-n+1,r)\varepsilon_r^{m-n+1})-(-1)^{m-1}\psi_\chi\\ &\qquad\left(c_{p,r}(m,r,1)f_p^1\varepsilon_r^{m-1}+(-1)^m c_{r,p}(m,r,m-1)\varepsilon_r^{m-1}f_p^1\right)\\ &=(\chi\otimes 1-1\otimes\chi)\sum_{u+v=m}c_{i,j}(m,r,u)\varepsilon_i^u\otimes_\Lambda\varepsilon_j^v.\end{aligned}$$

The modules $\mathbb{K}$ are generated by one elements so we get

$$\begin{aligned}&b_{m,r}(m-n+1,r)c_{p,r}(m-n+1,r,1)f_p^1\varepsilon_r^{m-n}+(-1)^{m-n+1}b_{m,r}(m-n+1,r)\\ &c_{r,p}(m-n+1,r,m-n)\varepsilon_r^{m-n}f_p^1-(-1)^{m-1}b_{m-1,r}(m-n,r)c_{p,r}(m,r,1)f_p^1\varepsilon_r^{m-n}\\ &\qquad+b_{m-1,r}(m-n,r)c_{r,p}(m,r,m-1)\varepsilon_r^{m-n}f_p^1\\ &\qquad=c_{r,j}(m,r,n)f_r^1\varepsilon_j^{m-n}+(-1)^{n(m-n)}c_{i,r}(m,r,m-n)\varepsilon_i^{m-n}f_r^1.\end{aligned}$$

We would obtain the following expressions for the above equality to hold,

$$\begin{aligned}&b_{m,r}(m-n+1,r)c_{p,r}(m-n+1,r,1)-(-1)^{m-1}b_{m-1,r}(m-n,r)c_{p,r}(m,r,1)\\ &\qquad=c_{p,r}(m,r,n)\end{aligned}$$

and

$$\begin{aligned}&(-1)^{m-n+1}b_{m,r}(m-n+1,r)c_{r,p}(m-n+1,r,m-n)\\ &\qquad+b_{m-1,r}(m-n,r)c_{r,p}(m,r,m-1)\\ &\qquad=(-1)^{n(m-n)}c_{r,p}(m,r,m-n).\end{aligned}$$

The scalars $c_{p,r}(m-n+1,r,*)$ come from the differentials on the resolution $\mathbb{K}$, so they are not equal to 0 for all r. In case $c_{p,r}(m-n+1,r,*)\neq 0$ for all r, the first equality in the last expression yields

$$b_{m,r}(m-n+1,r)=\frac{(-1)^{m-1}b_{m-1,r}(m-n,r)c_{p,r}(m,r,1)+c_{p,r}(m,r,n)}{c_{p,r}(m-n+1,r,1)} \tag{3.9}$$

while the second one yields

$$\begin{aligned}&b_{m,r}(m-n+1,r)\\ &=\frac{b_{m-1,r}(m-n,r)c_{r,p}(m,r,m-1)+(-1)^{n(m-n)+1}c_{r,p}(m,r,m-n)}{(-1)^{m-n}c_{r,p}(m-n+1,r,m-n)}.\end{aligned} \tag{3.10}$$

We now present the Gerstenhaber bracket structure on Hochschild cohomology using these scalars.

THEOREM 3.4. *Let $\Lambda = kQ/I$ be a quiver algebra that is Koszul. Denote by $\{f_r^m\}_{r=0}^{t_m}$ elements of kQ defining a minimal projective resolution of Λ_0 as a right Λ-module. Let $\mathbb{K}$ be the projective bimodule resolution of Λ with $\mathbb{K}_m$ having basis $\{\varepsilon_r^m\}_{r=0}^{t_m}$. Assume that $\eta:\mathbb{K}_n\to\Lambda$ and $\theta:\mathbb{K}_m\to\Lambda$ represent elements in $\mathrm{HH}^*(\Lambda)$ and are given by $\eta(\varepsilon_i^n)=\lambda_i$ for $i=0,1,\ldots,t_n$ and $\theta(\varepsilon_j^m)=\beta_j$ for*

$j = 0, 1, \ldots, t_m$, with each λ_i and β_j paths of length of 1. Then the class of the bracket $[\eta, \theta] : \mathbb{K}_{n+m-1} \to \Lambda$ can be expressed on the r-th basis element ε_r^{m+n-1} as

$$[\eta, \theta](\varepsilon_r^{m+n-1}) = \sum_{i=0}^{t_n} \sum_{j=0}^{t_m} b_{m-n+1,r}(n, i)\lambda_i - (-1)^{(m-1)(n-1)}(b_{m-n+1,r}(m, j)\beta_j$$

for some scalars $b_{m-n+1,r}(n, i)$ and $b_{m-n+1,r}(m, j)$ associated with homotopy lifting maps $\psi_{\theta(j)}$ and $\psi_{\eta^{(i)}}$ respectively.

PROOF. This is same as [**13**, Theorem 3.15] and proved therein. □

4. Some computations and examples

In this section, we give examples in which the scalars $b_{m,r}(m - n + 1, *)$ are obtained from $b_{m-1,r}(m - n, **)$ using the recurrence relations of Theorem (3.2), Equations (3.9) and (3.10). In most of the examples, we described the scalars $c_{p,q}(m, r, n)$ which are also used in the recurrence relations.

EXAMPLE 4.1. Let's consider the following quiver

$Q :=$ 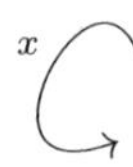 1 ,

and take $A = k[x]/(x^n)$ to be the truncated polynomial ring. The idempotent $f_0^0 = 1$, generates $\mathbb{K}_0$, $f_0^1 = x$ generates $\mathbb{K}_1$ and $f_0^n = x^n$ generates $\mathbb{K}_n$. Notice that we can write $f_0^n = f_0^1 f_0^{n-1} = f_0^{n-1} f_0^1$, so then $c_{0,0}(n, 0, 1) = c_{0,0}(n, 0, n-1) = 1$. If we **assume that** $n = 2$, that is, we mod out by the ideal $I = (x^2)$, some calculations show that using $\varepsilon_0^n = 1 \otimes f_0^n \otimes 1$,

$$\begin{aligned}
d(\varepsilon_0^1) &= d(1 \otimes x \otimes 1) = x(1 \otimes 1) - (1 \otimes 1)x = x\varepsilon_0^0 - \varepsilon_0^0 x \\
d(\varepsilon_0^2) &= d(1 \otimes x \otimes x \otimes 1) = x(1 \otimes x \otimes 1) - 1 \otimes x^2 \otimes 1 + (1 \otimes x \otimes 1)x \\
&= x\varepsilon_0^1 + \varepsilon_0^1 x \\
d(\varepsilon_0^3) &= d(1 \otimes x \otimes x \otimes x \otimes 1) = x(1 \otimes x \otimes x \otimes 1) - (1 \otimes x \otimes x \otimes 1)x \\
&= x\varepsilon_0^1 - \varepsilon_0^1 x \quad \text{and more generally} \\
d(\varepsilon_0^n) &= x\varepsilon_0^{n-1} - (-1)^{n-1}\varepsilon_0^{n-1}x = f_0^1\varepsilon_0^{n-1} - (-1)^{n-1}\varepsilon_0^{n-1}f_0^1,
\end{aligned}$$

Let $\eta : \mathbb{K}_1 \to A$ be defined by $\eta(\varepsilon_0^1) = x$. Also let $\chi : \mathbb{K}_2 \to A$ be defined by $\chi(\varepsilon_0^2) = x$. A short calculation shows that η and χ are cocycles. A diagonal map for this particular resolution is given by $\Delta_{\mathbb{K}}(\varepsilon_0^m) = \sum_{i+j=m} \varepsilon_0^i \otimes \varepsilon_0^j$. It can be verified by direct evaluation of Equation (2.4) that the map

$$\psi_\eta : \mathbb{K}_m \to \mathbb{K}_m \qquad \text{defined by} \qquad \psi_\eta(\varepsilon_0^m) = m\varepsilon_0^m$$

is a homotopy lifting map for η that is

$$
\begin{aligned}
(4.1)\qquad (d\psi_\eta - (-1)^0\psi_\eta d)(\varepsilon_0^m) &= d(m\varepsilon_0^m) - \psi_\eta(x\varepsilon_0^{m-1} - (-1)^{m-1}\varepsilon_0^{m-1}x) \\
&= mx\varepsilon_0^{m-1} - (-1)^{m-1}m\varepsilon_0^{m-1}x - (m-1)x\varepsilon_0^{m-1} \\
&\quad + (-1)^{m-1}(m-1)\varepsilon_0^{m-1}x \\
&= x\varepsilon_0^{m-1} - (-1)^{m-1}\varepsilon_0^{m-1}x \qquad \text{is equal to} \\
(\eta\otimes 1 - 1\otimes\eta)\Delta_{\mathbb{K}}(\varepsilon_0^m) &= (\eta\otimes 1 - 1\otimes\eta)\sum_{i+j=m}\varepsilon_0^i\otimes\varepsilon_0^j \\
&= \eta\otimes 1(\varepsilon_0^1\otimes\varepsilon_0^{m-1}) - (-1)^{m-1}1\otimes\eta(\varepsilon_0^{m-1}\otimes\varepsilon_0^1) \\
&= x\varepsilon_0^{m-1} - (-1)^{m-1}\varepsilon_0^{m-1}x,
\end{aligned}
$$

where the Koszul sign convention has been employed in the expansion of

$$(1\otimes\eta)(\varepsilon_0^{m-1}\otimes\varepsilon_0^1) = (-1)^{degree(\eta)\cdot(m-1)}\varepsilon_0^{m-1}\eta(\varepsilon_0^1).$$

We note that by the general definition given in Theorem 3.2, the map $\psi_\eta : \mathbb{K}_{m-1} \to \mathbb{K}_{m-1}$ defined by $\psi_\eta(\varepsilon_0^{m-1}) = (m-1)\varepsilon_0^{m-1}$ implies that $b_{m-1,0}(m-1,0) = m-1$. The map η is a 1-cocycle so $n=1$, $r=p=0$.

We can use the expression of (3.10) to verify that

$$
\begin{aligned}
&b_{m,r}(m-n+1,r) \\
&= \frac{b_{m-1,r}(m-n,r)c_{r,p}(m,r,m-1) + (-1)^{n(m-n)+1}c_{r,p}(m,r,m-n)}{(-1)^{m-n}c_{r,p}(m-n+1,r,m-n)} \\
&b_{m,0}(m,0) = \frac{b_{m-1,0}(m-1,0)c_{0,0}(m,0,m-1) + (-1)^m c_{0,0}(m,0,m-1)}{(-1)^{m-1}c_{0,0}(m,0,m-1)} \\
&\qquad = \frac{m-1+(-1)^m}{1} = m, \qquad \text{when } m \text{ is even,}
\end{aligned}
$$

and the expression of (3.9) to verify that

$$
\begin{aligned}
b_{m,0}(m,0) &= \frac{(-1)^{m-1}b_{m-1,0}(m-1,0)c_{0,0}(m,0,1) + c_{0,0}(m,0,1)}{c_{0,0}(m,0,1)} \\
&= \frac{m-1+1}{1} = m, \qquad \text{when } m \text{ is odd.}
\end{aligned}
$$

Similarly, it is a straightforward calculation (same calculations as (4.1)) to verify that the map $\psi_\chi : \mathbb{K}_m \to \mathbb{K}_{m-1}$ defined by

$$\psi_\chi(\varepsilon_0^m) = b_{m,0}(m-1,0)\varepsilon_0^{m-1} = \begin{cases} \varepsilon_0^{m-1}, & \text{when m is even} \\ 0, & \text{when m is odd} \end{cases}$$

is a homotopy lifting map for χ. In this case $b_{m,0}(m-1,0) = 1$ when m is even and 0 when m is odd. But we can also use the expression of (3.9) to verify that when m is even,

$$
\begin{aligned}
b_{m+1,0}(m,0) &= \frac{(-1)^{m-1}b_{m,0}(m-1,0)c_{0,0}(m,0,1) + c_{0,0}(m,0,2)}{c_{0,0}(m-n+1,0,1)} \\
&= -1+1 = 0,
\end{aligned}
$$

and when m is odd,

$$\begin{aligned} b_{m+1,0}(m,0) &= \frac{(-1)^{m-1}b_{m,0}(m-1,0)c_{0,0}(m,0,1)+c_{0,0}(m,0,2)}{c_{0,0}(m-n+1,0,1)} \\ &= 0+1=1. \end{aligned}$$

EXAMPLE 4.2. The following example of a homotopy lifting map was first given in [**10**, Example 4.7.2]. We will now verify that the recurrence relations also hold. Let k be a field and $A = k[x]/(x^3)$. Consider the following projective bimodule resolution of A:

$$\mathbb{P}_\bullet : \qquad \cdots \to A\otimes A \xrightarrow{\cdot u} A\otimes A \xrightarrow{\cdot u} \cdots \xrightarrow{\cdot v} A\otimes A \xrightarrow{\cdot u} A\otimes A\ (\xrightarrow{\mu} A)$$

where $u = x\otimes 1 - 1\otimes x$ and $v = x^2\otimes 1 + x\otimes x + 1\otimes x^2$. We consider the following elements $e_m := 1\otimes 1$, $r = 0$ for all m in the m-th module $P_m := A\otimes A$. A diagonal map $\Delta_{\mathbb{P}} : \mathbb{P} \to \mathbb{P}\otimes_A \mathbb{P}$ for this resolution is given by

$$\Delta_{\mathbb{P}}(e_m) = \sum_{j+l=m} (-1)^j e_j \otimes e_l.$$

By comparing $\Delta_{\mathbb{P}}(e_m)$ with Equation (2.9), the scalars $c_{rr}(m,r,j) = (-1)^j$ for all m, r. Consider the Hochschild 1-cocycle $\alpha : \mathbb{P}_1 \to A$ defined by $\alpha(e_1) = x$ and $\alpha(e_m) = 0$ for all $m \neq 1$. With a slight change in notation, it was shown in [**10**, Example 4.7.2] that the following $\psi_\alpha : \mathbb{P}_{2m} \to \mathbb{P}_{2m}$ defined by $\psi_\alpha(e_{2m}) = -3m\cdot e_{2m}$ is a homotopy lifting map for α. We note that the map ψ_α was regarded as an A_∞-coderivation in [**10**]. It can be also verified that ψ_α is a homotopy lifting map. We can use the recurrence relations of Equation (3.10) to obtain $b_{2m+1,r}(2m+1,r)$ from $b_{2m,r}(2m,r) = -3m$. That is

$$\begin{aligned} &b_{2m+1,r}(2m+1,r) \\ &\qquad = \frac{b_{2m,r}(2m,r)c_{r,r}(2m,r,2m) + (-1)^{2m+1}c_{r,r}(2m+1,r,2m)}{(-1)^{2m}c_{r,r}(2m+1,r,2m)} \\ &\qquad = \frac{-3m(-1)^{2m} + (-1)^{2m+1}(-1)^{2m}}{(-1)^{2m}(-1)^{2m-1+1}} = \frac{-3m-1}{1}, \end{aligned}$$

so it follows that $\psi_\alpha : \mathbb{P}_{2m+1} \to \mathbb{P}_{2m+1}$ is defined by $\psi_\alpha(e_{2m+1}) = (-3m-1)e_{2m+1}$.

EXAMPLE 4.3. Let k be a field of characteristics different from 2. Consider the quiver algebra $A = kQ/I$ (also examined in [**3**, Example 5]) defined using the following finite quiver:

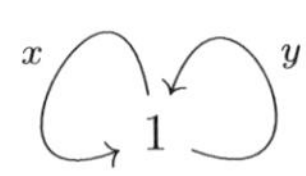

with one vertex and two arrows x, y. We denote by e_1 the idempotent associated with the only vertex. Let I, an ideal of the path algebra kQ be defined by $I = \langle x^2, xy + yx\rangle$. Since $\{x^2, xy+yx\}$ is a quadratic Grobner basis for the ideal generated by relations under the length lexicographic order with $x > y > 1$, the algebra is Koszul.

In order to define a comultiplicative structure, we take $t_0 = 0, t_n = 1$ for all n, $f_0^0 = e_1$, $f_1^0 = 0$, $f_0^1 = x$, $f_1^1 = y$, $f_0^2 = x^2$, $f_1^2 = xy + yx$, $f_0^3 = x^3$, $f_1^3 = x^2y + xyx + yx^2$, and in general $f_0^n = x^n$, $f_1^n = \sum_{i+j=n-1} x^i y x^j$. We also see that $f_0^n = f_0^r f_0^{n-r}$

and $f_1^n = f_0^r f_1^{n-r} + f_1^r f_0^{n-r}$ so $c_{00}(n,0,r) = c_{01}(n,1,r) = c_{10}(n,1,r) = 1$ and all other $c_{pq}(n,i,r) = 0$. With the above stated, we can construct the resolution $\mathbb{K}$ for the algebra A. A calculation shows that

$$d_1(\varepsilon_0^1) = x\varepsilon_0^0 - \varepsilon_0^0 x, \qquad d_1(\varepsilon_1^1) = y\varepsilon_0^0 - \varepsilon_0^0 y$$
$$d_2(\varepsilon_0^2) = x\varepsilon_0^1 + \varepsilon_0^1 x, \qquad d_2(\varepsilon_1^2) = y\varepsilon_0^1 + \varepsilon_0^1 y + x\varepsilon_1^1 + \varepsilon_1^1 x.$$

Consider the following map $\theta : \mathbb{K}_1 \to A$ defined by $\theta = (0 \ \ y)$. With the following calculations

$$\theta d_2(\varepsilon_0^2) = \theta(x\varepsilon_0^1 + \varepsilon_0^1 x) = 0$$

and

$$\theta d_2(\varepsilon_1^2) = \theta(y\varepsilon_0^1 + \varepsilon_0^1 y + x\varepsilon_1^1 + \varepsilon_1^1 x) = 0 + xy + yx = 0,$$

θ is a cocycle. The comultiplicative map $\Delta : \mathbb{K} \to \mathbb{K} \otimes_A \mathbb{K}$ on ε_0^1, $\varepsilon_1^1, \varepsilon_0^2, \varepsilon_1^2$ is given by

$$\begin{aligned}
\Delta(\varepsilon_0^1) &= c_{00}(1,0,0)\varepsilon_0^0 \otimes \varepsilon_0^1 + c_{00}(1,0,1)\varepsilon_0^1 \otimes \varepsilon_0^0 = \varepsilon_0^0 \otimes \varepsilon_0^1 + \varepsilon_0^1 \otimes \varepsilon_0^0,\\
\Delta(\varepsilon_1^1) &= \varepsilon_0^0 \otimes \varepsilon_1^1 + \varepsilon_1^1 \otimes \varepsilon_0^0,\\
\Delta(\varepsilon_0^2) &= \varepsilon_0^0 \otimes \varepsilon_0^2 + \varepsilon_0^1 \otimes \varepsilon_0^1 + \varepsilon_0^2 \otimes \varepsilon_0^0,\\
\Delta(\varepsilon_1^2) &= \varepsilon_0^0 \otimes \varepsilon_1^2 + \varepsilon_0^1 \otimes \varepsilon_1^1 + \varepsilon_1^1 \otimes \varepsilon_0^1 + \varepsilon_1^2 \otimes \varepsilon_0^0.
\end{aligned}$$

From Theorems (3.2), it can be verified by direct calculations using Equation (4.1) that the first, second and third degrees of the homotopy lifting maps ψ_θ associated θ is the following:

$$\psi_{\theta_0}(\varepsilon_i^0) = 0, \qquad \psi_{\theta_1}(\varepsilon_0^1) = 0, \qquad \psi_{\theta_1}(\varepsilon_1^1) = \varepsilon_1^1 \qquad \psi_{\theta_2}(\varepsilon_0^2) = 0.$$

The scalars $b_{1,1}(1,1) = 1$ and for other $(m,r) \neq (1,1),(2,1)$, $b_{m,r}(m,r) = 0$. Since θ is a 1-cocycle, $n = 1$. Also, $\theta = (0 \ y)$ and when compared with $\eta = \big(0 \ \cdots \ 0 \ (f_w^1)^{(i)} \ 0 \ \cdots \ 0\big)$ as given in Theorem (3.2), $f_w^1 = y$ and $i = 1$. To obtain $b_{2,1}(2,1)$ from $b_{1,r}(1,s)$ for some s, we take $m = 2, r = 1$, so that $m - n = 1$. Since $t_{m-n} = t_1$, we would have $r'' = 0$ or $r'' = 1$. From the statement of the theorem, we must have $i + r'' = r$, so $r'' = 0$ and $c_{r,r''}(m - n + 1, r', 1) = c_{10}(2, r', 1) = 1$. It then follows from the first recurrence relations in Theorem (3.2) that

$$b_{m,r}(m-n+1,r') = \frac{(-1)^{m-1} b_{m-1,\bar{r}}(m-n,r'')c_{r,\bar{r}}(m,r,1) + c_{i,r''}(m,r,n)}{c_{r,r''}(m-n+1,r',1)}.$$
$$b_{2,1}(2,1) = \frac{-b_{1,\bar{r}}(1,0)c_{1,\bar{r}}(2,1,1) + c_{1,0}(2,1,1)}{c_{1,0}(2,1,1)} = \frac{0+1}{1} = 1,$$

so $\psi_{\theta_2}(\varepsilon_1^2) = b_{2,1}(2,1)\varepsilon_1^2 = \varepsilon_1^2$.

5. Finding Maurer-Cartan elements

In this section, we find the Maurer-Cartan elements of a quiver algebra. We first recall the definition of a Maurer-Cartan element.

DEFINITION 5.1. A Hochschild 2-cocycle η is said to satisfy the Maurer-Cartan equation if

$$d(\eta) + \frac{1}{2}[\eta,\eta] = 0. \tag{5.1}$$

Applying the definition of the bracket using homotopy lifting, we obtain the following version of the Maurer-Cartan equation for the resolution $\mathbb{K}$. $d_3^*(\eta) + \frac{1}{2}(\eta\psi_\eta + \eta\psi_\eta) = d_3^*(\eta) + \eta\psi_\eta = 0$.

We begin with the following finite quiver:

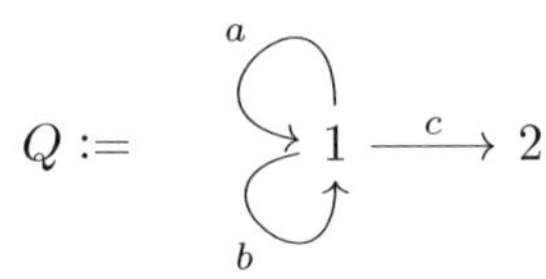

with two vertices and three arrows a, b, c. We denote by e_1 and e_2 the idempotents associated with vertices 1 and 2. Let kQ be the path algebra associated with Q and take for each $q \in k$, $I_q \subseteq kQ$ to be an admissible ideal of kQ generated by $I_q = \langle a^2, b^2, ab - qba, ac\rangle$ so that

$$\{A_q = kQ/I_q\}_{q\in k}$$

This family of quiver algebras have been well studied in [**12, 13**] and [**14**]. We simply recall the main tools needed to find Maurer-Cartan elements. To define a set of generators for the resolution $\mathbb{K}$ we start by letting kQ_0 to be the ideal of kQ generated by the vertices of Q with basis $f_0^0 = e_1, f_1^0 = e_2$. Next, set kQ_1 to be the ideal generated by paths with basis $f_0^1 = a, f_1^1 = b$ and $f_2^1 = c$. Set f_j^2, $j = 0, 1, 2, 3$ to be the set of paths of length 2 that generates the ideal I, that is $f_0^2 = a^2, f_1^2 = ab - qba, f_2^2 = b^2, f_3^2 = ac$, and define a comultiplicative equation on the paths of length $n > 2$ in the following way.

$$\begin{cases} f_0^n = a^n, \\ f_s^n = f_{s-1}^{n-1}b + (-q)^s f_s^{n-1}a, \quad (0 < s < n), \\ f_n^n = b^n, \\ f_{n+1}^n = a^{(n-1)}c, \end{cases}$$

The resolution $\mathbb{K} \to A_q$ has basis elements $\{\varepsilon_i^n\}_{i=0}^{t_n}$ such that for each i, we have

$$\varepsilon_i^n = (0, \dots, 0, o(f_i^n) \otimes_k t(f_i^n), 0, \dots, 0).$$

The differentials on $\mathbb{K}_n$ are given explicitly for this family by

$$\begin{aligned} d_1(\varepsilon_2^1) &= c\varepsilon_1^0 - \varepsilon_0^0 c \\ d_n(\varepsilon_r^n) &= (1 - \partial_{n,r})[a\varepsilon_r^{n-1}) + (-1)^{n-r}q^r\varepsilon_r^{n-1}a] \\ &\quad + (1 - \partial_{r,0})[(-q)^{n-r}b\varepsilon_{r-1}^{n-1} + (-1)^n\varepsilon_{r-1}^{n-1}b], \quad \text{for } r \leq n \\ d_n(\varepsilon_{n+1}^n) &= a\varepsilon_n^{n-1} + (-1)^n\varepsilon_0^{n-1}c, \quad \text{when } n \geq 2, \end{aligned}$$

where $\partial_{r,s} = 1$ when $r = s$ and 0 when $r \neq s$.

Calculations from [**12**] show that for this family, the comultiplicative map can be expressed in the following way

$$\Delta_{\mathbb{K}}(\varepsilon_s^n) = \begin{cases} \sum_{r=0}^{n} \varepsilon_0^r \otimes \varepsilon_0^{n-r}, & s=0 \\ \sum_{w=0}^{n} \sum_{j=\max\{0,s+w-n\}}^{\min\{w,s\}} (-q)^{j(n-s+j-w)} \varepsilon_j^w \otimes \varepsilon_{s-j}^{n-w}, & 0<s<n \\ \sum_{t=0}^{n} \varepsilon_t^t \otimes \varepsilon_{n-t}^{n-t}, & s=n \\ \varepsilon_0^0 \otimes \varepsilon_{n+1}^n + \left(\sum_{t=0}^{n} \varepsilon_0^t \otimes \varepsilon_{n-t+1}^{n-t} \right) + \varepsilon_{n+1}^n \otimes \varepsilon_0^0, & s=n+1. \end{cases}$$

EXAMPLE 5.2. Let $A_1 = kQ/I_1$ be a member of the family where $I = I_1 = \langle a^2, b^2, ab - ba, ac \rangle$. We now find Hochschild 2-cocycles that satisfy the Maurer-Cartan equation of 5.1. Suppose that the A_1^e-module homomorphism $\eta : \mathbb{K}_2 \to A_1$ defined by

$$\eta = \begin{pmatrix} \lambda_0 & \lambda_1 & \lambda_2 & \lambda_3 \end{pmatrix},$$

which is a cocycle. That is $d^*\eta = 0$, with $\lambda_i \in \Lambda_q$ for all i. Since $d^*\eta : \mathbb{K}_3 \to A_1$, we obtain using $d^*\eta(\varepsilon_i^3) = \eta d(\varepsilon_i^3)$,

$$\eta\left(\begin{cases} a\varepsilon_0^2 - \varepsilon_0^2 a \\ a\varepsilon_1^2 + \varepsilon_1^2 a + b\varepsilon_0^2 - \varepsilon_0^2 b \\ a\varepsilon_2^2 - \varepsilon_2^2 a - b\varepsilon_1^2 - \varepsilon_1^2 b \\ b\varepsilon_2^2 - \varepsilon_2^2 b \\ a\varepsilon_3^2 - \varepsilon_0^2 c \end{cases} \right) = \begin{cases} a\lambda_0 - \lambda_0 a, & \text{if } i=0 \\ a\lambda_1 + q\lambda_1 a + q^2 b\lambda_0 - \lambda_0 b & \text{if } i=1 \\ a\lambda_2 - q^2\lambda_2 a - qb\lambda_1 - \lambda_1 b & \text{if } i=2 \\ b\lambda_2 - \lambda_2 b & \text{if } i=3 \\ a\varepsilon_3^2 - \varepsilon_0^2 c & \text{if } i=4 \end{cases}$$

which will be equated to $\begin{pmatrix} 0 & 0 & 0 & 0 & 0 \end{pmatrix}$ and solved. We solve this system of equations with the following in mind. There is an isomorphism of A_1^e-modules $\mathrm{Hom}_{A_1^e}(A_1 o(f_i^n) \otimes_k t(f_i^n) A_1, A_1) \simeq o(f_i^n) A_1\, t(f_i^n)$ ensuring that

$$\begin{aligned} o(f_i^2)\lambda_i t(f_i^2) &= o(f_i^2)\eta(\varepsilon_i^2)t(f_i^2) = o(f_i^2)\eta(o(f_i^2) \otimes_k t(f_i^2))t(f_i^2) \\ &= \phi(o(f_i^2)^2 \otimes_k t(f_i^2)^2) = \phi(o(f_i^2) \otimes_k t(f_i^2)) = \lambda_i. \end{aligned}$$

This means that for $i = 0, 1, 2$ each λ_i should satisfy $e_1\lambda_i e_1 = \lambda_i$ since the origin and terminal vertex of f_0^2, f_1^2, f_2^2 is e_1 and $e_1\lambda_3 e_2 = \lambda_3$. We obtain 9 solutions presented in Table 1.

Now suppose that there is some $\phi : \mathbb{K}_1 \to A_1$ such that $\phi d_2(\varepsilon_i^2) = \eta(\varepsilon_i^2)$, i=0,1,2,3. If $\phi = \begin{pmatrix} 0 & \frac{1}{2}a & 0 \end{pmatrix}$, we get $\eta = \begin{pmatrix} 0 & 0 & ab & 0 \end{pmatrix}$, so $\eta = \begin{pmatrix} 0 & 0 & ab & 0 \end{pmatrix} \in$

solutions	1	2	3	4	5	6	7	8	9
λ_0	a	ab	0	0	0	0	0	0	0
λ_1	0	0	0	0	0	0	ab	0	0
λ_2	0	0	a	b	ab	e_1	0	0	0
λ_3	0	0	0	0	0	0	0	c	bc

TABLE 1. Possible values of $\eta(\varepsilon_r^2) = \lambda_r$ for different r.

$\operatorname{Im}(d_2^*)$. If ϕ is equal to $\begin{pmatrix}0 & \frac{1}{2}e_1 & 0\end{pmatrix}$, $\begin{pmatrix}e_1 & 0 & 0\end{pmatrix}$ or $\begin{pmatrix}b & 0 & 0\end{pmatrix}$, we obtain the following for η; $\begin{pmatrix}0 & 0 & b & 0\end{pmatrix}$, $\begin{pmatrix}0 & 0 & 0 & c\end{pmatrix}$ and $\begin{pmatrix}0 & 0 & 0 & bc\end{pmatrix}$ respectively. Therefore $\mathrm{HH}^2(A_1) = \frac{\operatorname{Ker} d_3^*}{\operatorname{Im} d_2^*}$ is generated as a k-vector space by $\langle \widetilde{\eta}, \bar{\eta}, \chi, \bar{\chi}, \sigma \rangle$ where $\widetilde{\eta} = \begin{pmatrix}a & 0 & 0 & 0\end{pmatrix}$, $\bar{\eta} = \begin{pmatrix}ab & 0 & 0 & 0\end{pmatrix}$, $\bar{\chi} = \begin{pmatrix}0 & ab & 0 & 0\end{pmatrix}$, $\chi = \begin{pmatrix}0 & 0 & a & 0\end{pmatrix}$ and $\sigma = \begin{pmatrix}0 & 0 & e_1 & 0\end{pmatrix}$.

Given in Table 2 are the first, second and third degree homotopy lifting maps associated to each the above elements of $\mathrm{HH}^2(A_1)$. It can be easily verified using the homotopy lifting equation in Definition (2.4) that these indeed are homotopy lifting maps.

The following Lemma follows immediately.

LEMMA 5.3. *Let $A_1 = kQ/I_1$ be a member of the family of quiver algebras where $I_1 = \langle a^2, b^2, ab - ba, ac \rangle$. The Hochschild 2-cocycles $\eta = \begin{pmatrix}a & 0 & 0 & 0\end{pmatrix}$, $\chi = \begin{pmatrix}0 & 0 & a & 0\end{pmatrix}$, $\bar{\eta} = \begin{pmatrix}ab & 0 & 0 & 0\end{pmatrix}$, $\bar{\chi} = \begin{pmatrix}0 & ab & 0 & 0\end{pmatrix}$ and $\sigma = \begin{pmatrix}0 & 0 & e_1 & 0\end{pmatrix}$ are Maurer-Cartan elements.*

PROOF. Let γ be any of those elements of $\mathrm{HH}^2(A_1)$. We make use of Equation (5.1). Since they are all cocycles, $d_3^*(\gamma) = 0$. Also observe that $\gamma\psi_{\gamma_3}(\varepsilon_i^3) = 0$ for all $\gamma \in \mathrm{HH}^2(A_1)$, therefore $d_3^*(\gamma) + \gamma\psi_\gamma = 0$. □

6. Deformation of algebras using reduction system

Let $A_1 = kQ/I_1$ be a member of family of quiver algebras introduced in Section 5, we now show using the combinatorial star product of Equation (2.13) that $\mathrm{HH}^2(A)$ has 5 elements satisfying the Maurer-Cartan equation.

EXAMPLE 6.1. Recall that for $A_1 = kQ/I$, $I = \langle a^2, b^2, ab - ba, ac \rangle$. If we use the set $\{(a^2, 0), (b^2, 0), (ab, ba), (ac, 0)\}$ as the reduction system, this system is reduction finite and reduction unique. All the one overlaps given by S_3 resolve to 0 uniquely. The reduction system

$$R = \{(a^2, 0), (b^2, 0), (ab, ba), (ac, 0)\}$$

satisfies the diamond condition ($\diamond$) where $\varphi(a^2) = 0, \varphi(b^2) = 0, \varphi(ab) = ba$ and $\varphi(ac) = 0$. The set S and the set Irr_S of irreducible paths in the algebra are given respectively by $S = \{a^2, b^2, ab, ac\}$ and $\mathrm{Irr}_S = \{e_1, e_2, a, b, c, ba, bc\}$, so $\dim(A_1) = 7$. The paths a^2 and ab overlap at a so $(aa)(ab) = a^2 b \in S_3$. The set of one-overlaps is given as

$$S_3 = \{a^3, b^3, a^2 b, ab^2, a^2 c\}.$$

Notice that in the quiver Q, the path $a^2, b^2, ab \in S$ are all parallel to the irreducible paths $e_1 = e, a, b, ba$ and the path $ac \in S$ is parallel to c and bc. Any element $\widetilde{\varphi} : kS \to A_1 \cong k\mathrm{Irr}_S$ viewed as $\widetilde{\varphi} \in \operatorname{Hom}(kS, A_1) \otimes (\tau)$ has the following general form

$$\begin{aligned}
\widetilde{\varphi}(a^2) &= (\lambda_e + \lambda_a a + \lambda_b b + \lambda_{ba} ba)\tau \\
\widetilde{\varphi}(b^2) &= (\mu_e + \mu_a a + \mu_b b + \mu_{ba} ba)\tau \\
\widetilde{\varphi}(ab) &= (\nu_e + \nu_a a + \nu_b b + \nu_{ba} ba)\tau \\
\widetilde{\varphi}(ac) &= (w_c c + w_{bc} bc)\tau
\end{aligned}$$

For $\eta = \begin{pmatrix} a & 0 & 0 & 0 \end{pmatrix}$, we get
$\psi_{\eta_1}(\varepsilon_i^1) = 0,\ i = 0,1,2,$ $\psi_{\eta_2}(\varepsilon_i^2) = \begin{cases} \varepsilon_0^1, & \text{if } i = 0 \\ 0, & \text{if } i = 1,2,3 \end{cases},\quad \psi_{\eta_3}(\varepsilon_i^3) = \begin{cases} 0, & \text{if } i = 0 \\ \varepsilon_1^2, & \text{if } i = 1 \\ 0, & \text{if } i = 2 \\ 0, & \text{if } i = 3 \\ \varepsilon_3^2, & \text{if } i = 4 \end{cases}$
For $\chi = \begin{pmatrix} 0 & 0 & a & 0 \end{pmatrix}$, we get
$\psi_{\chi_1}(\varepsilon_i^1) = 0,\ i = 0,1,2,$ $\psi_{\chi_2}(\varepsilon_i^2) = \begin{cases} 0, & \text{if } i = 0,1,3 \\ \varepsilon_0^1, & \text{if } i = 2 \end{cases},\quad \psi_{\chi_3}(\varepsilon_i^3) = \begin{cases} 0, & \text{if } i = 0 \\ 0, & \text{if } i = 1 \\ 0, & \text{if } i = 2 \\ \varepsilon_1^2, & \text{if } i = 3 \\ 0, & \text{if } i = 4 \end{cases}$
For $\bar{\eta} = \begin{pmatrix} ab & 0 & 0 & 0 \end{pmatrix}$, we get
$\psi_{\bar{\eta}_1}(\varepsilon_i^1) = 0,\ i = 0,1,2,$ $\psi_{\bar{\eta}_2}(\varepsilon_i^2) = \begin{cases} a\varepsilon_1^1 + \varepsilon_0^1 b & \text{if } i = 0 \\ 0, & \text{if } i = 1 \\ 0 & \text{if } i = 2 \\ 0 & \text{if } i = 3 \end{cases},\quad \psi_{\bar{\eta}_3}(\varepsilon_i^3) = \begin{cases} -a\varepsilon_1^2, & \text{if } i = 0 \\ 0, & \text{if } i = 1 \\ 0, & \text{if } i = 2 \\ 0, & \text{if } i = 3 \\ b\varepsilon_3^2 + \varepsilon_1^2 c, & \text{if } i = 4 \end{cases}$
For $\bar{\chi} = \begin{pmatrix} 0 & ab & 0 & 0 \end{pmatrix}$, we get
$\psi_{\bar{\chi}_1}(\varepsilon_i^1) = 0,\ i = 0,1,2,$ $\psi_{\bar{\chi}_2}(\varepsilon_i^2) = \begin{cases} 0 & \text{if } i = 0 \\ a\varepsilon_1^1 + \varepsilon_0^1 b, & \text{if } i = 1 \\ 0 & \text{if } i = 2 \\ 0 & \text{if } i = 3 \end{cases},\quad \psi_{\bar{\chi}_3}(\varepsilon_i^3) = \begin{cases} 0, & \text{if } i = 0 \\ a\varepsilon_1^2 - 2\varepsilon_0^2 b, & \text{if } i = 1 \\ \varepsilon_1^2 b, & \text{if } i = 2 \\ 0, & \text{if } i = 3 \\ 0, & \text{if } i = 4 \end{cases}.$
For $\sigma = \begin{pmatrix} 0 & 0 & e_1 & 0 \end{pmatrix}$, we get
$\psi_{\sigma_1}(\varepsilon_i^1) = 0,\ i = 0,1,2,$ $\psi_{\sigma_2}(\varepsilon_i^2) = 0,\ i = 0,1,2,3$ $\psi_{\sigma_3}(\varepsilon_i^3) = 0,\ i = 0,1,2,3,4$

TABLE 2. Homotopy lifting maps associated to some cocycles in degrees 1,2,3.

for scalars $\lambda_e, \lambda_a, \cdots, w_c, w_{bc} \in k$. By [**1**, Corollary 7.37], $\widetilde{\varphi}$ is a Maurer-Cartan element if and only if for each $uvw \in S_3$ with $uv, vw \in S$, Equation (2.13) holds. That is

$$(\pi(u) \star \pi(v)) \star \pi(w) = \pi(u) \star (\pi(v) \star \pi(w)) (\text{mod } \tau^2)$$

since we are considering first order deformations. We now check conditions on the scalars for the associativity of the star product. This product defined for example for $a, b \in A_1$ is given by

$$a \star b = \varphi(ab) + \widetilde{\varphi}(ab)\tau.$$

We check for all elements of S_3. For instance, the calculations involved in using a^3 to check that $a \star (a \star a) = (a \star a) \star a$ are the following.

$$\begin{aligned} a \star (a \star a) &= a \star (\varphi(a^2) + \widetilde{\varphi}(a^2)\tau) = a \star (\lambda_e + \lambda_a a + \lambda_b b + \lambda_{ba} ba)\tau \\ &= (\lambda_e \varphi(a) + \lambda_a \varphi(a^2) + \lambda_b \varphi(ab) + \lambda_{ba}\varphi(aba))\tau \\ &\quad + (\lambda_e \widetilde{\varphi}(a) + \lambda_a \widetilde{\varphi}(a^2) + \lambda_b \widetilde{\varphi}(ab) + \lambda_{ba}\widetilde{\varphi}(aba))\tau^2 \\ &= (\lambda_e a + \lambda_b ba)\tau \end{aligned}$$

and it is equal to

$$\begin{aligned} (a \star a) \star a &= (\varphi(a^2) + \widetilde{\varphi}(a^2)\tau) \star a = (\lambda_e + \lambda_a a + \lambda_b b + \lambda_{ba} ba)\tau \star a \\ &= (\lambda_e \varphi(a) + \lambda_a \varphi(a^2) + \lambda_b \varphi(ba) + \lambda_{ba}\varphi(ba^2))\tau \\ &\quad + (\lambda_e \widetilde{\varphi}(a) + \lambda_a \widetilde{\varphi}(a^2) + \lambda_b \widetilde{\varphi}(ba) + \lambda_{ba}\widetilde{\varphi}(ba^2))\tau^2 \\ &= (\lambda_e a + \lambda_b ba)\tau. \end{aligned}$$

This then implies that $\lambda_e = \lambda_e$ and $\lambda_b = \lambda_b$. For $a^2 b = a \star (a \star b) = (a \star a) \star b$, we obtain $a \star (a \star b) = (\nu_e a + \nu_b ba)\tau$ and $(a \star a) \star b = (\lambda_e b + \lambda_a ba)\tau$, so we get $\nu_e = \lambda_e = 0$ and $\nu_b = \lambda_a$. Equivalent calculations for ab^2 and $a^2 c$ yield $\mu_e = \nu_e = 0$ and $\mu_b = \nu_a$ and $\lambda_e = \lambda_b = 0$. We can now rewrite

$$\begin{aligned} \widetilde{\varphi}(a^2) &= (\lambda_a a + \lambda_{ba} ba)\tau \\ \widetilde{\varphi}(b^2) &= (\mu_a a + \mu_b b + \mu_{ba} ba)\tau \\ \widetilde{\varphi}(ab) &= (\mu_b a + \lambda_a b + \nu_{ba} ba)\tau \\ \widetilde{\varphi}(ac) &= (w_c c + w_{bc} bc)\tau \end{aligned}$$

so the Maurer-Cartan elements of $\mathrm{HH}^2(A_1)$ are parametrized by

$$\widetilde{\varphi} = (\lambda_a, \lambda_{ba}, \mu_a, \mu_b, \mu_{ba}, \nu_{ba}, w_c, w_{bc}) \in k^8.$$

Our next goal is to show that three of these parameters can be eliminated by a coboundary so that $\widetilde{\varphi} \in k^5$ and thus $\dim(\mathrm{HH}^2(A_1)) = 5$. Let $\widetilde{\varphi}'$ be defined by

$$\begin{aligned} \widetilde{\varphi}'(a^2) &= (\lambda'_a a + \lambda'_{ba} ba)\tau, & \widetilde{\varphi}'(b^2) &= (\mu'_a a + \mu'_b b + \mu'_{ba} ba)\tau \\ \widetilde{\varphi}'(ab) &= (\mu'_b a + \lambda'_a b + \nu'_{ba} ba)\tau, & \widetilde{\varphi}'(ac) &= (w'_c c + w'_{bc} bc)\tau \end{aligned}$$

From [**1**, Corollary 7.44], two cocycles $\widetilde{\varphi}$ and $\widetilde{\varphi}'$ are cohomologous or satisfy $\widetilde{\varphi} - \widetilde{\varphi} = \langle \Theta \rangle$, $\Theta \in \mathrm{Hom}(kQ_1, k\mathrm{Irr}_S)$ if

$$T(\varphi(s)) + \widetilde{\varphi}'(s) = T(s_1) \star \cdots \star T(s_m) (mod\ \tau^2)$$

for some $T : k\mathrm{Irr}_S[\tau]/(\tau^2) \to \mathrm{Irr}_S[\tau]/(\tau^2)$ defined by $T(x) = x + \Theta(x)\tau$ with $s = s_1 s_2 \cdots s_m$ a path of length m. Any $\Theta \in \mathrm{Hom}(kQ_1, k\mathrm{Irr}_S)$ has a general form

$$\begin{aligned} \Theta(a) &= \alpha_e + \alpha_a a + \alpha_b b + \alpha_{ba} ba \\ \Theta(b) &= \beta_e + \beta_a a + \beta_b b + \beta_{ba} ba \\ \Theta(c) &= \gamma_c c + \gamma_{bc} bc, \end{aligned}$$

where $(a \star \Theta(a)) = \alpha_e a + \alpha_a \varphi(a^2) + \alpha_b \varphi(ab) + \alpha_{ba} \varphi(aba)) + (\alpha_e \widetilde{\varphi}(a) + \alpha_a \widetilde{\varphi}(a^2) + \alpha_b \widetilde{\varphi}(ab) + \alpha_{ba} \widetilde{\varphi}(aba))\tau$. Whenever $s = a^2$, then $T(\varphi(a^2)) + \widetilde{\varphi}'(a^2) = T(a) \star T(a)$ yields the following:

$$T(\varphi(a^2)) + \widetilde{\varphi}'(a^2) = (\lambda'_a a + \lambda'_{ba} ba)\tau. \tag{6.1}$$

$$\begin{aligned} T(a) \star T(a) &= (a + \Theta(a)\tau) \star (a + \Theta(a)\tau) \\ &= a \star a + (a \star \Theta(a))\tau + (\Theta(a) \star a)\tau + (\Theta(a) \star \Theta(a))\tau^2 \\ &= (\lambda_a a + \lambda_{ba} ba)\tau + (\alpha_e a + \alpha_b ba)\tau + (\alpha_e a + \alpha_b ba)\tau + 0 \\ &= (\lambda_a a + \lambda_{ba} ba + 2\alpha_e a + 2\alpha_b ba)\tau \end{aligned}$$

and comparing with Equation (6.1) we arrive at $\lambda'_a - \lambda_a = 2\alpha_e$ and $\lambda'_{ba} - \lambda_{ba} = 2\alpha_b$. With other similar equivalent calculations on s being b^2, ab, ac, we get

$$\begin{array}{lll} (a^2) : \lambda'_a - \lambda_a = 2\alpha_e & \lambda'_{ba} - \lambda_{ba} = 2\alpha_b & \\ (b^2) : \mu'_a - \mu_a = 0 & \mu'_b - \mu_b = 2\beta_e & \mu'_{ba} - \mu_{ba} = 2\beta_a \\ (ab) : \mu'_b - \mu_b = \beta_e & \lambda'_a - \lambda_a = \alpha_e & \nu'_{ba} - \nu_{ba} = \alpha_a + \beta_b \\ (ac) : w'_c - w_c = \alpha_e & w'_{bc} - w_{bc} = \alpha_b & \end{array}$$

This implies that three variables in the parametric definition of $\widetilde{\varphi}$ can be eliminated or simply $\widetilde{\varphi} = (\lambda_a, \lambda_{ba}, \mu_a, \mu_b, \mu_{ba}, \nu_{ba}, w_c, w_{bc}) \in k^8$ is cohomologous to $\widetilde{\varphi} = (\lambda_a, \lambda_{ba}, \mu_a, \mu_b, 0, \nu_{ba}, 0, 0) \in k^8$. Therefore $\widetilde{\varphi}$ is in k^5 or equivalently the dimension of $\mathrm{HH}^2(A_1)$ is equal to 5.

References

[1] S. Barmeier, Z. Wang, *Deformation of path algebras of quivers with relations,* ArXiv: 2002.10001v4.

[2] G. M. Bergman, *The diamond lemma for ring theory,* Adv. in Math. **29** (1978), no. 2, 178–218, DOI 10.1016/0001-8708(78)90010-5. MR506890

[3] R.-O. Buchweitz, E. L. Green, N. Snashall, and Ø. Solberg, *Multiplicative structures for Koszul algebras,* Q. J. Math. **59** (2008), no. 4, 441–454, DOI 10.1093/qmath/ham056. MR2461267

[4] S. Chouhy and A. Solotar, *Projective resolutions of associative algebras and ambiguities,* J. Algebra **432** (2015), 22–61, DOI 10.1016/j.jalgebra.2015.02.019. MR3334140

[5] E. L. Green and Ø. Solberg, *An algorithmic approach to resolutions,* J. Symbolic Comput. **42** (2007), no. 11-12, 1012–1033, DOI 10.1016/j.jsc.2007.05.002. MR2368070

[6] E. L. Green, G. Hartman, E. N. Marcos, and Ø. Solberg, *Resolutions over Koszul algebras,* Arch. Math. (Basel) **85** (2005), no. 2, 118–127, DOI 10.1007/s00013-005-1299-9. MR2161801

[7] T. Karadağ, D. McPhate, P. S. Ocal, T. Oke, and S. Witherspoon, *Gerstenhaber brackets on Hochschild cohomology of general twisted tensor products,* J. Pure Appl. Algebra **225** (2021), no. 6, Paper No. 106597, 14, DOI 10.1016/j.jpaa.2020.106597. MR4168969

[8] R. Martínez-Villa, *Introduction to Koszul algebras,* Rev. Un. Mat. Argentina **48** (2007), no. 2, 67–95 (2008). MR2388890

[9] C. Negron and S. Witherspoon, *An alternate approach to the Lie bracket on Hochschild cohomology,* Homology Homotopy Appl. **18** (2016), no. 1, 265–285, DOI 10.4310/HHA.2016.v18.n1.a14. MR3498646

[10] C. Negron, Y. Volkov, and S. Witherspoon, *A_∞-coderivations and the Gerstenhaber bracket on Hochschild cohomology,* J. Noncommut. Geom. **14** (2020), no. 2, 531–565, DOI 10.4171/jncg/372. MR4130838

[11] N. Snashall, *Support varieties and the Hochschild cohomology ring modulo nilpotence,* Proceedings of the 41st Symposium on Ring Theory and Representation Theory, Symp. Ring Theory Represent. Theory Organ. Comm., Tsukuba, 2009, pp. 68–82. MR2512413

[12] T. Oke, *Cup product and Gerstenhaber bracket on Hochschild cohomology of a family of quiver algebras*, Comm. Algebra **50** (2022), no. 5, 1821–1841, DOI 10.1080/00927872.2021.1990941. MR4401117

[13] T. Oke, *Bracket structure on Hochschild cohomology of Koszul quiver algebras using homotopy liftings,* arXiv:2103.12331.

[14] T. N. Oke, *On the Lie Algebra Structure on Hochschild Cohomology of Koszul Quiver Algebras*, ProQuest LLC, Ann Arbor, MI, 2021. Thesis (Ph.D.)–Texas A&M University. MR4435716

[15] S. J. Witherspoon, *Hochschild cohomology for algebras*, Graduate Studies in Mathematics, vol. 204, American Mathematical Society, Providence, RI, [2019] ©2019, DOI 10.1090/gsm/204. MR3971234

[16] Y. Volkov, *Gerstenhaber bracket on the Hochschild cohomology via an arbitrary resolution*, Proc. Edinb. Math. Soc. (2) **62** (2019), no. 3, 817–836, DOI 10.1017/s0013091518000901. MR3974969

Department of Mathematics & Statistics, Wake Forest University, North Carolina
Email address: oket@wfu.edu

Contemporary Mathematics
Volume **791**, 2024
https://doi.org/10.1090/conm/791/15871

Centers and automorphisms of PI quantum matrix algebras

Jason Gaddis and Thomas Lamkin

Abstract. We study PI quantum matrix algebras and their automorphisms using the noncommutative discriminant. In the multi-parameter case at $n = 2$, we show that all automorphisms are graded when the center is a polynomial ring. In the single-parameter case, we determine a presentation of the center and show that the automorphism group is not graded, though we are able to describe certain families of automorphisms in this case, as well as those of certain subalgebras.

1. Introduction

In [**37**], Yakimov resolved the Launois-Lenagan conjecture [**23**], computing the automorphism group of a generic single-parameter quantum matrix algebra. In this work we are interested in a related question, but in the non-generic case. That is, we study automorphisms of single- and multi-parameter quantum matrix algebras at roots of unity.

Throughout, $\Bbbk$ is an algebraically closed field of characteristic zero. All algebras are $\Bbbk$-algebras and all tensor products should be regarded as over $\Bbbk$. For $\xi \in \Bbbk^\times$, we write $\operatorname{ord}(\xi)$ to denote the order of ξ as an element of the multiplicative group $\Bbbk^\times$. The center of an algebra A will be denoted $\mathcal{Z}(A)$.

Let $\lambda \in \Bbbk^\times$ and let $\mathbf{p}$ be a multipliciatively antisymmetric matrix over $\Bbbk$ (i.e., $p_{ij} = p_{ji}^{-1}$ for all i, j). The *$(n \times n)$ multi-parameter quantum matrix algebra*, $\mathcal{O}_{\lambda,\mathbf{p}}(M_n(\Bbbk))$, is generated by $\{x_{ij}\}_{1 \leq i,j \leq n}$ subject to the relations

$$x_{lm}x_{ij} = \begin{cases} p_{li}p_{jm}x_{ij}x_{lm} + (\lambda - 1)p_{li}x_{im}x_{lj} & l > i, m > j \\ \lambda p_{li}p_{jm}x_{ij}x_{lm} & l > i, m \leq j \\ p_{jm}x_{ij}x_{lm} & l = i, m > j. \end{cases}$$

Let $q \in \Bbbk^\times$. One can recover the single-parameter quantum matrix algebra $\mathcal{O}_q(M_n(\Bbbk))$ from the above definition by setting $p_{ij} = q$ for $i > j$ and $\lambda = q^{-2}$. The single-paramter and multi-parameter quantum matrix algebras are also related through cocycle twisting, which we review in the next section.

For $q \in \Bbbk^\times$ a nonroot of unity, Alev and Chamarie showed that $\operatorname{Aut}(\mathcal{O}_q(M_2(\Bbbk)))$ is a semidirect product $(\Bbbk^\times)^3 \rtimes \{\tau\}$ where $(\Bbbk^\times)^3$ is isomorphic to the group of scalar

---

2020 *Mathematics Subject Classification.* 16W20,16T20,16S38,16W50.

Key words and phrases. Quantum matrix algebras, noncommutative discriminant, automorphisms.

Lamkin was partially supported by the Miami University Dean's Scholar program.

automorphisms and τ is the transposition given by $\tau(x_{ij}) = x_{ji}$ for all $1 \le i,j \le 2$ [**1**, Theorem 2.3]. Launois and Lenagan proved that $\operatorname{Aut}(\mathcal{O}_q(M_3(\Bbbk)))$ has a similar description, thus motivating their conjecture that such a result should hold for $\operatorname{Aut}(\mathcal{O}_q(M_n(\Bbbk)))$. As mentioned above, this conjecture was resolved by Yakimov [**37**]. See also [**24**] for results in the case of *rectangular* quantum matrix algebras.

Automorphisms of some quantum algebras at roots of unity have been considered previously, for example by Alev and Dumas [**2**]. Our primary tool for studying automorphisms in the non-generic case will be noncommutative discriminants, as developed by Ceken, Palmieri, Wang, and Zhang [**7, 8**]. However, the rank of $\mathcal{O}_{\lambda,\mathbf{p}}(M_n(\Bbbk))$ over its center is large and so it is difficult to compute this discriminant directly.

There are several ways one may overcome this obstruction. One could view $\mathcal{O}_{\lambda,\mathbf{p}}(M_n(\Bbbk))$ as a specialization and consider the Poisson structure on the center so as to employ the Poisson techniques of Nguyen, Trampel, and Yakimov [**28**]. On the other hand, one could view $\mathcal{O}_{\lambda,\mathbf{p}}(M_n(\Bbbk))$ as an iterated Ore extension and in this way invoke results of Kirkman, Moore, and the first author [**12**]. More recently, Chan, Won, Zhang and the first author introduced and utilized the *reflexive hull discriminant* in cases where an algebra is not free over its center [**6**]. Finally, one could use a property $\mathcal{P}$-discriminant as introduced in [**25**]. Our strategy will be a mixture of these.

In Section 2, we review necessary prerequisites on quantum matrix algebras and noncommutative discriminants. Section 3 studies automorphisms in the multiparameter case.

THEOREM 1 (Theorem 3.5). *Let $M = \mathcal{O}_{\lambda,\mathbf{p}}(M_2(\Bbbk))$ be PI. If $\mathcal{Z}(M)$ is polynomial, then* $\operatorname{Aut}(M) = \operatorname{Aut}_{\mathrm{gr}}(M)$.

In addition we give explicit conditions for the center to be polynomial and compute the automorphism group, showing that the result aligns with that of Launois and Lenagan.

For the $n = 3$ case, we do not know if there exist examples where the center is polynomial. However, under this hypothesis it is possible to show that all automorphisms are graded. We refer the interested reader to the appendix of the arXiv version of this paper for that argument. Note that, in general, centers of PI AS regular algebras are quite complicated even for basic examples such as PI skew polynomial rings. See [**5, 8**] for partial progress on this question.

In the remainder of the paper, we study the single-parameter case $\mathcal{O}_q(M_n(\Bbbk))$ assuming q is a root of unity. For much of this we will assume $\operatorname{ord}(q) = m \ge 3$ is an odd positive integer, but offer some general results as well.

Section 4 is a study on the center of $\mathcal{O}_q(M_n(\Bbbk))$. In our setting the generators of the center were computed by Jakobsen and Zhang [**18**]. We give a presentation of the center using Gröbner basis techniques (Theorem 4.4). We are further able to derive several properties of the center.

THEOREM 2 (Theorem 4.5). *Assume that q is a root of unity such that* $\operatorname{ord}(q) =$ *$m \ge 3$ is odd. Then $Z = \mathcal{Z}(\mathcal{O}_q(M_n(\Bbbk)))$ is not Gorenstein and $\mathcal{O}_q(M_n(\Bbbk))$ is not projective over Z.*

In Section 5 we study automorphisms of $\mathcal{O}_q(M_n(\Bbbk))$. We show that the Launois-Lenagan conjecture does not extend to this setting.

THEOREM 3 (Theorem 5.1). *Assume that q is a root of unity. Then $\mathcal{O}_q(M_n(\Bbbk))$ has non-graded automorphisms. Moreover,* $\mathrm{Aut}(\mathcal{O}_q(M_n(\Bbbk)))$ *contains a free product on two generators.*

We show additionally that $\mathrm{Aut}(\mathcal{O}_q(M_2(\Bbbk)))$ contains an automorphism which restricts to a wild automorphism on a polynomial subalgebra (Theorem 5.2). Using discriminants, we establish a certain ideal that is fixed by every automorphism (Proposition 5.3). We conclude by studying certain subalgebras of $\mathcal{O}_q(M_3(\Bbbk))$.

2. Background on quantum matrix algebras and discriminants

In this section we provide some basic background and references for our main results and computations below. For an algebra A, we denote its Gelfand-Kirillov (GK) by $\mathrm{GKdim}(A)$ and its global dimension by $\mathrm{gldim}(A)$. If A is commutative, we denote its Krull dimension by $\mathrm{Kdim}(A)$.

2.1. Quantum matrix algebras. Notice that if $\lambda = 1$, then $\mathcal{O}_{\lambda,\mathbf{p}}(M_n(\Bbbk))$ is isomorphic to some quantum affine space. On the other hand, when $\lambda = -1$, then $\mathcal{O}_{\lambda,\mathbf{p}}(M_n(\Bbbk))$ does not have the same Hilbert series as a polynomial ring in n^2 variables. Hence, we assume throughout $\lambda^2 \neq 1$.

Recall that if $\Gamma = (\gamma_{ij})$ is a multiplicatively antisymmetric matrix, then the *quantum affine space* $\mathcal{O}_\Gamma(\Bbbk^n)$ is the algebra generated by $x_1, \ldots, x_n$ with relations $x_i x_j = \gamma_{ij} x_j x_i$ for all i, j. Quantum matrix algebras are related to quantum affine spaces in a way explained in the next proposition.

We begin by reviewing several well-known properties of $\mathcal{O}_{\lambda,\mathbf{p}}(M_n(\Bbbk))$ and its center. Recall that a prime affine algebra A which is finitely generated over its center is *homologically homogeneous* if all simple A-modules have the same projective dimension over the center.

PROPOSITION 2.1. *Assume $\lambda^2 \neq 1$. Let $M = \mathcal{O}_{\lambda,\mathbf{p}}(M_n(\Bbbk))$. Then the following hold.*

(1) *M is Artin-Schelter regular with* $\mathrm{gldim}(M) = \mathrm{GKdim}(M) = n^2$.
(2) *M is PI if and only if λ and the p_{ij}, $1 \leq i, j \leq n$, are roots of unity.*
(3) *If M is PI, then $\mathcal{Z}(M)$ is a Cohen-Macaulay (CM) normal Krull domain.*

PROOF. (1) This follows from [**3**, Theorem 2].

(2) This is clear by noting that $\mathrm{Frac}\, M$ is isomorphic to the quotient division ring of $\mathcal{O}_\Gamma(\Bbbk^{n^2})$ where Γ is the multiplicatively antisymmetric matrix whose entries are products of the p_{ij} and λ ([**16**, Corollary 4.7]).

(3) Let $m = \mathrm{lcm}(\mathrm{ord}(p_{ij}), \mathrm{ord}(\lambda p_{ji}))$. Then it is straightforward to see that $C = \Bbbk[x_{ij}^m \mid 1 \leq i, j \leq n]$ is a central subalgebra of M. Moreover, M is a free C-module with basis $\{\prod_{i,j=1}^n x_{ij}^{r_{ij}} \mid 0 \leq r_{ij} < m\}$. Since C is CM, M is a CM C-module. Therefore, M is homologically homogeneous by [**34**, Lemma 2.4]. That $\mathcal{Z}(M)$ is a normal CM domain then follows by [**34**, Theorem 2.3(d)]. Since M is a homologically homogeneous Noetherian PI domain, Z is a Krull domain by [**35**, Theorem 1.4(iii)]. □

The quantum determinant of $\mathcal{O}_{\lambda,\mathbf{p}}(M_n(\Bbbk))$ is defined as

$$D_{\lambda,\mathbf{p}} = \sum_{\pi \in S_n} \left(\prod_{\substack{1 \le i < j \le n \\ \pi(i) > \pi(j)}} (-p_{\pi(i),\pi(j)}) \right) x_{1,\pi(1)} x_{2,\pi(2)} \cdots x_{n,\pi(n)}, \tag{2.1}$$

where $l(\pi)$ is the number of inversions in π. By [**3**, Theorem 3(a)],

$$D_{\lambda,\mathbf{p}} x_{ij} = \lambda^{j-i} \left(\prod_{\ell=1}^{n} p_{j\ell} p_{\ell i} \right) x_{ij} D_{\lambda,\mathbf{p}}. \tag{2.2}$$

That is, $D_{\lambda,\mathbf{p}}$ is a normal element in $\mathcal{O}_{\lambda,\mathbf{p}}(M_n(\Bbbk))$.

In the single-parameter case, (2.1) reduces to

$$D_q = \sum_{\pi \in S_n} (-q)^{l(\pi)} x_{1,\pi(1)} x_{2,\pi(2)} \cdots x_{n,\pi(n)}, \tag{2.3}$$

where $l(\pi)$ is the number of inversions in π. By (2.2) or [**30**, Theorem 4.6.1], D_q is a central element in the single-parameter case. By [**30**, Corollary 4.4.4], we have the following quantum Laplace expansion formula in the single-parameter case,

$$D_q = \sum_{j=1}^{n} (-q)^{j-i} x_{ij} A(i,j) = \sum_{j=1}^{n} (-q)^{j-i} x_{ji} A(j,i). \tag{2.4}$$

We caution the reader that our convention differs from Parshall and Wang (we replace q^{-1} by q).

For subsets $I, J \subset \{1,\dots,n\}$ with $|I| = |J| \ge 1$, the subalgebra of $\mathcal{O}_{\lambda,\mathbf{p}}(M_n(\Bbbk))$ generated by $\{x_{ij} \mid i \in I, j \in J\}$ is another quantum matrix algebra and the quantum determinant of this subalgebra, denoted $D(I,J)$, is a *quantum minor* of $\mathcal{O}_{\lambda,\mathbf{p}}(M_n(\Bbbk))$.

Let $I, J \subset \{1,\dots,n\}$ with $|I| = |J| = k \ge 1$. Set $\widetilde{I} = \{1,\dots,n\} \backslash I$ and $\widetilde{J} = \{1,\dots,n\} \backslash J$. For simplicity, let $\widetilde{i} = \widetilde{\{i\}}$ in the case of a singleton. Set $A(I,J) = D(\widetilde{I},\widetilde{J})$ (and $A(i,j)$ in the case of singletons). For $1 \le t \le n$, let

$$D(t) := D(\{1,\dots,t\}, \{n-t+1,\dots,n\}). \tag{2.5}$$

Cocycle twisting for $\mathcal{O}_{\lambda,\mathbf{p}}(M_n(\Bbbk))$ was first established by Artin, Schelter, and Tate, though our review follows from Goodearl and Lenagan [**13**].

The algebra $\mathcal{O}_{\lambda,\mathbf{p}}(M_n(\Bbbk))$ has a bigrading by $\mathbb{Z}^n \times \mathbb{Z}^n$ in which x_{ij} has degree (ϵ_i, ϵ_j). Choose q such that $\lambda = q^{-2}$. We write $\mathbf{a}, \mathbf{b} \in \mathbb{Z}^n$ for $\mathbf{a} = (a_i)$ and $\mathbf{b} = (b_i)$. Define a map $c : \mathbb{Z}^n \times \mathbb{Z}^n \to \Bbbk^\times$ by

$$c(\mathbf{a},\mathbf{b}) = \prod_{i>j} (q p_{ji})^{a_i b_j}.$$

Then

$$c(\mathbf{a}+\mathbf{a}',\mathbf{b}) = c(\mathbf{a},\mathbf{b}) c(\mathbf{a}',\mathbf{b}) \quad \text{and} \quad c(\mathbf{a},\mathbf{b}+\mathbf{b}') = c(\mathbf{a},\mathbf{b}) c(\mathbf{a},\mathbf{b}'),$$

so that c is a multiplicative bicharacter on $\mathbb{Z}^n$ and thus a 2-cocycle. Then

$$c(\epsilon_i, \epsilon_j) = \begin{cases} q p_{ji} & \text{if } i > j \\ 1 & \text{if } i \le j. \end{cases}$$

Let $A_q = \mathcal{O}_q(M_n(\Bbbk))$. We use c to twist the multiplication on A_q to obtain an algebra A'_q. There is an isomorphism of graded vector spaces $A_q \to A'_q$ given by $a \mapsto a'$. The multiplication on A'_q is given by

$$a'b' = c(u_1, v_1)^{-1}c(u_2, v_2)(ab)'$$

for homogeneous $a, b \in A_q$ of bidegrees (u_1, u_2) and (v_1, v_2), respectively. Then it is easily checked that $A'_q \cong \mathcal{O}_{\lambda,\mathbf{p}}(M_n(\Bbbk))$.

Using the method of cocycle twists above we can easily establish a multi-parameter version of (2.4).

2.2. Discriminants. Applications of discriminants to the study of automorphism groups of noncommutative algebras were studied by Ceken, Palmieri, Wang, and Zhang [**7, 8**]. Since then, it has been implemented by various authors and many tools have been developed to aid in their computation, thus resolving a host of problems in noncommutative algebra and noncommutative invariant theory.

Let A be a $\Bbbk$-algebra and C a central subalgebra which is a domain and such that A is finitely generated over C. In general, A may not be free over C, so we pass to a localization F of C such that $A_F := A \otimes_C F$ is free (and finitely generated) over F. Set $w = \operatorname{rank}_F(A_F) < \infty$. Then left-multiplication gives a natural embedding $\operatorname{lm} : A \to A_F \to \operatorname{End}_F(A_F)$. By the hypotheses, $\operatorname{End}_F(A_F) \cong M_w(F)$ and we let $\operatorname{tr}_{\text{int}}$ denote the usual matrix trace in $M_w(F)$. The *regular trace* is then the composition

$$\operatorname{tr}_{\text{reg}} : A \xrightarrow{\operatorname{lm}} M_w(F) \xrightarrow{\operatorname{tr}_{\text{int}}} F.$$

Here, we only consider the case that the algebra A is free over C of rank $w < \infty$, and the image of $\operatorname{tr}_{\text{reg}}$ is in C. In this case, we can take a C-algebra basis $\{c_1, \dots, c_w\}$ of A. Then the discriminant is defined as

$$d(A/C) = \det(\operatorname{tr}_{\text{reg}}(c_i c_j)_{i,j=1}^{w}) \in C.$$

Note that the discriminant is unique up to a unit in C. For any automorphism ϕ of A which preserves C, the discriminant is preserved by ϕ up to a unit in C.

Now let A be a prime $\Bbbk$-algebra and let Z be the center of A so that A is free over Z and the image of tr is in Z. Suppose that $X = \operatorname{Spec} Z$ is an affine normal $\Bbbk$-variety. Let U be an open subset of X such that $X \backslash U$ has codimension 2 in X. If if there exists an element $d \in Z$ such that the principal ideal (d) of Z agrees with $d(A/C)$ on U, then [**6**, Lemma 2.3(2)] implies that $d(A/Z) =_{Z^\times} d$. In [**6**], this was called the *reflexive hull discriminant* and defined more generally in the context of the *modified discriminant.* However, freeness implies that the modified discriminant and the discriminant agree. Consequently, the reflexive hull discriminant and the discriminant agree.

One can also define discriminants relative to some (algebraic) property. Let A be an algebra and let $Z = \mathcal{Z}(A)$. Let $\mathcal{P}$ be a property defined for $\Bbbk$-algebras that is invariant under algebra isomorphisms. Following [**25**], we define the $\mathcal{P}$*-locus* of A to be

$$L_{\mathcal{P}}(A) := \{\mathfrak{m} \in \operatorname{MaxSpec}(Z) \mid Z_{\mathfrak{m}} \text{ has property } \mathcal{P}\},$$

and the $\mathcal{P}$*-discriminant ideal* to be

$$I_{\mathcal{P}}(A) := \bigcap_{\mathfrak{m} \in L_{\mathcal{P}}(A)} \mathfrak{m} \subseteq Z.$$

By [**25**, Lemma 3.5], if $\phi : A \to B$ is an algebra isomorphism, then

$$\phi(L_{\mathcal{P}}(A)) = L_{\mathcal{P}}(B) \quad \text{and} \quad \phi(I_{\mathcal{P}}(A)) = I_{\mathcal{P}}(B).$$

Our definition of the $\mathcal{P}$-locus differs slightly from that of [**25**] in that we study localizations of the center instead of quotients of the algebra. This choice reflects the fact that, in the case of $\mathcal{O}_q(M_2(\Bbbk))$, it is both easier and sufficient to determine useful information about the automorphism group using the above definition.

3. The 2×2 multi-parameter case

In this section we compute the automorphism group for the 2×2 quantum matrix algebra in the multi-parameter setting. The relations in $\mathcal{O}_{\lambda,\mathbf{p}}(M_2(\Bbbk))$, explicitly, are

$$\begin{aligned} x_{12}x_{11} &= p_{12}x_{11}x_{12} & x_{21}x_{11} &= (\lambda p_{21})x_{11}x_{21} \\ x_{22}x_{12} &= (\lambda p_{21})x_{12}x_{22} & x_{22}x_{21} &= p_{12}x_{21}x_{22} \\ x_{21}x_{12} &= (\lambda p_{21})p_{21}x_{12}x_{21} & x_{22}x_{11} &= x_{11}x_{22} + (\lambda - 1)p_{21}x_{12}x_{21}. \end{aligned}$$

The subalgebra R of $\mathcal{O}_{\lambda,\mathbf{p}}(M_2(\Bbbk))$ generated by $\{x_{11}, x_{12}, x_{21}\}$ is a quantum affine 3-space. Then $\mathcal{O}_{\lambda,\mathbf{p}}(M_2(\Bbbk)) = R[x_{22}; \sigma, \delta]$ where σ is an automorphism of R and δ a σ-derivation of R, and these are determined by

$$\begin{aligned} \sigma(x_{11}) &= x_{11} & \sigma(x_{12}) &= \lambda p_{21}x_{12} & \sigma(x_{21}) &= p_{12}x_{21} \\ \delta(x_{11}) &= (\lambda - 1)p_{21}x_{12}x_{21} & \delta(x_{12}) &= 0 & \delta(x_{21}) &= 0. \end{aligned}$$

Note that

$$\sigma(\delta(x_{11})) = \sigma((\lambda - 1)p_{21}x_{12}x_{21}) = \lambda(\lambda - 1)p_{21}(x_{12}x_{21}) = \lambda^{-1}\delta(x_{11}) = \lambda^{-1}\delta(\sigma x_{11}).$$

So $\delta(\sigma(x_{11})) = \lambda\sigma(\delta(x_{11}))$. This holds trivially with x_{11} replaced by x_{12} or x_{21}, and so (σ, δ) is a (left) λ-skew derivation of R.

We will assume the following hypothesis throughout this section.

HYPOTHESIS 3.1. Let $M = \mathcal{O}_{\lambda,\mathbf{p}}(M_2(\Bbbk))$ as above where p_{12} and λp_{12} are roots of unity such that $\operatorname{ord}(p_{12})$ and $\operatorname{ord}(\lambda p_{21})$ are relatively prime. We set $\ell = \operatorname{lcm}(\operatorname{ord}(p_{12}), \operatorname{ord}(\lambda p_{21}))$ and let $y_{ij} = x_{ij}^{\ell}$ for $i, j \in \{1, 2\}$. Throughout, let $Z = \mathcal{Z}(M)$ and let $D = D_{\lambda,\mathbf{p}}$.

LEMMA 3.2. *Assume Hypothesis* 3.1. *Then* $Z = \Bbbk[y_{11}, y_{12}, y_{21}, y_{22}]$.

PROOF. It is easy to show that x_{12}^{ℓ} and x_{21}^{ℓ} are central. To check x_{22}^{ℓ} is central, we apply [**14**, Lemma 6.2]. Note that $\delta^2(x_{11}) = 0$, and so we have

$$x_{22}^{\ell}x_{11} = x_{11}x_{22}^{\ell} + [\ell]_{\lambda}\sigma^{\ell-1}\delta(x_{11})x_{22}^{\ell-1} = x_{11}x_{22}^{\ell}$$

where the last equality holds because $\operatorname{ord}(\lambda)$ divides ℓ. A similar proof shows that x_{11}^{ℓ} is central.

On the other hand, suppose $z \in Z$ and write $z = \sum f_{ij}x_{11}^{i}x_{22}^{j}$ where each f_{ij} is an element in the subalgebra generated by x_{12} and x_{21}. Let $f_{uv}x_{11}^{u}x_{22}^{v}$ be the element of highest degree in z according to the ordering $x_{22} > x_{11}$. Hence,

$$0 = [x_{11}, z] = [x_{11}, f_{uv}]x_{11}^{u}x_{22}^{v} + (\text{lower degree terms in } x_{22}).$$

Thus $[x_{11}, f_{uv}] = 0$ and similarly $[f_{uv}, x_{22}] = 0$. Write $f_{uv} = \sum \gamma_{ij} x_{12}^i x_{21}^j$ with $\gamma_{ij} \in \Bbbk$. Then

$$0 = [x_{11}, f_{uv}] = x_{11} \sum \gamma_{ij}(1 - p_{12}^i(\lambda p_{21})^j) x_{12}^i x_{21}^j$$
$$0 = [f_{uv}, x_{22}] = \sum \gamma_{ij}((\lambda p_{21})^i p_{12}^j - 1) x_{12}^i x_{21}^j x_{22}.$$

By our hypothesis, for every i, j with $\gamma_{ij} \neq 0$, the first equation gives that $p_{12}^i = 1$ and $(\lambda p_{21})^j = 1$. Similarly, the second equation gives $(\lambda p_{21})^i = 1$ and $p_{12}^j = 1$. Thus, for every i with $\gamma_{ij} \neq 0$, $\operatorname{ord}(p_{12})$ and $\operatorname{ord}(\lambda p_{21})$ divide i, so $\ell \mid i$. Similarly, $\ell \mid j$. That is, f_{uv} is a polynomial in x_{12}^ℓ and x_{21}^ℓ. Now,

$$0 = [x_{12}, f_{uv} x_{11}^u x_{22}^v] = (p_{12}^u - (\lambda p_{21})^v) f_{uv} x_{11}^u x_{12} x_{22}^v$$
$$0 = [x_{21}, f_{uv} x_{11}^u x_{22}^v] = ((\lambda p_{21})^u - p_{12}^v) f_{uv} x_{11}^u x_{21} x_{22}^v.$$

A similar argument to the above shows that $\ell \mid u$ and $\ell \mid v$. □

Our strategy will then be the following. We will consider a localization of M which is isomorphic to a localization of a quantum affine space. The discriminant of this quantum affine space over its center is known (see [**8**]). We compute this on "enough" localizations, then patch them together using the method discussed in Section 2. We remark that using localizations to compute discriminants has also been employed in [**9, 26**].

LEMMA 3.3. *Assume Hypothesis* 3.1. *Let* $Y = \{y_{11}^k : k \geq 0\}$. *Then* Y *is an Ore set in* M *and* $MY^{-1} = M[y_{11}^{-1}]$ *is isomorphic to a localization of a quantum affine space.*

PROOF. It is clear that Y is an Ore set in M. Recall that R is the subalgebra of M generated by $\{x_{11}, x_{12}, x_{21}\}$. Note that Y may also be regarded as an Ore set in R. Let $R' = RY^{-1} = R[y_{11}^{-1}]$. The automorphism σ and the σ-derivation δ introduced above have a unique extension to R' which, by an abuse of notation, we also call σ and δ, respectively. Hence we have

$$M' = (R[x_{22}; \sigma, \delta])Y^{-1} = (R')[x_{22}; \sigma, \delta].$$

The σ-derivation δ is inner if there exists ω such that $\delta(r) = \omega r - \sigma(r)\omega$ for all $r \in R'$. Note that x_{11} is a unit in R' since $x_{11}(x_{11}^{m-1}(x_{11}^m)^{-1}) = 1$. Set $\omega = p_{21} x_{11}^{-1} x_{12} x_{21} \in R'$. Then we have

$$\begin{aligned}
\omega x_{11} - \sigma(x_{11})\omega &= (p_{21} x_{11}^{-1} x_{12} x_{21}) x_{11} - x_{11}(p_{21} x_{11}^{-1} x_{12} x_{21}) \\
&= p_{21}(\lambda - 1) x_{12} x_{21} = \delta(x_{11}) \\
\omega x_{12} - \sigma(x_{12})\omega &= (p_{21} x_{11}^{-1} x_{12} x_{21}) x_{12} - \lambda p_{21} x_{12}(p_{21} x_{11}^{-1} x_{12} x_{21}) \\
&= p_{21}(\lambda p_{21}^2 - (\lambda p_{21}) p_{21})(x_{11}^{-1} x_{12}^2 x_{21}) = 0 \\
\omega x_{21} - \sigma(x_{21})\omega &= (p_{21} x_{11}^{-1} x_{12} x_{21}) x_{21} - p_{12} x_{21}(p_{21} x_{11}^{-1} x_{12} x_{21}) \\
&= p_{21}(1 - p_{12}(\lambda^{-1} p_{12})(\lambda p_{21}^2))(x_{11}^{-1} x_{12} x_{21}^2) x_{21} = 0.
\end{aligned}$$

It follows that for all $r \in R$,

$$x_{22} r = \sigma(r) x_{22} + \delta(r) = \sigma(r) x_{22} + \omega r - \sigma(r)\omega \Rightarrow (x_{22} - \omega) r = \sigma(r)(x_{22} - \omega).$$

Thus, $M' = R'[x_{22}; \sigma, \delta] = R'[x_{22} - \omega; \sigma]$. □

Next we compute the discriminant of M over Z, which agrees with the formula for the single-parameter quantum matrix algebra case considered in [**28**].

LEMMA 3.4. *Assume Hypothesis* 3.1. *Let* $\Omega := D^\ell = y_{11}y_{22} - y_{12}y_{21}$. *Then*

$$d(M/Z) =_{\Bbbk^\times} (x_{12}x_{21}D)^{\ell^4(\ell-1)} =_{\Bbbk^\times} (y_{12}y_{21}\Omega)^{\ell^3(\ell-1)}.$$

PROOF. Let $S = \Bbbk_{\mathbf{q}}[x_1, x_2, x_3, x_4]$ be the quantum affine space whose localization is isomorphic to $M' = M[y_{11}^{-1}]$ by Lemma 3.3. Here, x_1, x_2, x_3, x_4 represent the images of $x_{11}, x_{12}, x_{21}, x_{22} - \omega$, respectively, in S. Note that S is free over the central subalgebra $C = \Bbbk[x_1^\ell, x_2^\ell, x_3^\ell, x_4^\ell]$ and so by [**8**, Proposition 2.8],

$$d(S/C) =_{\Bbbk^\times} (x_1x_2x_3x_4)^{\ell^4(\ell-1)}.$$

Set $S' = S[y_1^{-1}]$ and let $C' = C[y_1^{-1}]$. Then $d(S'/C') =_{(C')^\times} d(S/C)$ by [**9**, Lemma 1.3]. Recall that $M' \cong S'$. Set $Z' = Z[y_{11}^{-1}]$. Then

$$d(M'/Z') =_{(Z')^\times} (x_{12}x_{21}D)^{\ell^4(\ell-1)}.$$

By a completely symmetric argument, localizing at powers of y_{22} in place of y_{11} gives the same result. It is left to show that this agrees with $d(M/Z)$.

Clearly M is a prime $\Bbbk$-algebra and $X = \operatorname{Spec} Z$ is an affine normal $\Bbbk$-variety. Thus, (M, Z) satisfies [**6**, Hypothesis 2.1]. Let U_1 and U_2 be the open subsets of X with $y_{11} \neq 0$ and $y_{22} \neq 0$, respectively. Let I be the ideal in Z generated by y_{11} and y_{22}. Then we have

$$\begin{aligned}\operatorname{codim}(X\backslash(U_1 \cup U_2)) &= \operatorname{codim}((X\backslash U_1) \cap (X\backslash U_2)) = \operatorname{codim}(V(y_{11}) \cap V(y_{22})) \\ &= \operatorname{codim}(V(y_{11}, y_{22})) = \operatorname{codim}(I) = \operatorname{height}(I) = 2.\end{aligned}$$

Thus, by [**6**], the discriminant on $U_1 \cup U_2$ extends to a discriminant over X. □

We conclude this section by computing the automorphism group of M.

THEOREM 3.5. *Assume Hypothesis* 3.1. *Then* $\operatorname{Aut}(M) = (\Bbbk^\times)^3 \rtimes \{\tau\}$.

PROOF. The same argument as in [**28**, Proposition 5.10] shows that $d(M/Z)$ is locally dominating. Now by [**7**, Theorem 2.7], $\operatorname{Aut}(M)$ is affine. It follows easily that $\operatorname{Aut}(M)$ is graded (see [**24**, Proposition 4.2]). That is, $\operatorname{Aut}(M) = \operatorname{Aut}_{\mathrm{gr}}(M)$.

Let $\phi \in \operatorname{Aut}(M)$. By [**11**, Theorem 3.4], the (homogeneous) degree one normal elements of M are of the form $\alpha_1 x_{12} + \alpha_2 x_{21}$ for $\alpha_i \in \Bbbk$. By the same argument as in [**11**, Lemma 5.5], the ideals (x_{12}) and (x_{21}) are either fixed or swapped by any automorphism. As the automorphism τ swaps them, then up to conjugation by τ we may assume ϕ fixes these ideals. Hence $\phi(x_{12}) = bx_{12}$ and $\phi(x_{21}) = cx_{21}$ for some $b, c \in \Bbbk^\times$. Write,

$$\begin{aligned}\phi(x_{11}) &= a_1x_{11} + a_2x_{12} + a_3x_{21} + a_4x_{22} \\ \phi(x_{22}) &= d_1x_{11} + d_2x_{12} + d_3x_{21} + d_4x_{22},\end{aligned}$$

for some $a_i, d_i \in \Bbbk$. Then

$$\begin{aligned}0 &= \phi(x_{12}x_{11} - p_{12}x_{11}x_{12}) \\ &= b(a_2(1 - p_{12})x_{12}^2 + a_3p_{12}(1 - \lambda p_{21})x_{12}x_{21} + a_4(1 - p_{21}^2)x_{12}x_{22}).\end{aligned}$$

Hence, $a_2 = a_3 = 0$. Now

$$0 = \phi(x_{21}x_{11} - \lambda p_{21}x_{11}x_{21}) = ca_4(1 - \lambda)x_{12}x_{22}.$$

Since we assume that $\lambda \neq 1$, this implies that $a_4 = 0$. Similarly one shows that $d_1 = d_2 = d_3 = 0$. Set $a = a_1$ and $d = d_1$. Then we have

$$\begin{aligned} bc(\lambda - 1)p_{21}x_{12}x_{21} &= \phi(x_{22}x_{11} - x_{11}x_{22}) = (dx_{22})(ax_{11}) - (ax_{11})(dx_{22}) \\ &= ad(\lambda - 1)p_{21}x_{12}x_{21}. \end{aligned}$$

Thus, we let H denote the automorphisms of M satisfying $\phi(x_{11}) = ax_{11}$, $\phi(x_{12}) = bx_{12}$, $\phi(x_{21}) = cx_{21}$, and $\phi(x_{22}) = dx_{22}$ satisfying $ad = bc$. Then $\mathrm{Aut}(M) = H \rtimes \{\tau\}$. □

4. The center of $\mathcal{O}_q(M_n(\Bbbk))$

From here on, we study single-parameter quantum matrix algebras $\mathcal{O}_q(M_n(\Bbbk))$ at roots of unity. Recall that $\mathcal{O}_q(M_n(\Bbbk))$ is generated by $\{x_{ij}\}_{1 \leq i,j \leq n}$ subject to the relations

$$x_{ij}x_{kl} = \begin{cases} x_{kl}x_{ij} + (q - q^{-1})x_{il}x_{kj} & k > i, l > j \\ qx_{kl}x_{ij} & k > i, l = j \\ qx_{kl}x_{ij} & k = i, l > j \\ x_{ij}x_{lm} & k > i, l < j. \end{cases}$$

In the generic case, the center of $\mathcal{O}_q(M_n(\Bbbk))$ is $\Bbbk[D_q]$ ([**29**, Theorem 1.6], [**31**]). Our interest is in the non-generic case. For most of this section we work under the following hypothesis.

HYPOTHESIS 4.1. Assume that q is a root of unity such that $\mathrm{ord}(q) = m \geq 3$ is odd. Let $Z = \mathcal{Z}(\mathcal{O}_q(M_n(\Bbbk)))$.

Our goal will be to give a full presentation for Z in the setting of Hypothesis 4.1. Recall that τ is the automorphism of $\mathcal{O}_q(M_n(\Bbbk))$ defined by $x_{ij} \mapsto x_{ji}$ and that the notation $D(t)$ was introduced in (2.5). Assuming Hypothesis 4.1, a result of Jakobsen and Zhang [**18**, Theorem 6.2] shows that Z is generated by

$$\{x_{ij}^m, D_q, D(t)^r\tau(D(n-t)^{m-r}) \mid i, j, t = 1, \dots, n \text{ and } r = 0, \dots, m\}. \tag{4.1}$$

Jakobsen and Zhang also computed a generating set in the case of m even. However, as it is significantly more complicated, we do not consider that problem here.

We begin in earnest by setting up some notation, and proving a technical lemma relating quantum minors. For a positive integer N, we use the notation $[N] = \{1, \dots, N\}$. We will rely on some results of Scott [**32**]. In the notation of that paper, $D(t) = \Delta_{[t],[n]-[n-t]}$ for $1 \leq t \leq n-1$. Hence, $\tau(D(t)) = \Delta_{[n]-[n-t],[t]}$. Given $I, J \subseteq [n]$, we write $I \prec J$ if $i < j$ for all $i \in I$ and $j \in J$.

LEMMA 4.2. *Let* $1 \leq t, t' \leq n-1$. *Then*

(1) $D(t)\tau(D(t')) = \tau(D(t'))D(t)$,
(2) $D(t)D(t') = D(t')D(t)$.

PROOF. (1) This is trivial if $t' \leq n-t$ as then each term of $D(t)$ and each term of $\tau(D(t'))$ commute, so we may assume $t' > n-t$. Let

$$\begin{aligned} I &:= \{t'+1, \dots, n, n+1, \dots, n+t'\} \text{ and} \\ J &:= \{1, \dots, n-t, 2n-t+1, \dots, 2n\}. \end{aligned}$$

In the notation of [**32**], $I = S([n]-[n-t'],[t'])$ and $J = S([t],[n]-[n-t])$. We claim that I and J are weakly separated ([**32**, Definition 2]).

Since $|I| = |J| = n$ and $t' > n - t$, we have

$$I - J = \{t' + 1, \dots, 2n - t\}, \text{ and}$$
$$J - I = \{1, \dots, n - t, n - t' + 1, \dots, 2n\} = [n - t] \sqcup \{n + t' + 1, \dots, 2n\}.$$

Denoting $J' = [n-t]$ and $J'' = \{n + t' + 1, \dots, 2n\}$, we see $J' \prec I - J \prec J''$ once again using $t' > n - t$. Hence by [**32**, Theorem 1], $D(t)$ and $\tau(D(t'))$ skew-commute by a factor of q^c for some $c \in \mathbb{Z}$. By [**32**, Theorem 2], the integer c can be computed as

$$c = |J''| - |J'| + \Big|[n] - [n - t']\Big| - \Big|[t]\Big| = n - t' - n + t + t' - t = 0.$$

The result follows.

(2) WLOG, assume $t' \leq t$. Since D_q is a central element in each $\mathcal{O}_q(M_n(\Bbbk))$, it follows that $D(t)$ is a central element in the subalgebra generated by the x_{ij} with $1 \leq i \leq t, n - t + 1 \leq j \leq n$. In particular, $D(t)$ commutes with $D(t')$. $\square$

Our next goal is to determine the ideal of relations in Z. In order to simplify notation, denote

$$Z_{ij} := x_{ij}^m, \quad D := D_q, \quad Y_{tr} := D(t)^r \tau(D(n-t)^{m-r})$$

for $1 \leq i, j, t \leq n$ and $0 \leq r \leq m$. Hence, Z_{ij}, Y_{tr}, and D generate Z by (4.1). By [**19**, Proposition 2.12], we have the relation $D^m = \det(Z_{ij})$. More generally, $D(t)^m = \det(A_t)$ and $\tau(D(t))^m = \det(B_t)$, where $A_t = (Z_{ij})_{1 \leq i \leq t, n-t+1 \leq j \leq n}$ and $B_t = (Z_{ij})_{n-t+1 \leq i \leq n, 1 \leq j \leq t}$, for each $1 \leq t \leq n$. Consequently the Y_{t0} and Y_{tm} are superfluous, and since $D(n) = D$ and $D(0) = 1$, we only need the Y_{tr} with $1 \leq t \leq n - 1$ and $1 \leq r \leq m - 1$ to generate Z.

We now define a monomial ordering on these elements. Given a word w in the alphabet

$$\{Z_{ij}, Y_{tr}, D \mid 1 \leq i, j \leq n, 1 \leq t \leq n - 1, 1 \leq r \leq m - 1\},$$

let w_Y denote the subword consisting of the Y_{tr} in w, and let w_Z be the quotient w/w_Y. For example, if $w = Y_{11}Y_{21}^2 D Z_{12} Z_{13} Z_{14}$, then $w_Y = Y_{11}Y_{21}^2$ and $w_Z = D Z_{12} Z_{13} Z_{14}$. Now, given two such words w and w', we will say $w > w'$ if

- $w_Y >_{deglex} w'_Y$ with respect to the ordering $Y_{11} > \cdots > Y_{1,m-1} > Y_{21} > \cdots > Y_{n-1,m-1}$, or
- $w_Y = w'_Y$ and $w_Z >_{lex} w'_Z$ with respect to the ordering $D > Z_{11} > \cdots > Z_{1n} > Z_{21} > \cdots > Z_{nn}$.

LEMMA 4.3. *Assume Hypothesis 4.1. Let $\leq$ denote the monomial ordering defined above. With respect to $\leq$, the following families of elements form a Gröbner basis of the ideal they generate:*

(1) $D^m - \det(Z_{ij})$,
(2) $Y_{ti}Y_{tj} - Y_{t,i+j} \det(B_{n-t})$ *if* $i + j < m$,
(3) $Y_{ti}Y_{tj} - \det(A_t) \det(B_{n-t})$ *if* $i + j = m$,
(4) $Y_{ti}Y_{tj} - Y_{t,i+j-m} \det(A_t)$ *if* $i + j > m$,

for $1 \leq t \leq n - 1$.

PROOF. The relations hold by Lemma 4.2. We use the diamond lemma to show they form a Gröbner basis. Indeed, with respect to the monomial ordering $\leq$, one computes the following leading terms:

- $\mathrm{LT}(D^m - \det(Z_{ij}) = D^m$,

- $\operatorname{LT}(Y_{ti}Y_{tj} - Y_{t,i+j}\det(B_{n-t})) = Y_{ti}Y_{tj}$, for $i+j<m$,
- $\operatorname{LT}(Y_{ti}Y_{tj} - \det(A_t)\det(B_{n-t})) = Y_{ti}Y_{tj}$, for $i+j=m$,
- $\operatorname{LT}(Y_{ti}Y_{tj} - Y_{t,i+j-m}\det(A_t)) = Y_{ti}Y_{tj}$, for $i+j>m$.

There are no inclusion ambiguities and the only overlap ambiguities are of the form $Y_{ti}Y_{tj}Y_{tk}$ for some $1 \le i,j,k \le m-1$. Suppose $i+j+k<m$ and $m<i+j, j+k$. Then both $Y_{t,i+j}Y_{t,k}\det(B_{n-t})$ and $Y_{t,i}Y_{t,j+k}\det(B_{n-t})$ can be reduced to $Y_{t,i+j+k-m}\det(A_t)\det(B_{n-t})$. A similar reduction occurs in the case $i+j+k \le m$. The other cases are left to the reader. □

We are now in a position to determine a presentation for Z. We keep the notation of Lemma 4.3 and let I denote the ideal generated by the Gröbner basis computed therein.

THEOREM 4.4. *Assume Hypothesis* 4.1. *Let*

$$T = \Bbbk[Z_{ij}, Y_{tr}, D \mid 1 \le i,j \le n, 1 \le t \le n-1, 1 \le r \le m-1]$$

and let $R = T/I$. Then the following hold:

(1) $\operatorname{Kdim}(R) = n^2$,
(2) *R is an integral domain, and*
(3) $Z \cong R$.

PROOF. Since $\mathcal{O}_q(M_n(\Bbbk))$ is a finitely generated module over Z, $\operatorname{GKdim}(Z) = \operatorname{GKdim}(\mathcal{O}_q(M_n(\Bbbk))) = n^2$ ([**21**, Proposition 5.5]). Therefore, Z is an integral domain of Krull dimension n^2 ([**21**, Theorem 4.5]), and there is an obvious surjective homomorphism from R onto Z. Thus, (3) follows from (1) and (2).

Throughout we abuse notation and identify elements in $\Bbbk[Z_{ij}, Y_{tr}, D]$ with their respective images in the quotient R.

(1) By [**22**, Theorem 5.6.36 and Corollary 5.7.10(a)], it suffices to compute the Krull dimension of the graded ring $\widetilde{R} = T/\operatorname{LT}(I)$, which we do by computing the degree of the Hilbert polynomial of $\widetilde{R}$.

By Lemma 4.3, we need to count the number of (monic) monomials in the variables Z_{ij}, Y_{tr}, D of a given degree containing at most $m-1$ copies of D, and containing at most one Y_{tr} for each $1 \le t \le n-1$. A simple counting argument shows that the total number of such monomials containing no Z_{ij} is

$$m\sum_{i=0}^{n-1}\binom{n-1}{i}(m-1)^i = m^n,$$

with the equality being the binomial theorem applied to $((m-1)+1)^{n-1}$. Thus for sufficiently large degrees N, the number of "good" monomials of degree N is asymptotically equal to the number of monomials in the Z_{ij} of degree N. This is well-known to be $\binom{N+n^2-1}{N}$, which is a degree n^2-1 polynomial in N. In other words, the Hilbert polynomial of $\widetilde{R}$ has degree n^2-1, and thus $\operatorname{Kdim}(\widetilde{R}) = n^2$ ([**22**, Theorem 5.4.15(b)]).

(2) Before we prove that R is a domain, first note that any nonzero polynomial in the Z_{ij} is a regular element of R. Indeed, suppose $f(Z_{ij})g \in I$ for some $g \in T$. Then by Lemma 4.3,

$$\operatorname{LT}(fg) = \operatorname{LT}(f)\operatorname{LT}(g) \in \operatorname{LT}(I) = (D^m, Y_{ti}Y_{tj} \mid 1 \le t \le n-1, 1 \le i,j \le m-1).$$

Since $LT(f)$ is a monomial in the Z_{ij}, it must be that $\mathrm{LT}(g) \in LT(I)$, so that $\mathrm{LT}(g) = \mathrm{LT}(g')$ for some $g' \in I$. Then $f(g-g') \in I$ with $\mathrm{LT}(g-g') < \mathrm{LT}(g)$. Since $<$ is a well-order, we can repeat this argument until we have written g as a sum of elements in I, proving $g \in I$ and f is not a zero divisor in R.

Let $S = \Bbbk[Z_{ij} \mid i,j \in 1,\dots,n]\backslash\{0\}$. By the preceding discussion, S consists of regular elements and so it suffices to prove that the localization $Q = RS^{-1}$ is a domain.

First note that for $1 \le t \le n-1$ and $1 \le k \le m-2$, we have the relation $Y_{tk} = Y_{t,k+1}Y_{t,m-1}\det(A_t)^{-m+k+1}$ in Q. By repeatedly applying this relation, we obtain $Y_{tk} = Y_{t,m-1}^{m-k}\det(A_t)^{-m+k+1}$. Conversely, these relations imply $Y_{ti}Y_{tj} = Y_{t,i+j}\det(A_t)$ for $i+j<m$. Moreover, the relations $Y_{t1}Y_{t,m-1} = \det(A_t)\det(B_{n-t})$ can be rewritten as $Y_{t,m-1}^m = \det(A_t)^{m-1}\det(B_{n-t})$, from which the rest of the relations of the form $Y_{ti}Y_{tj} = \det(A_t)\det(B_{n-t})$ with $i+j=m$ follow. In addition, the above formulas for $Y_{t,m-1}^m$ and Y_{tk} for $k<m-2$ imply $Y_{t,i}Y_{t,j} = Y_{t,i+j}\det(B_{n-t})$ for $i+j<m$. Next, note that since $\det(B_{n-t}) = Z_{nt}f+g$ for some $f \in S$ and some $g \in S \cup \{0\}$, we can rewrite the relations $Y_{t,m-1}^m = \det(A_t)^{m-1}\det(B_{n-t})$ as $(Y_{t,m-1}^m\det(A_t)^{1-m} - g)f^{-1} = Z_{nt}$. Similarly, we can rewrite the relation $D^m = \det(Z_{ij})$ to solve for Z_{nn}.

It follows that Q is (isomorphic to) a localization of the polynomial ring

$$\Bbbk[X_{ij}, Y_{t,m-1}, D \mid 1 \le i,t \le n-1, 1 \le j \le n].$$

Thus, Q is a domain. □

Having determined a presentation for the center Z of $\mathcal{O}_q(M_n(\Bbbk))$, we are now ready to discuss several properties of Z.

THEOREM 4.5. *Assume Hypothesis* 4.1. *Then Z is not Gorenstein. Moreover, $\mathcal{O}_q(M_n(\Bbbk))$ is not projective over Z.*

PROOF. We show the localization of Z at $\mathfrak{m} = (Z_{ij}, Y_{tr}, D)$ is not Gorenstein. Since $\{Z_{ij} \mid 1 \le i,j \le n\}$ is a system of parameters for $Z_{\mathfrak{m}}$, it suffices to show $A = Z_{\mathfrak{m}}/(Z_{ij})$ is not Gorenstein by [**4**, Theorem 2.1.2(d) and Proposition 3.1.19(b)]. As $\operatorname{depth}(A) = 0$, this is equivalent to showing $\dim_{\Bbbk}\operatorname{Hom}_A(\Bbbk, A) \ge 2$ by [**4**, Lemma 1.2.19 and Theorem 3.2.10]. But this is simple to see as

$$\phi : 1 \mapsto D^{m-1}\prod_{t=1}^{n-1} Y_{t1} \quad \text{and} \quad \psi : 1 \mapsto D^{m-1}\prod_{t=1}^{n-1} Y_{t2}$$

are clearly $\Bbbk$-linearly independent A-algebra homomorphisms from $\Bbbk$ to A.

Suppose $\mathcal{O}_q(M_n(\Bbbk))$ is projective over Z, so that there is a projective basis $\{z_i \mid i \in I\}$ with corresponding Z-module homomorphisms $f_i : \mathcal{O}_q(M_n(\Bbbk)) \to Z$. Then, in particular, $x_{1n} = \sum_{i\in I} f_i(x_{1n})z_i$. Since x_{1n} is not a term in any central element, a simple graded argument shows that some $f_j(x_{1n})$ is a nonzero scalar. However, since $x_{1n}\tau(D(n-1))^m = Y_{11}\tau(D(n-1))$, we have

$$\begin{aligned}\tau(D(n-1)^m)f_j(x_{1n}) &= f_j(x_{1n}\tau(D(n-1))^m) = f_j(Y_{11}\tau(D(n-1)))\\ &= Y_{11}f_j(\tau(D(n-1))).\end{aligned}$$

But this is impossible, since x_{1n} divides the (possibly 0) RHS, but not the LHS. □

As a consequence of Theorem 4.5, $\mathcal{O}_q(M_n(\Bbbk))$ is not Azumaya over its center.

5. Automorphisms of $\mathcal{O}_q(M_n(\Bbbk))$

As mentioned earlier, in the case where q is not a root of unity, every automorphism of $\mathcal{O}_q(M_n(\Bbbk))$ is graded. Similarly, as shown in Theorem 3.5, one can find hypotheses under which every automorphism of $\mathcal{O}_{\lambda,\mathbf{p}}(M_2(\Bbbk))$ is graded. In this section, we will show that in the non-generic single-parameter case, the automorphism group of $\mathcal{O}_q(M_n(\Bbbk))$ contains a free group on two generators. Moreover, if $n = 2$, we construct an automorphism of $\mathcal{O}_q(M_2(\Bbbk))$ which induces a wild automorphism of a three variable polynomial ring. Finally, in the setting of Hypothesis 4.1, we use discriminants to compute an ideal of $\mathcal{O}_q(M_2(\Bbbk))$ fixed by all automorphisms, and we investigate some natural subalgebras of $\mathcal{O}_q(M_3(\Bbbk))$, determining whether they have non-graded automorphisms as well. We do not assume Hypothesis 4.1 in general for this section. Several results indeed hold for any root of unity q and, in one case, for any parameter q.

The next result establishes both that $\mathcal{O}_q(M_n(\Bbbk))$ contains non-graded automorphisms and that its automorphism group contains a free product on two generators.

THEOREM 5.1. *Let q be any root of unity, and let* $\operatorname{ord}(q) = m$. *Then there are automorphisms ϕ and ψ of $\mathcal{O}_q(M_n(\Bbbk))$ given by*

$$\phi : x_{ij} \mapsto \begin{cases} x_{11} + A(1,1)^{m-1} & i = j = 1 \\ x_{ij} & (i,j) \neq (1,1), \end{cases}$$

$$\psi : x_{ij} \mapsto \begin{cases} x_{ij} & (i,j) \neq (n,n) \\ x_{nn} + A(n,n)^{m-1} & i = j = n. \end{cases}$$

Moreover, the proper subgroup of $\operatorname{Aut}(\mathcal{O}_q(M_n(\Bbbk)))$ *generated by ϕ and ψ is isomorphic to a free group on two generators.*

PROOF. We only show ϕ is an automorphism, as the proof for ψ is similar. The map

$$\rho : x_{ij} \mapsto \begin{cases} x_{11} - A(1,1)^{m-1} & i = j = 1 \\ x_{ij}, & (i,j) \neq (1,1), \end{cases}$$

is an inverse to ϕ, so ϕ is bijective. To show that ϕ is a homomorphism, notice that we only have to show $\phi(x_{11}x_{ij} - qx_{ij}x_{11}) = 0$, for either $i = 1$ or $j = 1$, and $\phi(x_{11}x_{ij} - x_{ij}x_{11} - (q - q^{-1})x_{i1}x_{1j}) = 0$, for $2 \leq i,j \leq n$, as the other relations trivially map to 0. For the latter type, note that x_{ij} commutes with $A(1,1)$ for each $2 \leq i,j \leq n$ as, by definition, $A(1,1)$ is the quantum determinant of the quantum matrix subalgebra generated by these x_{ij}. On the other hand, for the former relation type, [**30**, Lemma 4.5.1] shows that $A(1,1)^{m-1}x_{ij} = qx_{ij}A(1,1)^{m-1}$ for exactly one of i,j equal to 1. Therefore, both relation types indeed map to zero.

We now prove that the subgroup generated by ϕ and ψ is isomorphic to the free group $\langle\phi\rangle * \langle\psi\rangle \cong \mathbb{Z} * \mathbb{Z}$. We will show that every word $w = \psi^{m_r}\phi^{n_r}\cdots\psi^{m_1}\phi^{n_1}$ is not the identity automorphism, where $r \geq 1$ and where the $n_i, m_i \in \mathbb{Z}$ are not all 0. Let $\leq_1$ denote the lexicographical order with respect to the ordering $x_{11} < \cdots < x_{1n} < x_{21} \cdots < x_{nn}$, and let $\leq_2$ denote the reverse lexicographical order with respect to the same ordering of the x_{ij}.

First, notice that for $r, s \in \mathbb{Z}$,

$$\phi^r : x_{ij} \mapsto \begin{cases} x_{11} + rA(1,1)^{m-1} & i = j = 1 \\ x_{ij} & (i,j) \neq (1,1), \end{cases}$$

$$\psi^s : x_{ij} \mapsto \begin{cases} x_{ij} & (i,j) \neq (n,n) \\ x_{nn} + sA(n,n)^{m-1} & i = j = n. \end{cases}$$

Now, the key observation is that if $\mathrm{LT}_{\leq_2}(f) = c\prod_{i,j=1}^n x_{ij}^{a_{ij}}$ for $f \in \mathcal{O}_q(M_n(\Bbbk))$, $c \in \Bbbk$, and $a_{ij} \in \mathbb{N}$, then

$$\mathrm{LT}_{\leq_1}(\phi^r(f)) = cr \left(\prod_{\substack{i,j=1 \\ (i,j)\neq(1,1)}}^n x_{ij}^{a_{ij}} \right) \prod_{k=2}^n x_{kk}^{a_{11}(m-1)}.$$

Likewise, if $\mathrm{LT}_{\leq_1}(g) = d\prod_{i,j=1}^n x_{ij}^{b_{ij}}$ for $g \in \mathcal{O}_q(M_n(\Bbbk))$, $d \in \Bbbk$, and $b_{ij} \in \mathbb{N}$, then

$$\mathrm{LT}_{\leq_2}(\psi^s(g)) = ds \left(\prod_{\substack{i,j=1 \\ (i,j)\neq(n,n)}}^n x_{ij}^{b_{ij}} \right) \prod_{k=1}^{n-1} x_{kk}^{b_{11}(m-1)}.$$

It follows by induction that $w(x_{11}) \neq x_{11}$ if w is not a power of ψ, and that $w(x_{nn}) \neq x_{nn}$ if w is not a power of ϕ. Since ϕ and ψ are both infinite order automorphisms, w is not the identity on $\mathcal{O}_q(M_n(\Bbbk))$, as desired.

Finally, note that the subgroup generated by ϕ and ψ is proper since it does not contain, for instance, the transpose automorphism $\tau : x_{ij} \to x_{ji}$. □

5.1. Automorphisms of $\mathcal{O}_q(M_2(\Bbbk))$. In this subsection alone, we restrict to the case $n = 2$ and adopt the following notation changes:

(1) $x_{11}, x_{12}, x_{21}, x_{22} := a, b, c, d$, respectively;
(2) $Z_{11}, Z_{12}, Z_{21}, Z_{22} := u, v, w, z$, respectively;
(3) $Y_{1r} = b^r c^{m-r} := t_r$ for $1 \leq r \leq m-1$.

Though we continue to assume that q is a root of unity, we do not immediately assume that $\mathrm{ord}(q) \geq 3$ is an odd integer.

Before proceeding, first recall that an automorphism of a polynomial ring $A = F[x_1, \dots, x_t]$ over a field F is *elementary* if it is of the form

$$(x_1, \dots, x_{i-1}, x_i, x_{i+1}, \dots, x_n) \mapsto (x_1, \dots, x_{i-1}, \alpha x_i + f, x_{i+1}, \dots, x_n),$$

for some $0 \neq \alpha \in F$ and $f \in F[x_1, \dots, \hat{x}_i, \dots, x_n]$. An automorphism of A is then *tame* if it is a composition of elementary automorphisms. An automorphism which is not tame is called *wild*.

If $n = 1, 2$, then it is well-known that every automorphism of A is tame ([**20**, **36**]). When $n = 3$, Umirbaev and Shestakov proved the *Nagata automorphism*, constructed by Nagata in [**27**, §2.1], is wild in the characteristic zero case ([**33**]). The following theorem can be seen as an analogue of this result for $\mathcal{O}_q(M_2(\Bbbk))$, as the automorphism therein restricts to an automorphism of $k[b, c, u]$ inspired by the Nagata automorphism, which is then shown to be wild using the methods of [**33**].

THEOREM 5.2. *Let q be a root of unity of with* $\operatorname{ord}(q) = m$. *Let* $\Omega = cu + b^{m+1}$. *Then the map*

$$\begin{aligned}\sigma : a &\mapsto a\\ b &\mapsto b + \Omega u^2\\ c &\mapsto c - \sum_{i=1}^{m+1} \binom{m+1}{i} b^{m+1-i}\Omega^i u^{2i-1}\\ d &\mapsto d + qa^{m-1}\left(c\Omega u^2 - \sum_{i=1}^{m+1}\binom{m+1}{i} b^{m+1-i}\Omega^i u^{2(i-1)}(b+\Omega u^2)\right),\end{aligned}$$

extends to an automorphism of $\mathcal{O}_q(M_2(\Bbbk))$. Moreover, the restriction $\sigma|_{\Bbbk[b,c,u]}$ is a wild automorphism.

PROOF. That σ is injective is easy to see, while surjectivity follows from the observation that $\sigma(\Omega) = \Omega$. Moreover, that σ sends all relations to zero is simple to check, proving σ is an automorphism.

For an element $f \in \Bbbk[b,c,u]$, denote the highest homogeneous part of f with respect to the standard grading of $\Bbbk[b,c,u]$ by $\overline{f}$. We show that none of $\overline{\sigma(b)}, \overline{\sigma(c)}, \overline{\sigma(u)}$ are contained in the subalgebra generated by the other two; that $\sigma|_{\Bbbk[b,c,u]}$ is wild then follows by [**33**, Corollary 8]. Indeed, we have that

$$\overline{\sigma(b)} = b^{m+1}u^2, \quad \overline{\sigma(c)} = b^{(m+1)^2}u^{2m+1}, \quad \overline{\sigma(u)} = u,$$

from which it is clear none of these elements are contained in the subalgebra generated by the other two. □

The next result requires our presentation of the center from Section 4. Hence, we reinvoke Hypothesis 5.1. That is, we assume that $\operatorname{ord}(q) = m \geq 3$ is an odd integer. Using discriminants, we determine an ideal of $\mathcal{O}_q(M_2(\Bbbk))$ that is fixed by every automorphism.

Recall the Jacobian criterion provides a simple way to determine whether the localization of an affine $\Bbbk$-algebra $R = k[x_1, \ldots, x_r]/I$ at a maximal ideal is regular. In the course of proving the criterion however, even more is established: if $J_{\mathfrak{m}}$ denotes the Jacobian matrix of I evaluated mod $\mathfrak{m}$, then

$$\operatorname{rank}(J_{\mathfrak{m}}) = r - \dim_{\Bbbk}(M/M^2) = r - \operatorname{edim}(R_{\mathfrak{m}}),$$

where M is the maximal ideal of $R_{\mathfrak{m}}$ and $\operatorname{edim}(R_{\mathfrak{m}})$ is the embedding dimension of $R_{\mathfrak{m}}$ [**15**, Theorem 5.6.12 and its proof]. It follows that $\operatorname{rank}(J_{\mathfrak{m}})$ is an easy to compute invariant of the algebras $R_{\mathfrak{m}}$. This motivates the choice of $\mathcal{P}$ in the following proposition.

PROPOSITION 5.3. *Assume Hypothesis 4.1. Let $\mathcal{P}$ be the property of being a local ring with embedding dimension $m+3$. Then the $\mathcal{P}$-discriminant ideal of $\mathcal{O}_q(M_2(\Bbbk))$ is $(v, w, t_1, \ldots, t_{m-1})$. Consequently, if $\sigma \in \operatorname{Aut}(\mathcal{O}_q(M_2(\Bbbk)))$, then $\sigma((b,c)) = (b,c)$.*

PROOF. By the above discussion and Theorem 4.4, the embedding dimension condition on a localization $Z_{\mathfrak{m}}$ of Z at some $\mathfrak{m} \in \operatorname{MaxSpec}(Z)$ is equivalent to $\operatorname{rank}(J_{\mathfrak{m}}) = 1$. Again by Theorem 4.4, the ideal of relations I of Z is generated by the following families of elements:

(1) $D^m - uz + vw$,
(2) $t_it_j - t_{i+j}w$ if $i + j < m$,
(3) $t_it_j - vw$ if $i + j = m$, and
(4) $t_it_j - t_{i+j-m}v$ if $i + j > m$.

Since $\Bbbk$ is algebraically closed, to compute $J_{\mathfrak{m}}$ we need only compute the Jacobian matrix of the generators of I, then evaluate at the point corresponding to $\mathfrak{m}$ in the variety X whose coordinate ring is Z.

Let P be a point in X and denote the coordinates of P simply by their respective variables; e.g., the v-coordinate of P is denoted just v. Suppose that $t_i = 0$ for some i. Then we may assume $i = 1$ as if $i > 1$, we have $t_1t_{i-1} = t_iw = 0$ by the second relation family, from which $t_1 = 0$ follows by induction. Now, if $w \neq 0$, then $t_i = 0$ for all i by induction and the second family of relations. On the other hand, if $w = 0$, then $t_i^2 = 0$ for all $i < \frac{m}{2}$ by the second family. Hence, by the fourth family and induction, $t_i^2 = 0$ for all $i > \frac{m}{2}$ so that each $t_i = 0$. Therefore, either all t_i are 0, or no t_i is 0. In these cases, we have $vw = 0$ or $v, w \neq 0$, respectively, by the third relation family.

Now, suppose P corresponds to a maximal ideal $\mathfrak{m}$ with $\operatorname{rank}(J_{\mathfrak{m}}) = 1$. Then by the above discussion, each $t_i = 0$, and so $vw = 0$. Next, notice that if $w \neq 0$, then $J_{\mathfrak{m}}$ contains the rank 2 submatrix

$$\begin{bmatrix} -z & w & 0 & -u & mD^{m-1} & 0 & 0 & 0 & \dots & 0 \\ 0 & 0 & 0 & 0 & 0 & 0 & -w & 0 & \dots & 0 \end{bmatrix}.$$

Similarly, if $v \neq 0$, then $\operatorname{rank}(J_{\mathfrak{m}}) \geq 2$. Finally, if $u = z = 0$, then $D^m = uz - vw = 0$, so that $\operatorname{rank}(J_{\mathfrak{m}}) = 0$. We conclude that $\operatorname{rank}(J_{\mathfrak{m}}) = 1$ if and only if $\mathfrak{m} = (u - a, v, w, z - b, D - c, t_1, t_2, \dots, t_{m-1})$ for any $(a, b) \neq (0, 0)$ with $c^m = ab$. Taking the intersection of these maximal ideals, we see that the $\mathcal{P}$-discriminant ideal of $\mathcal{O}_q(M_2(\Bbbk))$ is $(v, w, t_1, \dots, t_{m-1})$.

Let $\sigma \in \operatorname{Aut}(\mathcal{O}_q(M_2(\Bbbk)))$. We claim σ fixes the ideal (b, c). By [**25**, Lemma 3.5], σ fixes the ideal $(v, w, t_1, \dots, t_{m-1})$. Notice that

$$(b, c)^m = (v, w, t_1, \dots, t_{m-1}).$$

Moreover, since $\mathcal{O}_q(M_2(\Bbbk))/(b, c) \cong k[a, d]$, the ideal (b, c) is completely prime. Hence, if $x \in (b, c)$, then

$$\sigma(x)^m = \sigma(x^m) \in (v, w, t_1, \dots, t_{m-1}) \subseteq (b, c),$$

from which it follows that $\sigma(x) \in (b, c)$. □

5.2. Automorphisms of certain subalgebras of $\mathcal{O}_q(M_3(\Bbbk))$. In this section we study automorphisms of certain subalgebras of $\mathcal{O}_q(M_3(\Bbbk))$. For a subset a $S \subseteq \mathcal{O}_q(M_3(\Bbbk))$, we let $\langle S \rangle$ denote the subalgebra of $\mathcal{O}_q(M_3(\Bbbk))$ generated by S. The particular subalgebras we study are listed below:

$$B_1 = \langle x_{ij} \rangle_{1 \leq i \leq 2, 1 \leq j \leq 3}, \quad B_2 = \langle B_1 \cup x_{31} \rangle, \quad B_3 = \langle B_2 \cup x_{32} \rangle, \quad C = \langle x_{ij} \rangle_{i+j \geq 4}.$$

The subalgebra B_1 is isomorphic to $\mathcal{O}_q(M_{2,3}(\Bbbk))$, i.e., the 2×3 quantum matrix algebra [**24**]. Below we show that B_1 has polynomial center. Using results of Nguyen, Trampel, and Yakimov [**28**], we are then able to conclude that all automorphisms of B_1 are graded.

The algebra B_3 appears in [**17**], therein denoted $A_{n,r}$, in the authors' study of the algebras obtained by factoring out the ideal of $\mathcal{O}_q(M_n(\Bbbk))$ generated all $(r+1)\times(r+1)$ quantum subdeterminants. The algebra B_2 acts as an intermediary between B_1 and B_3. We will show that both B_2 and B_3 have non-graded automorphisms.

Like B_1, C is a 6-generated subalgebra of $\mathcal{O}_q(M_3(\Bbbk))$ that has polynomial center under our standard hypotheses, as shown below, such that all automorphisms are graded. We do not know if there are any 7- or 8-generated subalgebras of $\mathcal{O}_q(M_3(\Bbbk))$ which have polynomial centers, though we suspect not.

The following proof is motivated in part by [**18**].

PROPOSITION 5.4. *Assume Hypothesis* 4.1. *The center of* B_1 *is polynomial. Consequently,* $\mathrm{Aut}(B_1) = \mathrm{Aut}_{\mathrm{gr}}(B_1)$.

PROOF. We will prove that $\mathcal{Z}(B_1) = \Bbbk[Z_{ij}]$. By [**28**, Proposition 5.10], the discriminant of B_1 over $\Bbbk[Z_{ij}]$ is locally dominating. Then [**7**, Theorem 2.7] implies that all all automorphisms of B_1 are graded.

Let $\overline{B_1}$ denote the associated quasipolynomial algebra of B_1 [**10**]; that is, if we write

$$x_{ij}x_{kl} = q^{a_{(i,j),(k,l)}}x_{kl}x_{ij} + P_{(i,j),(k,l)}$$

in B_1, then $\overline{B_1}$ is the skew-polynomial algebra generated by w_{ij} with $1 \le i \le 2, 1 \le j \le 3$, subject to the relations

$$w_{ij}w_{kl} = q^{a_{(i,j),(k,l)}}w_{kl}w_{ij}.$$

Now, if we order the basis elements of B_1 via the lexicographical order with respect to $x_{11} > x_{12} > x_{13} > x_{21} > x_{22} > x_{23}$, then we can define a (nonlinear) map $T : B_1 \to \overline{B_1}$ by $P \mapsto \mathrm{LT}(P)$. By our choice of order, T is multiplicative, from which it follows that T maps central elements to central elements. Now, a simple computation reveals that $\overline{B_1}$ has polynomial center $\mathcal{Z}(\overline{B_1}) = \Bbbk[w_{ij}^m]$. Therefore, for any element $Y \in \mathcal{Z}(B_1)$, there is a polynomial $f \in \Bbbk[Z_{ij}]$ such that $\mathrm{LT}(f) = \mathrm{LT}(Y)$. Hence $\mathrm{LT}(Y - f) < \mathrm{LT}(Y)$, so that $Y - f \in \Bbbk[Z_{ij}]$ by induction. □

Non-graded automorphisms are introduced in extending B_1 to B_2 and B_3. An example of such an automorphism is given in the next proposition. This result does not require q to be a root of unity, but that hypothesis will be necessary in the latter part of the subsequent remark.

PROPOSITION 5.5. *Define* $\phi : B_3 \to B_3$ *by*

$$x_{ij} \mapsto \begin{cases} x_{31} + x_{12}x_{23} - qx_{13}x_{22} & i = 3, j = 1 \\ x_{ij} & \textit{otherwise.} \end{cases}$$

Then $\phi \in \mathrm{Aut}(B_3)$ *and* $\phi|_{B_2} \in \mathrm{Aut}(B_2)$.

PROOF. Clearly, we need only prove ϕ is an automorphism, as the second part follows trivially. That ϕ is a bijection is simple to see, so we need only show it maps the relations of B_3 to 0. The only nontrivial relation is then $x_{11}x_{31} - qx_{31}x_{11}$:

$$\begin{aligned}\phi(x_{31}x_{11}) &= x_{31}x_{11} + x_{12}x_{23}x_{11} - qx_{13}x_{22}x_{11} \\ &= q^{-1}x_{11}x_{31} + x_{12}(x_{11}x_{23} - (q-q^{-1})x_{13}x_{21}) \\ &\quad - qx_{13}(x_{11}x_{22} - (q-q^{-1})x_{12}x_{21}) \\ &= q^{-1}(x_{11}x_{31} + x_{11}x_{12}x_{23} - qx_{11}x_{13}x_{22}) \\ &\quad - (q-q^{-1})x_{12}x_{13}x_{21} + (q-q^{-1})x_{12}x_{13}x_{21} \\ &= \phi(q^{-1}x_{11}x_{31}),\end{aligned}$$

as desired. □

REMARK 5.6. Let $\phi \in \mathrm{Aut}(B_3)$ be as in Proposition 5.5. Since B_3 is fixed by the transpose automorphism τ, one can prove in a manner similar to that of Theorem 5.1 that $\langle\phi\rangle * \langle\tau\phi\tau\rangle$ is a proper subgroup of $\mathrm{Aut}(B_3)$. While $\tau\phi\tau$ does not restrict to an automorphism of $\mathrm{Aut}(B_2)$, if q is a root of unity with $\mathrm{ord}(q) = m$, then one can still show $\mathrm{Aut}(B_2)$ (properly) contains a free group on two generators using ϕ and

$$\psi : x_{ij} \mapsto \begin{cases} x_{11} + x_{21}x_{22}^{m-1}x_{23}^{m-1}x_{31} & i=j=1 \\ x_{12} + Z_{22}x_{23}^{m-1}x_{31} & i=1, j=2 \\ x_{ij} & \text{otherwise.}\end{cases}$$

We finish by showing that all automorphisms of C are graded. To do this, we will compute the discriminant $d(C/\mathcal{Z}(C))$ using [**12**, Theorem 6.1], and show it is locally dominating. We first need to set up our notation.

Let C'' denote the subalgebra of C generated by $\{x_{22}, x_{23}, x_{32}, x_{33}\}$. Clearly $C'' \cong \mathcal{O}_q(M_2(\Bbbk))$. Then C is the iterated Ore extension $C''[x_{13};\sigma_1][x_{31};\sigma_2]$, where

$$\sigma_1(x_{22}) = x_{22},\ \sigma_1(x_{23}) = qx_{23},\ \sigma_1(x_{32}) = x_{32},\ \sigma_1(x_{33}) = qx_{33},$$
$$\sigma_2(x_{22}) = x_{22},\ \sigma_2(x_{23}) = x_{23},\ \sigma_2(x_{32}) = qx_{32},\ \sigma_2(x_{33}) = qx_{33},\ \sigma_2(x_{13}) = x_{13}.$$

As in [**12**], we say an automorphism σ of an algebra A is *inner* if there exists an $a \in A$ such that $xa = a\sigma(x)$ for all $x \in A$.

PROPOSITION 5.7. *Assume Hypothesis* 4.1.

(1) *The center of C is the polynomial ring*

$$\mathcal{Z}(C) = \Bbbk[Z_{13}, Z_{22}, Z_{23}, Z_{31}, Z_{32}, Z_{33}].$$

(2) *The automorphisms* $\sigma_1^r \in \mathrm{Aut}(C''), \sigma_2^r \in \mathrm{Aut}(C''[x_{13};\sigma_1])$ *are not inner for each* $1 \le r \le m-1$.

(3) *The discriminant of C over its center is*

$$d(C/\mathcal{Z}(C)) =_{\mathcal{Z}(C)^\times} \left(x_{13}^{m^6}x_{23}^{m^5}x_{32}^{m^5}x_{31}^{m^6}(x_{22}x_{33} - qx_{23}x_{32})^{m^5}\right)^{m-1}.$$

(4) *The discriminant $d(C/\mathcal{Z}(C))$ is locally dominating.*

(5) $\mathrm{Aut}(C) = \mathrm{Aut}_{\mathrm{gr}}(C)$.

PROOF. (1) That the center is polynomial follows as in Proposition 5.4. The details are left to the reader.

(2) If some power σ_1^r, $1 \le r \le m-1$, was an inner automorphism of C'', then we would have an element $0 \ne a \in \mathcal{O}_q(M_2(\Bbbk))$ so that $x_{23}a = q^r ax_{23}$ and $x_{32}a = ax_{32}$, which is clearly impossible.

Let $C' = C'''[x_{13}; \sigma_1]$. We claim that no power σ_2^r is inner for any $1 \le r \le m-1$. Suppose there is an $0 \ne a \in C'$ such that $Xa = a\sigma_2^r(X)$ for all $X \in C'$. Write a as a sum of basis elements: $a = \sum \gamma x_{13}^{b_1} x_{22}^{b_2} x_{23}^{b_3} x_{32}^{b_4} x_{33}^{b_5}$. From $X = x_{23}, x_{32}$ respectively, we obtain

$$b_1 + b_2 \equiv b_5 \bmod m;$$
$$r - b_5 \equiv -b_2 \bmod m.$$

for all terms of a. On the other hand, if $X = x_{22}$, then by comparing the leading terms with respect to the lexicographical ordering of the basis elements such that $x_{13} > x_{22} > x_{23} > x_{32} > x_{33}$, we find that $b_3 + b_4 \equiv 0 \bmod m$ for the leading term of a; a similar argument for $X = x_{33}$ gives $b_1 + b_3 + b_4 \equiv m - r \bmod m$ for the leading term of a. Putting these equations together, it follows that $2r \equiv 0 \bmod m$ for the leading term of a, which is impossible since m is odd.

(3) Let $S' = \mathcal{Z}(C) \cap (C')^{\sigma_2}$. By part (1), $S' = \Bbbk[Z_{13}, Z_{22}, Z_{23}, Z_{32}, Z_{33}]$. Moreover, by part (2), we can apply [**12**, Theorem 6.1] to obtain

$$d(C/\mathcal{Z}(C)) =_{\mathcal{Z}(C)^\times} (d(C'/S'))^m (x_{31}^{m-1})^{m^6}.$$

Denoting $S = S' \cap (C'')^{\sigma_2} = \Bbbk[Z_{22}, Z_{23}, Z_{32}, Z_{33}]$, another application of [**12**, Theorem 6.1] gives

$$d(C'/S') =_{(S')^\times} (d(C/S))^m (x_{13}^{m-1})^{m^5}.$$

Next, observe that $d(C/S)$ is one of the discriminants computed in [**28**, Theorem 5.7]. Hence

$$\begin{aligned} d(C/\mathcal{Z}(C)) &=_{\mathcal{Z}(C)^\times} \left(\left(\left(x_{32}^{m^3} x_{23}^{m^3} (x_{22}x_{33} - qx_{23}x_{32})^{m^3} \right)^m x_{13}^{m^5} \right)^m x_{31}^{m^6} \right)^{m-1} \\ &=_{\mathcal{Z}(C)^\times} \left(x_{13}^{m^6} x_{23}^{m^5} x_{32}^{m^5} x_{31}^{m^6} (x_{22}x_{33} - qx_{23}x_{32})^{m^5} \right)^{m-1}. \end{aligned}$$

(4) This follows from the same argument as [**28**, Proposition 5.10].

(5) This follows by [**7**, Theorem 2.7]. □

References

[1] J. Alev and M. Chamarie, *Dérivations et automorphismes de quelques algèbres quantiques* (French), Comm. Algebra **20** (1992), no. 6, 1787–1802, DOI 10.1080/00927879208824431. MR1162608

[2] J. Alev and F. Dumas, *Rigidité des plongements des quotients primitifs minimaux de $U_q(\mathrm{sl}(2))$ dans l'algèbre quantique de Weyl-Hayashi* (French), Nagoya Math. J. **143** (1996), 119–146, DOI 10.1017/S002776300000595X. MR1413010

[3] M. Artin, W. Schelter, and J. Tate, *Quantum deformations of* GL_n, Comm. Pure Appl. Math. **44** (1991), no. 8-9, 879–895, DOI 10.1002/cpa.3160440804. MR1127037

[4] W. Bruns and J. Herzog, *Cohen-Macaulay rings*, Cambridge Studies in Advanced Mathematics, vol. 39, Cambridge University Press, Cambridge, 1993. MR1251956

[5] K. Chan, J. Gaddis, R. Won, J. J. Zhang, *Ozone groups and centers of skew polynomial rings*, to appear in Int. Math. Res. Not. (2023).

[6] K. Chan, J. Gaddis, R. Won, and J. J. Zhang, *Reflexive hull discriminants and applications*, Selecta Math. (N.S.) **28** (2022), no. 2, Paper No. 40, 35 pp., DOI 10.1007/s00029-021-00755-x. MR4370373

[7] S. Ceken, J. H. Palmieri, Y.-H. Wang, and J. J. Zhang, *The discriminant controls automorphism groups of noncommutative algebras*, Adv. Math. **269** (2015), 551–584, DOI 10.1016/j.aim.2014.10.018. MR3281142

[8] S. Ceken, J. H. Palmieri, Y.-H. Wang, and J. J. Zhang, *The discriminant criterion and automorphism groups of quantized algebras*, Adv. Math. **286** (2016), 754–801, DOI 10.1016/j.aim.2015.09.024. MR3415697

[9] K. Chan, A. A. Young, and J. J. Zhang, *Discriminant formulas and applications*, Algebra Number Theory **10** (2016), no. 3, 557–596, DOI 10.2140/ant.2016.10.557. MR3513131

[10] C. De Concini and C. Procesi, *Quantum groups*, D-modules, representation theory, and quantum groups (Venice, 1992), Lecture Notes in Math., vol. 1565, Springer, Berlin, 1993, pp. 31–140, DOI 10.1007/BFb0073466. MR1288995

[11] J. Gaddis, *The isomorphism problem for quantum affine spaces, homogenized quantized Weyl algebras, and quantum matrix algebras*, J. Pure Appl. Algebra **221** (2017), no. 10, 2511–2524, DOI 10.1016/j.jpaa.2016.12.036. MR3646315

[12] J. Gaddis, E. Kirkman, and W. F. Moore, *On the discriminant of twisted tensor products*, J. Algebra **477** (2017), 29–55, DOI 10.1016/j.jalgebra.2016.12.019. MR3614142

[13] K. R. Goodearl and T. H. Lenagan, *Quantum determinantal ideals*, Duke Math. J. **103** (2000), no. 1, 165–190, DOI 10.1215/S0012-7094-00-10318-3. MR1758243

[14] K. R. Goodearl, *Prime ideals in skew polynomial rings and quantized Weyl algebras*, J. Algebra **150** (1992), no. 2, 324–377, DOI 10.1016/S0021-8693(05)80036-5. MR1176901

[15] G.-M. Greuel and G. Pfister, *A* **Singular** *introduction to commutative algebra*, Springer-Verlag, Berlin, 2002. With contributions by Olaf Bachmann, Christoph Lossen and Hans Schönemann; With 1 CD-ROM (Windows, Macintosh, and UNIX), DOI 10.1007/978-3-662-04963-1. MR1930604

[16] H. Haynal, *PI degree parity in q-skew polynomial rings*, J. Algebra **319** (2008), no. 10, 4199–4221, DOI 10.1016/j.jalgebra.2008.01.036. MR2407896

[17] H. P. Jakobsen and S. Jøndrup, *Quantized rank R matrices*, J. Algebra **246** (2001), no. 1, 70–96, DOI 10.1006/jabr.2001.8902. MR1872614

[18] H. P. Jakobsen and H. Zhang, *The center of the quantized matrix algebra*, J. Algebra **196** (1997), no. 2, 458–474, DOI 10.1006/jabr.1997.7121. MR1475119

[19] H. P. Jakobsen and H. Zhang, *Cyclic representations of the quantum matrix algebras*, Comm. Algebra **27** (1999), no. 2, 493–510, DOI 10.1080/00927879908826445. MR1671922

[20] H. W. E. Jung, *Über ganze birationale Transformationen der Ebene* (German), J. Reine Angew. Math. **184** (1942), 161–174, DOI 10.1515/crll.1942.184.161. MR8915

[21] G. R. Krause and T. H. Lenagan, *Growth of algebras and Gelfand-Kirillov dimension*, Revised edition, Graduate Studies in Mathematics, vol. 22, American Mathematical Society, Providence, RI, 2000, DOI 10.1090/gsm/022. MR1721834

[22] M. Kreuzer and L. Robbiano, *Computational commutative algebra. 2*, Springer-Verlag, Berlin, 2005. MR2159476

[23] S. Launois and T. H. Lenagan, *Automorphisms of quantum matrices*, Glasg. Math. J. **55** (2013), no. A, 89–100, DOI 10.1017/S0017089513000529. MR3110806

[24] S. Launois and T. H. Lenagan, *Primitive ideals and automorphisms of quantum matrices*, Algebr. Represent. Theory **10** (2007), no. 4, 339–365, DOI 10.1007/s10468-007-9059-0. MR2333441

[25] D.-M. Lu, Q.-S. Wu, and J. J. Zhang, *A Morita cancellation problem*, Canad. J. Math. **72** (2020), no. 3, 708–731, DOI 10.4153/s0008414x1900004x. MR4098598

[26] J. Levitt and M. Yakimov, *Quantized Weyl algebras at roots of unity*, Israel J. Math. **225** (2018), no. 2, 681–719, DOI 10.1007/s11856-018-1675-3. MR3805662

[27] M. Nagata, *On automorphism group of* $k[x, y]$, Kinokuniya Book Store Co., Ltd., Tokyo, 1972. Department of Mathematics, Kyoto University, Lectures in Mathematics, No. 5. MR0337962

[28] B. Nguyen, K. Trampel, and M. Yakimov, *Noncommutative discriminants via Poisson primes*, Adv. Math. **322** (2017), 269–307, DOI 10.1016/j.aim.2017.10.018. MR3720799

[29] M. Noumi, H. Yamada, and K. Mimachi, *Finite-dimensional representations of the quantum group* $\mathrm{GL}_q(n;\mathbf{C})$ *and the zonal spherical functions on* $\mathrm{U}_q(n-1)\backslash\mathrm{U}_q(n)$, Japan. J. Math. (N.S.) **19** (1993), no. 1, 31–80, DOI 10.4099/math1924.19.31. MR1231510

[30] B. Parshall and J. P. Wang, *Quantum linear groups*, Mem. Amer. Math. Soc. **89** (1991), no. 439, vi+157, DOI 10.1090/memo/0439. MR1048073

[31] N. Yu. Reshetikhin, L. A. Takhtadzhyan, and L. D. Faddeev, *Quantization of Lie groups and Lie algebras* (Russian), Algebra i Analiz **1** (1989), no. 1, 178–206; English transl., Leningrad Math. J. **1** (1990), no. 1, 193–225. MR1015339

[32] J. Scott, *Quasi-commuting families of quantum minors*, J. Algebra **290** (2005), no. 1, 204–220, DOI 10.1016/j.jalgebra.2001.12.001. MR2154990

[33] I. P. Shestakov and U. U. Umirbaev, *The tame and the wild automorphisms of polynomial rings in three variables*, J. Amer. Math. Soc. **17** (2004), no. 1, 197–227, DOI 10.1090/S0894-0347-03-00440-5. MR2015334

[34] J. T. Stafford and M. Van den Bergh, *Noncommutative resolutions and rational singularities: Special volume in honor of Melvin Hochster*, Michigan Math. J. **57** (2008), 659–674, DOI 10.1307/mmj/1220879430. MR2492474

[35] J. T. Stafford and J. J. Zhang, *Homological properties of (graded) Noetherian* PI *rings*, J. Algebra **168** (1994), no. 3, 988–1026, DOI 10.1006/jabr.1994.1267. MR1293638

[36] W. van der Kulk, *On polynomial rings in two variables*, Nieuw Arch. Wisk. (3) **1** (1953), 33–41. MR54574

[37] M. Yakimov, *The Launois-Lenagan conjecture*, J. Algebra **392** (2013), 1–9, DOI 10.1016/j.jalgebra.2013.07.001. MR3085018

Miami University, Department of Mathematics, Oxford, Ohio 45056
Email address: gaddisj@miamioh.edu

Miami University, Department of Mathematics, Oxford, Ohio 45056
Current address: Department of Mathematics, University of California San Diego, La Jolla, California 92903
Email address: tlamkin@ucsd.edu

Contemporary Mathematics
Volume **791**, 2024
https://doi.org/10.1090/conm/791/15872

On automorphisms of quantum Schubert cells

Garrett Johnson and Hayk Melikyan

Abstract. Automorphisms of the quantum Schubert cell algebras $\mathcal{U}_q^{\pm}[w]$ of De Concini, Kac, Procesi and Lusztig and their restrictions to some key invariant subalgebras are studied. We develop some general rigidity results and apply them to completely determine the automorphism group in several cases.

We focus primarily on those cases when the underlying Lie algebra $\mathfrak{g}$ is finite dimensional and simple with rank $r > 1$, and w is a parabolic element of the Weyl group, say $w = w_o^J w_o$, for some nonempty subset J of simple roots. Here, $\mathcal{U}_q^{\pm}[w]$ is a deformation of the universal enveloping algebra of the nilradical of a parabolic subalgebra of $\mathfrak{g}$. In this setting we conjecture that, with the exception of two specific low rank cases, the automorphism group of $\mathcal{U}_q^{\pm}[w]$ is the semidirect product of an algebraic torus of rank r with the group of Dynkin diagram symmetries that preserve J. This conjecture is a more general form of the Launois-Lenagan and Andruskiewitsch-Dumas conjectures regarding the automorphism groups of the algebras of quantum matrices and the algebras $\mathcal{U}_q^{+}(\mathfrak{g})$, respectively. We completely determine the automorphism group in several instances, including all cases when $\mathfrak{g}$ is of type F_4 or G_2, as well as those cases when the quantum Schubert cell algebras are the algebras of quantum symmetric matrices.

1. Introduction and summary of the results

Quantum Schubert cell algebras $\mathcal{U}_q^{\pm}[w]$ were introduced by De Concini, Kac, Procesi [**7**] and Lusztig [**26**]. They are a family of subalgebras of the Drinfeld-Jimbo quantized enveloping algebra $\mathcal{U}_q(\mathfrak{g})$ indexed by the elements w of the Weyl group, and have appeared in several contexts, including ring theory [**28, 32**], crystal basis theory [**22, 27**], and cluster algebras [**11, 13**]. Several important cases arise when the Weyl group element w is a parabolic element, say $w_o^J w_o$, for some nonempty subset J of simple roots. Here, the corresponding algebra $\mathcal{U}_q^{\pm}[w]$ can be viewed as a deformation of the universal enveloping algebra of the nilradical $\mathfrak{n}_J$ of a parabolic subalgebra of $\mathfrak{g}$. In such a setting, we denote the quantum Schubert cell algebra by $\mathcal{U}_q(\mathfrak{n}_J)$ and we refer to it as a *quantized nilradical* for short.

Throughout, the underlying base field for all algebras will be denoted by $\mathbb{K}$. We do not need to assume that $\mathbb{K}$ is algebraically closed or that it is of characteristic zero. The role of the Lie algebra $\mathfrak{g}$ in defining the $\mathbb{K}$-algebras $\mathcal{U}_q^{\pm}[w]$, $\mathcal{U}_q(\mathfrak{n}_J)$, and

2020 *Mathematics Subject Classification.* Primary 17B37; Secondary 16T20, 16W20.
Key words and phrases. quantum Schubert cell algebras, automorphisms, quantum algebras, nilradicals.
The authors were supported by NSF grant DMS-1900823.

$\mathcal{U}_q(\mathfrak{g})$ can be viewed as purely symbolic. We will denote the multiplicative group of nonunits by $\mathbb{K}^\times$.

We will turn our attention towards studying the automorphisms of these algebras. We assume that the deformation parameter $q \in \mathbb{K}^\times$ is not a root of unity. Basically, there is a dichotomy in the structure of quantum algebras depending on whether or not q is a root of unity. When q is a root of unity, these algebras more closely resemble deformations of modular Lie algebras (Lie algebras over fields of positive characteristic). These algebras have large centers, and are therefore closer to being commutative. This, in effect, gives less control over automorphisms. In such situations, various types of noncommutative discriminants [**4**–**6**] have been developed as tools to study automorphisms (see e.g. [**9**, **10**]).

Automorphism groups of quantized nilradicals (when q is not a root of unity) have already been studied in several cases. For instance, when the underlying Lie algebra is $\mathfrak{sl}(n)$ and J is a singleton, say $J = \{\alpha_k\}$, the quantized nilradical $\mathcal{U}_q(\mathfrak{n}_J)$ is isomorphic to the algebra of quantum $k \times (n-k)$ matrices. Launois and Lenagan prove in [**23**] that the automorphism group is $(\mathbb{K}^\times)^{n-1}$ whenever $k \neq n-k$ (and when $(n,k) \notin \{(4,1),(4,3)\}$) by using certain properties of height one prime ideals. These techniques do not apply when $k = n-k$, yet they conjecture that the automorphism group in this remaining case is $(\mathbb{K}^\times)^{n-1} \rtimes \mathbb{Z}_2$. Their conjecture was already known to be true in the 2×2 case by the work of Alev and Chamarie [**1**]. Launois and Lenagan later proved their conjecture for the 3×3 case in [**24**]. Finally, the Launois-Lenagan conjecture was proved in the remaining cases by Yakimov in [**35**].

An interesting phenomenon regarding automorphisms arises when $k = 1$ or $n-k = 1$. In this setting $\mathcal{U}_q(\mathfrak{n}_J)$ is isomorphic to $(n-1)$-dimensional quantum affine space $\mathcal{A}_q(\mathbb{K}^{n-1})$. As an algebra, $\mathcal{A}_q(\mathbb{K}^{n-1})$ is generated by elements $x_1, \ldots, x_{n-1}$ and has defining relations $x_i x_j = q x_j x_i$ whenever $i < j$. In [**1**], Alev and Chamarie studied automorphisms of several types of noncommutative algebras, including multiparameter and uniparameter quantum affine space. Their work predates the Launois-Lenagan conjecture. Interestingly, automorphisms of $\mathcal{A}_q(\mathbb{K}^{n-1})$ send each generator x_i to a scalar multiple of itself whenever $n \neq 4$. Alev and Chamarie proved in [**1**, Theorem 1.4.6] that every automorphism ϕ of $\mathcal{A}_q(\mathbb{K}^3)$ has the form

$$\phi(x_1) = a_1 x_1, \qquad \phi(x_2) = a_2 x_2 + b x_1 x_3, \qquad \phi(x_3) = a_3 x_3,$$

where $a_1, a_2, a_3 \in \mathbb{K}^\times$ and $b \in \mathbb{K}$. Hence, the automorphism group of $\mathcal{A}_q(\mathbb{K}^3)$ is isomorphic to the semidirect product $(\mathbb{K}^\times)^3 \rtimes \mathbb{K}$. On the other hand, if $n \neq 4$, they proved that the automorphism group of $\mathcal{A}_q(\mathbb{K}^{n-1})$ is isomorphic to $(\mathbb{K}^\times)^{n-1}$. Here, every automorphism sends x_i to a nonzero multiple of itself.

Automorphism groups of quantized nilradicals have also been determined in all cases when J is chosen to be the full set of simple roots. In particular, we assume now that $\mathfrak{g}$ is an arbitrarily chosen finite dimensional complex simple Lie algebra with $\operatorname{rank}(\mathfrak{g}) = r > 1$, and J is the full set of simple roots. Here, $\mathcal{U}_q(\mathfrak{n}_J)$ is the entire positive part of $\mathcal{U}_q(\mathfrak{g})$. The Chevalley generators $E_1, \ldots, E_r$ generate $\mathcal{U}_q^+(\mathfrak{g})$ as an algebra and satisfy the q-Serre relations. With this, it is not too difficult to observe that for every r-tuple $(a_1, \ldots, a_r) \in (\mathbb{K}^\times)^r$, there is an algebra automorphism ϕ of $\mathcal{U}_q^+(\mathfrak{g})$ such that $\phi(E_i) = a_i E_i$ $(i = 1, \ldots, r)$. Furthermore, for every symmetry ψ of the underlying Dynkin diagram, there is an algebra automorphism of $\mathcal{U}_q^+(\mathfrak{g})$

given by the rule $E_i \mapsto E_{\psi(i)}$. Andruskiewitsch and Dumas [**2**] conjectured that the automorphism group of $\mathcal{U}_q^+(\mathfrak{g})$ is generated by only these types of automorphisms. That is to say, they conjectured that

$$\operatorname{Aut}\left(\mathcal{U}_q^+(\mathfrak{g})\right) \cong \left(\mathbb{K}^\times\right)^{\operatorname{rank}(\mathfrak{g})} \rtimes \text{Dynkin-Aut}(\mathfrak{g}),$$

where Dynkin-Aut$(\mathfrak{g})$ is the automorphism group of the Dynkin diagram of $\mathfrak{g}$. Yakimov proved this conjecture in [**34**] using a rigidity result involving quantum tori.

In describing the automorphism group $\mathcal{U}_q(\mathfrak{n}_J)$ (for arbitrarily chosen $\mathfrak{g}$ and J), we need to introduce the subgroup of Dynkin diagram symmetries that fixes J,

$$\text{Dynkin-Aut}_J(\mathfrak{g}) := \{\psi \in \text{Dynkin-Aut}(\mathfrak{g}) \mid \psi(J) = J\}.$$

We conjecture the following result regarding the automorphism groups of quantized nilradicals.

CONJECTURE 1.1. *Let $\mathfrak{g}$ be a finite dimensional complex simple Lie algebra with* $\operatorname{rank}(\mathfrak{g}) > 1$. *Suppose J is a nonempty subset of simple roots, and let $\mathcal{U}_q(\mathfrak{n}_J)$ be the corresponding quantized nilradical. Then*

$$\operatorname{Aut}\left(\mathcal{U}_q(\mathfrak{n}_J)\right) \cong \left(\mathbb{K}^\times\right)^{\operatorname{rank}(\mathfrak{g})} \rtimes \text{Dynkin-Aut}_J(\mathfrak{g}).$$

provided $\mathcal{U}_q(\mathfrak{n}_J) \ncong \mathcal{A}_q(\mathbb{K}^3)$.

We remark that $\mathcal{U}_q(\mathfrak{n}_J)$ is isomorphic to $\mathcal{A}_q(\mathbb{K}^3)$ in only two situations: (1) $\mathfrak{g} = \mathfrak{sl}(4)$ and $J = \{\alpha_1\}$, or (2) $\mathfrak{g} = \mathfrak{sl}(4)$ and $J = \{\alpha_3\}$.

Conjecture 1.1 above has been resolved in several cases. As mentioned above, the proof of the Launois-Lenagan conjecture covers the situation when $\mathfrak{g}$ is of type A_n and J is a singleton, whereas the Andruskiewitsch-Dumas conjecture handles the case when $\mathfrak{g}$ is arbitrary and J is the full set of simple roots. We prove Conjecture 1.1 in some other situations, including when the underlying Lie algebra $\mathfrak{g}$ is of type F_4 or G_2.

THEOREM 1.2. *If $\mathfrak{g}$ is the Lie algebra of type F_4 and J is a nonempty subset of simple roots of $\mathfrak{g}$, then* $\operatorname{Aut}(\mathcal{U}_q(\mathfrak{n}_J)) \cong (\mathbb{K}^\times)^4$.

THEOREM 1.3. *If $\mathfrak{g}$ is the Lie algebra of type G_2 and J is a nonempty subset of simple roots of $\mathfrak{g}$, then* $\operatorname{Aut}(\mathcal{U}_q(\mathfrak{n}_J)) \cong (\mathbb{K}^\times)^2$.

We also develop some general theorems (Theorems 4.1, 4.2, 4.3, 4.4, 4.5) regarding automorphisms of quantum Schubert cell algebras that can be applied to help determine the automorphism groups of several other quantized nilradicals. More generally, quantum Schubert cell algebras belong to a larger family of algebras called Cauchon-Goodearl-Letzter (CGL) extensions, which originated in the works [**3, 12**]. Some general techniques have been developed in [**14**] to study automorphisms of CGL extensions. These techniques utilize properties of some key subalgebras of a CGL extension R, namely the normal subalgebra $\mathcal{N}(R)$ (the subalgebra generated by the normal elements), and the core $\mathcal{C}(R)$. Basically, the larger the core $\mathcal{C}(R)$, the more control one has over automorphisms [**15**, Theorem 4.2]. One has the most control over automorphisms when the core coincides with the entire algebra. Most quantized nilradicals appear to have this property.

Several other instances of quantized nilradicals appear in the literature, particularly when the nilradical $\mathfrak{n}_J$ is abelian. For example, when the underlying Lie algebra $\mathfrak{g}$ is of type C_n and $J = \{\alpha_n\}$, the corresponding quantized nilradical $\mathcal{U}_q(\mathfrak{n}_J)$ is the algebra of quantum $n \times n$ symmetric matrices [**21, 30**]. When $\mathfrak{g}$ is the Lie algebra of type D_n and $J = \{\alpha_{n-1}\}$ or $J = \{\alpha_n\}$, $\mathcal{U}_q(\mathfrak{n}_J)$ is the algebra of quantum antisymmetric matrices [**31**]. If $\mathfrak{g}$ is of type B_n and $J = \{\alpha_1\}$, $\mathcal{U}_q(\mathfrak{n}_J)$ is the odd-dimensional quantum Euclidean space, which was introduced by Faddeev, Reshetikhin, and Takhtadzhyan [**8**, Definition 12]. Simplified relations for this algebra appear in [**29**, Sections 2.1-2.2]. If $\mathfrak{g}$ is of type D_n and $J = \{\alpha_1\}$, $\mathcal{U}_q(\mathfrak{n}_J)$ is the even-dimensional quantum Euclidean space [**8, 29**]. The automorphism groups of even and odd-dimensional quantum Euclidean space are already known to satisfy Conjecture 1.1 [**14**, Example 4. 10]. We prove Conjecture 1.1 holds when $\mathcal{U}_q(\mathfrak{n}_J)$ is the algebra of quantum symmetric matrices.

THEOREM 1.4. *If $\mathfrak{g}$ is the Lie algebra of type C_n with $n > 1$ and $J = \{\alpha_n\}$ (i.e. $\mathcal{U}_q(\mathfrak{n}_J)$ is the algebra of $n \times n$ quantum symmetric matrices), then*

$$\mathrm{Aut}(\mathcal{U}_q(\mathfrak{n}_J)) \cong (\mathbb{K}^\times)^n .$$

Other examples of quantized nilradicals $\mathcal{U}_q(\mathfrak{n}_J)$ have been studied for cases when $\mathfrak{n}_J$ is non-abelian. For instance, the quantized nilradicals when $\mathfrak{g}$ is of type A_n and J is an arbitrary set of simple roots were studied in [**20**], where it was shown that $\mathcal{U}_q(\mathfrak{n}_J)$ is isomorphic to an algebra of coinvariants. With this, $\mathcal{U}_q(\mathfrak{n}_J)$ can be viewed as a deformation of the coordinate ring of a unipotent subgroup of a parabolic subgroup of $SL(n+1)$.

Each algebra $\mathcal{U}_q(\mathfrak{n}_J)$ can be equipped with a $\mathbb{N}$-grading such that, with respect to this grading, $\mathcal{U}_q(\mathfrak{n}_J)$ is connected and locally finite. We apply the results developed in Theorems 4.1, 4.2, 4.3, 4.4, 4.5 to illustrate that, for certain cases of $\mathfrak{g}$ and J, every automorphism of $\mathcal{U}_q(\mathfrak{n}_J)$ that preserves the $\mathbb{N}$-grading acts diagonally on the graded component of degree one (see Proposition 4.7).

We choose the first case listed in Proposition 4.7, namely when $\mathfrak{g}$ is of type B_6 and $J = \{\alpha_2, \alpha_5\}$, and completely determine the automorphism group of the corresponding quantized nilradical. We show here that every automorphism preserves the $\mathbb{N}$-grading by applying the results of [**15**, Theorem 4.2] involving the core of $\mathcal{U}_q(\mathfrak{n}_J)$. The same steps can be applied to other cases listed in Proposition 4.7.

THEOREM 1.5. *If $\mathfrak{g}$ is the Lie algebra of type B_6 and $J = \{\alpha_2, \alpha_5\}$, then*

$$\mathrm{Aut}(\mathcal{U}_q(\mathfrak{n}_J)) \cong (\mathbb{K}^\times)^6 .$$

Ideally, we would like to eventually develop a theory sufficient to completely determine the automorphism group of $\mathcal{U}_q(\mathfrak{n}_J)$ in all cases. A more general endeavor is to develop a theory sufficient to describe the automorphism groups of quantum Schubert cell algebras $\mathcal{U}_q^{\pm}[w]$. Interestingly, Ceken, Palmieri, Wang, and Zhang [**4**] describe a family of algebras such that the automorphism group of each algebra in this family is isomorphic to the semidirect product of an algebraic torus and a finite group. While quantum Schubert cells don't belong to this family of algebras, in many instances their automorphism groups seem to have this form. In a related work, one could attempt to find necessary and sufficient conditions on w so that $\mathrm{Aut}\left(\mathcal{U}_q^{\pm}[w]\right)$ is isomorphic to $(\mathbb{K}^\times)^n \rtimes G$ for some natural number $n \in \mathbb{N}$ and finite group G.

2. The algebra $\mathcal{U}_{\mathbf{q}}(\mathfrak{g})$

Let $\mathfrak{g}$ be a finite dimensional complex simple Lie algebra of rank r. Define the index set $\mathbf{I} := \{1, 2, \dots r\}$, and let $\Pi = \{\alpha_i\}_{i \in \mathbf{I}}$ be a set of simple roots of $\mathfrak{g}$ with respect to a fixed Cartan subalgebra $\mathfrak{h} \subset \mathfrak{g}$ such that the labelling of the simple roots agrees with the labelling in [**17**, Section 12.1].

The root system of $\mathfrak{g}$ will be denoted by Δ, and the sets of positive and negative roots will be denoted by Δ_+ and Δ_-, respectively. The corresponding triangular decomposition of $\mathfrak{g}$ will be denoted by

$$\mathfrak{g} = \mathfrak{n}^- \oplus \mathfrak{h} \oplus \mathfrak{n}^+,$$

where

$$\mathfrak{n}^\pm := \bigoplus_{\alpha \in \Delta_\pm} \mathfrak{g}_\alpha, \qquad \mathfrak{g}_\alpha := \{x \in \mathfrak{g} \mid [h, x] = \alpha(h)x \text{ for all } h \in \mathfrak{h}\}.$$

As usual, let $Q = \oplus_{i \in \mathbf{I}} \mathbb{Z}\alpha_i$ be the root lattice of $\mathfrak{g}$, and let $\langle\ ,\ \rangle : Q \times Q \to \mathbb{Z}$ be a symmetric nondegenerate ad-invariant $\mathbb{Z}$-bilinear form, normalized so that $\langle \alpha, \alpha \rangle = 2$ for short roots α. Define

$$c_{ij} := \frac{2\langle \alpha_i, \alpha_j \rangle}{\langle \alpha_i, \alpha_i \rangle}, \qquad (i, j \in \mathbf{I}),$$

and let $(c_{ij})_{i,j \in \mathbf{I}}$ be the associated Cartan matrix of $\mathfrak{g}$. We will denote the simple reflections in the Weyl group W of $\mathfrak{g}$ by

$$s_i, \qquad (i \in \mathbf{I}).$$

The corresponding generators of the braid group $\mathcal{B}_\mathfrak{g}$ of $\mathfrak{g}$ will be denoted by

$$T_i, \qquad (i \in \mathbf{I}).$$

Let $q \in \mathbb{K}$ be nonzero and not a root of unity, and let $q_i = q^{\langle \alpha_i, \alpha_i \rangle / 2}$ for $i \in \mathbf{I}$. As usual, define

$$\widehat{q} := q - q^{-1}.$$

For a natural number $n \in \mathbb{N}$, define

$$[n]_{q_i} := \frac{q_i^n - q_i^{-n}}{q_i - q_i^{-1}}, \qquad [n]_{q_i}! := [n]_{q_i}[n-1]_{q_i} \cdots [1]_{q_i}.$$

The quantized universal enveloping algebra $\mathcal{U}_q(\mathfrak{g})$ is an associative $\mathbb{K}$-algebra with standard Chevalley generators

$$K_\mu, E_i, F_i, \qquad (\mu \in Q, i \in \mathbf{I}).$$

There is a standard Q-gradation on the algebra $\mathcal{U}_q(\mathfrak{g})$,

$$\mathcal{U}_q(\mathfrak{g}) = \bigoplus_{\lambda \in Q} \mathcal{U}_q(\mathfrak{g})_\lambda, \quad \mathcal{U}_q(\mathfrak{g})_\mu := \left\{ u \in \mathcal{U}_q(\mathfrak{g}) : K_\lambda u = q^{\langle \lambda, \mu \rangle} u K_\lambda \text{ for all } \lambda \in Q \right\}.$$

With respect to this grading, the Chevalley generators are homogeneous elements. In particular,

$$E_i \in \mathcal{U}_q(\mathfrak{g})_{\alpha_i}, \quad F_i \in \mathcal{U}_q(\mathfrak{g})_{-\alpha_i}, \quad K_\mu \in \mathcal{U}_q(\mathfrak{g})_0, \qquad (i \in \mathbf{I}, \mu \in Q).$$

We will state the defining relations of $\mathcal{U}_q(\mathfrak{g})$, but first we find it convenient to introduce the abbreviation

$$[x, y] := xy - q^{\langle \mu, \eta \rangle} yx, \qquad x \in \mathcal{U}_q(\mathfrak{g})_\mu, \quad y \in \mathcal{U}_q(\mathfrak{g})_\eta, \tag{2.1}$$

for *q-commutators*. We will adopt this notation throughout. Next, for every homogeneous $x \in \mathcal{U}_q(\mathfrak{g})_\mu$, we define the linear operator $\mathrm{ad}_q\, x : \mathcal{U}_q(\mathfrak{g}) \to \mathcal{U}_q(\mathfrak{g})$ by the condition that

$$(\mathrm{ad}_q\, x)\,(y) = [x, y],$$

for every homogeneous element $y \in \mathcal{U}_q(\mathfrak{g})$. With this, the defining relations of $\mathcal{U}_q(\mathfrak{g})$ are

$$\begin{aligned} &K_0 = 1, \qquad K_\mu K_\lambda = K_{\lambda+\mu}, \\ &K_\mu E_i = q^{\langle \mu, \alpha_i \rangle} E_i K_\mu, \qquad K_\mu F_i = q^{-\langle \mu, \alpha_i \rangle} F_i K_\mu, \\ &E_i F_j - F_j E_i = \delta_{ij} \frac{K_{\alpha_i} - K_{-\alpha_i}}{q_i - q_i^{-1}} \end{aligned}$$

together with the q-Serre relations

$$\begin{aligned} (\mathrm{ad}_q\, E_i)^{1-c_{ij}}(E_j) &= 0, && (\text{for all } i \neq j), \\ (\mathrm{ad}_q\, F_i)^{1-c_{ij}}(F_j) &= 0, && (\text{for all } i \neq j). \end{aligned}$$

The algebra $\mathcal{U}_q(\mathfrak{g})$ has a triangular decomposition,

$$\mathcal{U}_q(\mathfrak{g}) \cong \mathcal{U}_q(\mathfrak{n}^-) \otimes \mathcal{U}_q(\mathfrak{h}) \otimes \mathcal{U}_q(\mathfrak{n}^+),$$

where $\mathcal{U}_q(\mathfrak{n}^-)$, $\mathcal{U}_q(\mathfrak{h})$, and $\mathcal{U}_q(\mathfrak{n}^+)$ are the subalgebras of $\mathcal{U}_q(\mathfrak{g})$ generated by the F's, K's, and E's respectively.

2.1. A $\mathbb{Z}$-grading on $\mathcal{U}_q(\mathfrak{g})$. The fundamental coweights of $\mathfrak{g}$ will be denoted by $\varpi_i^\vee$ ($i \in \mathbf{I}$). They are determined by the conditions $\langle \varpi_i^\vee, \alpha_j \rangle = \delta_{ij}$ for all $i, j \in \mathbf{I}$. We will let $P^\vee = \oplus_{i \in \mathbf{I}} \mathbb{Z}\varpi_i^\vee$ be the coweight lattice of $\mathfrak{g}$. For every integral coweight $\lambda \in P^\vee$ (i.e. $\langle \lambda, \alpha_i \rangle \in \mathbb{Z}$ for every $i \in \mathbf{I}$), there is an associated $\mathbb{Z}$-grading on $\mathcal{U}_q(\mathfrak{g})$ given by assigning the degree $\langle \lambda, \mu \rangle$ to every Q-homogeneous element of degree μ. To make the distinction between the Q-grading and the $\mathbb{Z}$-grading, we will write

$$\deg_Q(u) = \mu, \qquad \deg_{\mathbb{Z}}(u) = n,$$

respectively, to mean that $u \in \mathcal{U}_q(\mathfrak{g})$ is a Q-homogeneous element of degree $\mu \in Q$ and a $\mathbb{Z}$-homogeneous element of degree n. Technically, $\deg_{\mathbb{Z}}$ depends on the coweight λ. However, we adopt the notation $\deg_{\mathbb{Z}}$ rather than $\deg_\lambda$ whenever the choice of coweight λ is clear from the context.

2.2. Lusztig symmetries of $\mathcal{U}_\mathbf{q}(\mathfrak{g})$. In [**26**, Section 37.1.3], Lusztig defines an action of the braid group $\mathcal{B}_\mathfrak{g}$ via algebra automorphisms on $\mathcal{U}_q(\mathfrak{g})$. In fact, Lusztig defines the symmetries $T'_{i,1}$, $T'_{i,-1}$, $T''_{i,1}$, and $T''_{i,-1}$. By [**26**, Proposition 37.1.2], these are automorphisms of $\mathcal{U}_q(\mathfrak{g})$, while by [**26**, Theorem 39.4.3] they satisfy the braid relations. For short, we will adopt the abbreviation $T_i := T''_{i,1}$. With this convention, Lusztig's symmetries are given by the formulas

$$\begin{aligned} T_i(K_\mu) &= K_{s_i(\mu)}, \\ T_i(E_j) &= \begin{cases} -F_i K_{\alpha_i}, & (i = j), \\ (\mathrm{ad}_q\, E_i)^{(-c_{ij})}(E_j), & (i \neq j), \end{cases} \\ T_i(F_j) &= \begin{cases} -K_{-\alpha_i} E_i, & (i = j), \\ (-q_i)^{-c_{ij}} (\mathrm{ad}_q\, F_i)^{(-c_{ij})}(F_j), & (i \neq j), \end{cases} \end{aligned}$$

where, for a nonnegative integer n,

$$(\mathrm{ad}_q E_i)^{(n)} := \frac{(\mathrm{ad}_q E_i)^n}{[n]_{q_i}!}, \qquad (\mathrm{ad}_q F_i)^{(n)} := \frac{(\mathrm{ad}_q F_i)^n}{[n]_{q_i}!}.$$

If $w \in W$ has a reduced expression $w = s_{i_1} \cdots s_{i_N} \in W$, we write

$$T_w = T_{i_1} T_{i_2} \cdots T_{i_N}.$$

A key property of the braid symmetries is given in the following proposition (see e.g. [**19**, Proposition 8.20]).

PROPOSITION 2.1. *If $w \in W$ such that $w(\alpha_i) = \alpha_j$, then $T_w(E_i) = E_j$.*

3. The quantized nilradical $\mathcal{U}_q(\mathfrak{n}_J)$

For each nonempty set J of simple roots, let $\mathfrak{p}_J$ be the parabolic subalgebra of $\mathfrak{g}$ obtained by deleting the roots in J. The Levi decomposition of $\mathfrak{p}_J$ will be denoted by

$$\mathfrak{p}_J = \mathfrak{l}_J \ltimes \mathfrak{n}_J,$$

where $\mathfrak{l}_J$ is the Levi subalgebra and $\mathfrak{n}_J$ is the nilradical.

Let w_o denote the longest element of the Weyl group, and let $w_o^J \in W$ be the longest element in the subgroup $\langle s_i \mid i \notin J \rangle \subseteq W$. Define

$$w_J := w_o^J w_o \in W.$$

For a reduced expression,

$$w_J = s_{i_1} \cdot s_{i_2} \cdots s_{i_N} \in W,$$

where N is the length of w_J, define the roots

$$\beta_1 = \alpha_{i_1}, \quad \beta_2 = s_{i_1}\alpha_{i_2}, \ldots, \quad \beta_N = s_{i_1} \cdots s_{i_{N-1}} \alpha_{i_N},$$

and root vectors

$$X_{\beta_1} = E_{i_1}, \quad X_{\beta_2} = T_{s_{i_1}} E_{i_2}, \ldots, \quad X_{\beta_N} = T_{s_{i_1}} \cdots T_{s_{i_{N-1}}} E_{i_N}. \tag{3.1}$$

We will denote the set of radical roots by

$$\Delta_w := \{\beta_1, \ldots, \beta_N\}.$$

These roots are precisely the positive roots that get sent to negative roots by the action of w_J^{-1}. An analogous construction can be applied to obtain a list of negative roots by replacing the E's in (3.1) above with F's. The subalgebra of $\mathcal{U}_q(\mathfrak{g})$ generated by the root vectors $X_{\beta_1}, \ldots, X_{\beta_N}$ is contained in the positive part $\mathcal{U}_q(\mathfrak{n}^+)$ (see e.g. [**19**, Proposition 8.20]). This subalgebra will be denoted by $\mathcal{U}_q(\mathfrak{n}_J)$,

$$\mathcal{U}_q(\mathfrak{n}_J) := \langle X_{\beta_1}, \ldots X_{\beta_N} \rangle \subseteq \mathcal{U}_q(\mathfrak{n}^+),$$

and we refer to it as the *quantized nilradical of* $\mathfrak{p}_J$, or *quantized nilradical* for short. The subalgebra of $\mathcal{U}_q(\mathfrak{g})$ generated by the negative root vectors

$$X_{-\beta_1} = F_{i_1}, X_{-\beta_2} = T_{s_{i_1}} F_{i_2}, \ldots, X_{-\beta_N} = T_{s_{i_1}} \cdots T_{s_{i_{N-1}}} F_{i_N}.$$

is isomorphic to $\mathcal{U}_q(\mathfrak{n}_J)$.

Quantized nilradicals belong to a larger class of algebras called quantum Schubert cell algebras, which are indexed by elements w in the Weyl group. More generally, given a reduced expression of a Weyl group element w, the corresponding quantum Schubert cell algebra $\mathcal{U}_q^+[w]$ can be constructed in the same way as $\mathcal{U}_q(\mathfrak{n}_J)$

by replacing a reduced expression for w_J above with a reduced expression for w. De Concini, Kac, and Procesi [**7**, Proposition 2.2] proved that the algebra $\mathcal{U}_q^+[w]$ does not depend on the reduced expression for w. Furthermore, every quantum Schubert cell $\mathcal{U}_q^+[w]$ has a PBW basis

$$X_{\beta_1}^{m_1} \cdots X_{\beta_N}^{m_N}, \quad m_1, \ldots, m_N \in \mathbb{Z}_{\geq 0},$$

of standard monomials, and they have presentations as iterated Ore extensions,

$$\mathcal{U}_q^+[w] = \mathbb{K}[X_{\beta_1}][X_{\beta_2}; \sigma_2, \delta_2] \cdots [X_{\beta_N}; \sigma_N, \delta_N].$$

For $1 < i < j \leq N$, define the *interval subalgebra*

$$\mathbf{U}_{[i,j]} := \langle X_i, X_{i+1}, \ldots, X_j \rangle \subseteq \mathcal{U}_q^+[w] \tag{3.2}$$

as the subalgebra generated by $X_i, X_{i+1}, \ldots, X_j$. Standard monomials

$$X_{\beta_i}^{m_i} \cdots X_{\beta_j}^{m_j}, \quad m_i, \ldots, m_j \in \mathbb{Z}_{\geq 0}$$

form a basis of $\mathbf{U}_{[i,j]}$. The Levendorskii-Soibelmann straightening rule [**25**, Prop. 5.5.2] tells us that for all $1 \leq i < j \leq N$,

$$[X_{\beta_i}, X_{\beta_j}] \in \mathbf{U}_{[i+1,j-1]} \cap \mathcal{U}_q(\mathfrak{g})_{\beta_i+\beta_j},$$

(recall (2.1)). As a consequence of the straightening rule, we have the following corollary.

COROLLARY 3.1. *If $1 \leq i < j \leq N$ and there fails to exist a nonnegative integral combination of roots in $\{\beta_{i+1}, \ldots, \beta_{j-1}\}$ that sum to $\beta_i + \beta_j$, then $[X_{\beta_i}, X_{\beta_j}] = 0$.*

Furthermore, every quantum Schubert cell algebra $\mathcal{U}_q^+[w]$ has a quantum cluster algebra structure (provided the deformation parameter q satisfies some minor conditions) [**13**] with a set of frozen variables

$$\{\Theta_1, \ldots, \Theta_r\}, \quad r = \#\{s_i \in W : s_i < w \text{ w.r.t. the Bruhat order}\}.$$

The *normal* subalgebra of $\mathcal{U}_q^+[w]$ (the subalgebra generated by the normal elements) is generated by the frozen variables [**15**, Proposition 2.7].

4. Automorphisms of quantum Schubert cells

In this section, we assume that R is a quantum Schubert cell algebra, say $R = \mathcal{U}_q^+[w] \subseteq \mathcal{U}_q(\mathfrak{g})$. Fix a reduced expression

$$w = s_{i_1} s_{i_2} \cdots s_{i_N} \in W$$

and let

$$X_{\beta_1}, X_{\beta_2}, \ldots, X_{\beta_N} \in R$$

be the corresponding Lusztig root vectors. Recall that R can be written as an iterated Ore extension

$$R = \mathbb{K}[X_{\beta_1}][X_{\beta_2}; \sigma_2, \delta_2] \cdots [X_{\beta_N}; \sigma_N, \delta_N]. \tag{4.1}$$

Observe that the algebraic torus $\mathcal{H} = (\mathbb{K}^\times)^r$ of rank $r = \text{rank}(\mathfrak{g})$ acts canonically on $\mathcal{U}_q(\mathfrak{g})$ via algebra automorphisms. An element $h = (h_1, \ldots, h_r) \in \mathcal{H}$ acts by the rule

$$h.E_i = h_i E_i, \quad h.F_i = h_i^{-1} F_i, \quad h.K_\mu = K_\mu,$$

for all $1 \leq i \leq r$ and $\mu \in Q$. This action is preserved by R, and each Lusztig root vector X_{β_i} is an $\mathcal{H}$-eigenvector. In fact, every Q-homogeneous element is an $\mathcal{H}$-eigenvector. Furthermore, the iterated Ore extension presentation in (4.1) is a symmetric Cauchon-Goodearl-Letzter (CGL) extension presentation for R (see e.g. [**13**, Theorem 9.1.b]).

4.1. The function η. Following [**15**], to every iterated Ore extension presentation R, as in (4.1) above, we define the rank of R

$$\operatorname{rank}(R) = \#\{k \in \{1,\dots,N\} : \delta_k = 0\}.$$

Let S be a set of cardinality $\operatorname{rank}(R)$, and let $\eta : \{1,\dots,N\} \to S$ be a function such that $\eta(\{k \in \{1,\dots,N\} : \delta_k = 0\}) = S$. That is, we assign to each trivial derivation δ_k a unique element in S. We require also that, for every $k \in \{1,\dots,N\}$ such that $\delta_k \neq 0$,

$$\eta(k) = \eta\left(\max\{\ell \in \{1,\dots,k-1\} : \delta_k(X_{\beta_\ell}) \neq 0\}\right).$$

The existence of such a function η was proved in [**15**, Theorem 4.3], and it plays a key role in determining the homogeneous prime elements for any CGL extension R. When $R = \mathcal{U}_q^+[w]$, the rank of R agrees with the cardinality of the *support* of w,

$$\operatorname{supp}(w) := \{i \in \mathbf{I} : s_i < w \text{ w.r.t. the Bruhat order}\}.$$

In this setting, the function $\eta : \{1,\dots,N\} \to \operatorname{supp}(w)$ can be defined by the rule

$$\eta(k) = i_k, \qquad (1 \le k \le N), \tag{4.2}$$

(see e.g. [**13**, Theorem 9.5]).

4.2. The core of R. Following [**14**, Section 4.1], define $P_x(R)$ to be the set of those $i \in \{1,\dots,N\}$ such that X_{β_i} is prime. By [**14**, Proposition 2.6],

$$P_x(R) = \left\{i \in \{1,\dots,N\} : \{i\} = \eta^{-1}(\eta(i))\right\}. \tag{4.3}$$

For $1 \le j < k \le N$, the element $Q_{jk} := [X_{\beta_j}, X_{\beta_k}]$ can be written uniquely as a linear combination of monomials $X_{\beta_{j+1}}^{m_{j+1}} \cdots X_{\beta_{k-1}}^{m_{k-1}}$. Let $F_x(R)$ be the set of those $i \in P_x(R)$ such that X_{β_i} does not appear in any Q_{jk}. More precisely, no monomial in Q_{jk} with a nonzero coefficient contains a positive power of X_{β_i}. Define $C_x(R) := \{1,\dots,N\} \setminus F_x(R)$. The *core* of R, denoted by $\mathcal{C}(R)$, is defined as the subalgebra generated by the X_{β_i}'s with $i \in C_x(R)$,

$$\mathcal{C}(R) := \mathbb{K}\langle X_{\beta_i} : i \in C_x(R)\rangle.$$

4.3. Diagonal and graded automorphisms. An algebra automorphism $\phi : R \to R$ that sends every Lusztig root vector X_{β_i} $(1 \le i \le N)$ to a scalar multiple of itself will be called a *diagonal* automorphism. This notion is dependent upon the choice of reduced expression for w. Hence, whenever we refer to automorphisms of this type, we have a fixed reduced expression for w in mind. The set of diagonal automorphisms is a subgroup of the automorphism group $\operatorname{Aut}(R)$ of R. We will denote this subgroup by $\operatorname{Diag-Aut}(R)$. Thus, for an algebra automorphism $\phi : R \to R$,

$$\phi \in \operatorname{Diag-Aut}(R) \iff \phi(X_{\beta_i}) \in \mathbb{K}^\times X_{\beta_i}, \qquad (1 \le i \le N).$$

From Section 2.1, recall that every coweight $\lambda \in P^\vee$ induces a $\mathbb{Z}$-grading on $\mathcal{U}_q(\mathfrak{g})$. With this, the subalgebra $R = \mathcal{U}_q^+[w] \subseteq \mathcal{U}_q(\mathfrak{g})$ inherits this grading,

$$R = \bigoplus_{d\in\mathbb{Z}} R_d, \qquad R_d := \{u \in R \cap \mathcal{U}_q(\mathfrak{g})_\mu : \langle \mu, \lambda\rangle = d\}.$$

We assume throughout that $\lambda \in P^\vee$ is chosen so that the induced grading satisfies the following conditions:

(1) $R = R_0 \oplus R_1 \oplus R_2 \oplus \cdots$ (that is to say, $R_d = 0$ whenever $d < 0$),
(2) R_d is finite dimensional for every $d \geq 0$ (i.e. R is locally finite),
(3) $R_0 = \mathbb{K}$ (i.e. R is connected), and
(4) R is generated, as an algebra, by R_1.

These conditions mimic the standard grading on a commutative polynomial ring $\mathbb{K}[z_1, \ldots, z_N]$, where each variable z_i is assigned degree 1. It is always possible to choose λ so that the first three conditions above are satisfied. For example, $\lambda = \sum_{i \in \mathbf{I}} \varpi_i^\vee$ is one such choice. However, it is not always possible to select λ such that all four conditions are met. To give one example, it is not too difficult to verify that such a λ fails to exist for the case when the underlying Lie algebra is of type G_2 and $w = s_2 s_1 s_2$.

An algebra automorphism $\phi : R \to R$ is a *graded* algebra automorphism if it respects the $\mathbb{Z}_{\geq 0}$-grading. That is to say,

$$\phi(R_d) = R_d$$

for all $d \geq 0$. The set of graded automorphisms is a subgroup of the automorphism group of R. We denote the subgroup of graded automorphisms by Gr-Aut(R). Observe we have a chain of subgroups,

$$\text{Diag-Aut}(R) \subseteq \text{Gr-Aut}(R) \subseteq \text{Aut}(R).$$

Using (4.2) and (4.3), one can easily determine the set $P_x(R)$ from the reduced expression $w = s_1 \cdots s_N$. In many cases $P_x(R)$ is empty (and $\mathcal{C}(R) = R$), and in such situations we have the most control over the automorphisms of R (see e.g. [**14**]). The following theorem describes sufficient conditions on R to conclude that every automorphism of R is graded.

THEOREM 4.1. *Suppose $R = \mathcal{U}_q^+[w]$ is a quantum Schubert cell algebra with Lusztig root vectors $X_{\beta_1}, \ldots, X_{\beta_N}$. Suppose $\mathcal{C}(R) = R$. Suppose also that R is connected graded, locally finite, and generated by R_1. For every radical root $\beta_i \in \Delta_w$ with $X_{\beta_i} \in R_1$, suppose there exists $\beta_j \in \Delta_w$ such that $X_{\beta_i} X_{\beta_j} = \kappa X_{\beta_j} X_{\beta_i}$ for some scalar $\kappa \neq 1$. Then every algebra automorphism of R is a graded automorphism. In other words,*

$$\text{Aut}(R) = \text{Gr-Aut}(R).$$

PROOF. It was shown in [**14**, Theorem 4.2] that if R is a symmetric saturated CGL extension which is a connected graded algebra, then every unipotent automorphism restricted to $\mathcal{C}(R)$ is the identity. Since $\mathcal{C}(R) = R$, then the identity is the only unipotent automorphism of R. As a consequence of [**14**, Lemma 4.7] every automorphism ϕ is graded provided $\phi(R_d) \subseteq \oplus_{j \geq d} R_j$ for all $d \geq 0$. However, this condition was established in [**23**, Proposition 4.2]. □

4.4. The normal subalgebra $\mathcal{N}(\mathbf{R})$ and the sets $\mathbf{C}_\mathbf{d}^\mathbf{m}$ and $\gamma_{\mathbf{d},\ell}^\mathbf{m}$. Following [**15**], let $\mathcal{N}(R)$ be the *normal subalgebra* of R. It is the subalgebra generated by the normal elements of R. By [**15**, Theorem 4.3], $\mathcal{N}(R)$ is a generated by a finite set of Q-homogeneous prime elements

$$\{\Theta_i : i \in \text{supp}(w)\} \subseteq R.$$

We remark here that the element $\Theta_i \in R$ is written as $\Delta_{\varpi_i, w\varpi_i}$ in [**13**, Section 9.4]. We have the following commutation relations,

$$u\Theta_i = q^{-\langle (1+w)\varpi_i, \mu\rangle}\Theta_i u, \quad \deg_Q(u) = \mu, \quad i \in \operatorname{supp}(w), \tag{4.4}$$

(see e.g. [**33**, Eq. 3.30] and [**13**, Eq. 9.23]).

For $d \in \mathbb{Z}_{\geq 0}$, $m \in \mathbb{Z}$, and $x \in R_1$, define

$$\begin{aligned} C_d^m &:= \{x \in R_1 : xy = q^m yx \text{ for all } y \in \mathcal{N}(R)_d\}, \\ V_d^m(x) &:= \{y \in \mathcal{N}(R)_d : xy = q^m yx\} \end{aligned} \tag{4.5}$$

For $d, \ell \in \mathbb{Z}_{\geq 0}$ and $m \in \mathbb{Z}$, define the set

$$\gamma_{d,\ell}^m := \{x \in R_1 : \dim_{\mathbb{K}}(V_d^m(x)) = \ell\}. \tag{4.6}$$

The following proposition is an easy observation.

PROPOSITION 4.2. *Suppose $\phi \in$ Gr-Aut(R). Let C_d^m and $\gamma_{d,\ell}^m$ be as defined in (4.5) and (4.6). Then each C_d^m is a ϕ-invariant subspace of R_1, and each $\gamma_{d,\ell}^m$ is a ϕ-invariant subset of R_1.*

The following theorem gives us sufficient conditions to determine when a standard generator Θ_j of $\mathcal{N}(R)$ gets sent to a nonzero scalar multiple of itself by a graded algebra automorphism $\phi : R \to R$.

THEOREM 4.3. *Suppose $\phi \in$ Gr-Aut(R). For every $i \in \operatorname{supp}(w)$, let d_i be the degree of Θ_i. That is, $\Theta_i \in \mathcal{N}(R)_{d_i}$. If, for some $j \in \operatorname{supp}(w)$, there fails to exist a nonnegative integral combination of numbers in $\{d_i : i \in \operatorname{supp}(w)$ and $i \neq j\}$ that sum to d_j, then*

$$\phi(\Theta_j) \in \mathbb{K}^\times \Theta_j.$$

PROOF. Ordered monomials in the Θ_i's form a basis (over $\mathbb{K}$) of the normal subalgebra $\mathcal{N}(R)$ [**15**, Theorem 4.6]. As there fails to exist a nonnegative integral combination of numbers in $\{d_i : i \in \operatorname{supp}(w)$ and $i \neq j\}$ that sum to d_j, this implies that $\mathcal{N}(R)_{d_j}$ is a one-dimensional vector space over $\mathbb{K}$ spanned by the element Θ_j. Since $\mathcal{N}(R)$ is an invariant subalgebra of R under ϕ, then we have

$$\phi(\mathcal{N}(R)_{d_j}) = \mathcal{N}(R)_{d_j}.$$

Hence $\phi(\Theta_j) \in \mathbb{K}^\times \Theta_j$. □

The following theorem gives sufficient conditions to determine when a Lusztig root vector X_β gets sent to a nonzero scalar multiple of itself by a graded algebra automorphism $\phi : R \to R$.

THEOREM 4.4. *Suppose $\phi \in$ Gr-Aut(R) and $\phi(\Theta_i) \in \mathbb{K}^\times \Theta_i$ for some $i \in \operatorname{supp}(w)$. Suppose also that there exists a radical root $\beta \in \Delta_w$ with $X_\beta \in R_1$ such that*

$$\langle \beta - \beta', (1+w)\varpi_i\rangle \neq 0$$

for every radical root $\beta' \in \Delta_w \setminus \{\beta\}$ with $X_{\beta'} \in R_1$, then

$$\phi(X_\beta) \in \mathbb{K}^\times X_\beta.$$

PROOF. Suppose $x_1, \ldots, x_n$ is a list of the Lusztig root vectors in R_1. Without loss of generality, assume $x_1 = X_\beta$. There are integers $d_1, \ldots, d_n$ which can be computed explicitly using (4.4) such that $x_j\Theta_i = q^{d_j}\Theta_i x_j$. The given hypotheses imply that d_1 is not equal to any number in $\{d_2, \ldots, d_n\}$.

The automorphism ϕ sends x_1 to a linear combination of $x_1, \dots, x_n$, say $\phi(x_1) = \sum c_j x_j$, $(c_j \in \mathbb{K})$. Applying ϕ to the relation $x_1\Theta_i = q^{d_1}\Theta_i x_1$ yields

$$\sum c_j x_j \Theta_i = \sum c_j q^{d_1} \Theta_i x_j = \sum c_j q^{d_j} \Theta_i x_j.$$

Thus,

$$\sum c_j (q^{d_1} - q^{d_j})\Theta_i x_j = 0.$$

The elements $\Theta_i x_1, \dots, \Theta_i x_n$ are Q-homogeneous and have distinct degrees with respect to the Q-gradation. Hence, each of the coefficients $c_j(q^{d_1} - q^{d_j})$ equals zero. Since q is not a root of unity, $c_2 = \cdots = c_n = 0$. □

Using the same techniques in the proof of Theorem 4.4 above, we have a more general result.

THEOREM 4.5. *Suppose* $\phi \in \text{Gr-Aut}(R)$. *Define the set*

$$S(\phi, w) := \left\{ i \in \text{supp}(w) \mid \phi(\Theta_i) \in \mathbb{K}^\times \Theta_i \right\}.$$

Suppose that $\beta \in \Delta_w$ *is a radical root with* $X_\beta \in R_1$ *and satisfies the condition that for every radical root* $\beta' \in \Delta_w \backslash \{\beta\}$ *with* $X_{\beta'} \in R_1$, *there exists* $i \in S(\phi, w)$ *such that* $\langle \beta - \beta', (1 + w)\varpi_i \rangle \neq 0$, *then*

$$\phi(X_\beta) \in \mathbb{K}^\times X_\beta.$$

By applying the theorems above, we can prove that Conjecture 1.1 holds, for example, when the underlying Lie algebra $\mathfrak{g}$ is of type G_2.

THEOREM 4.6. *If* $\mathfrak{g}$ *is the Lie algebra of type* G_2 *and* J *is a nonempty subset of simple roots of* $\mathfrak{g}$, *then* $\text{Aut}(\mathcal{U}_q(\mathfrak{n}_J)) \cong (\mathbb{K}^\times)^2$.

PROOF. We consider the reduced expression $w_o = s_1 s_2 s_1 s_2 s_1 s_2$ for the longest element of the Weyl group of $\mathfrak{g}$. The corresponding radical roots and root vectors associated to this reduced expression will be denoted by $\beta_1, \dots, \beta_6$ and $x_1, \dots, x_6$, respectively. In other words, we define $x_i := X_{\beta_i}$ $(i = 1, \dots, 6)$. The positive part $\mathcal{U}_q^+(\mathfrak{g})$ of $\mathcal{U}_q(\mathfrak{g})$ is generated by $x_1, \dots, x_6$. The defining relations appear in the work of Hu and Wang [**16**], where they index the root vectors by Lyndon words. The correspondence between our notation and their notation is

$$x_1 \leftrightarrow E_1, [3]_q! x_2 \leftrightarrow E_{1112}, [2]_q x_3 \leftrightarrow E_{112}, [3]_q! x_4 \leftrightarrow E_{11212}, x_5 \leftrightarrow E_{12}, x_6 \leftrightarrow E_2.$$

From [**16**, Eqns. 2.2 - 2.7 and Lemma 3.1], the defining relations in $\mathcal{U}_q^+(\mathfrak{g})$ are

$$\begin{aligned}
x_1x_2 &= q^3 x_2 x_1, & x_2x_6 &= q^{-3}x_6x_2 + \widehat{q}x_3x_5 + \zeta x_4,\\
x_1x_3 &= q x_3 x_1 + [3]_q x_2, & x_3x_4 &= q^3 x_4 x_3,\\
x_1x_4 &= x_4x_1 + q\widehat{q}x_3^2, & x_3x_5 &= q x_5 x_3 + [3]_q x_4,\\
x_1x_5 &= q^{-1}x_5x_1 + [2]_q x_3, & x_3x_6 &= x_6x_3 + q\widehat{q}x_5^2,\\
x_1x_6 &= q^{-3}x_6x_1 + x_5, & x_4x_5 &= q^3x_5x_4,\\
x_2x_3 &= q^3x_3x_2, & x_4x_6 &= q^3x_6x_4 + \eta x_5^3\\
x_2x_4 &= q^3x_4x_2 + \eta x_3^3, & x_5x_6 &= q^3x_6x_5,\\
x_2x_5 &= x_5x_2 + q\widehat{q}x_3^2, & &
\end{aligned}$$

where $\zeta := q^{-3} - q^{-1} - q \in \mathbb{K}$ and $\eta = q^3\frac{\widehat{q}^2}{[3]_q} \in \mathbb{K}$.

Observe that when J is a singleton, the parabolic element $w_o^J w_o$ has a unique reduced expression, and it appears as a substring of the reduced expression for the longest element w_o. Hence, each quantized nilradical $\mathcal{U}_q(\mathfrak{n}_J)$ is isomorphic to an *interval* subalgebra of $\mathcal{U}_q^+(\mathfrak{g})$. In particular, when $J = \{\alpha_1\}$, $\mathcal{U}_q(\mathfrak{n}_J)$ is isomorphic to the subalgebra generated by $x_1, \dots, x_5$, whereas if $J = \{\alpha_2\}$, the corresponding quantized nilradical is isomorphic to the subalgebra generated by $x_2, \dots, x_6$. For convenience, we will identify each quantized nilradical with an appropriate interval subalgebra.

First consider the case when $J = \{\alpha_1\}$. Here we choose the coweight $\lambda = \varpi_1 \in P^\vee$ to equip $\mathcal{U}_q(\mathfrak{n}_J)$ with a $\mathbb{N}$-gradation. With respect to this grading, the degree one generators are x_1 and x_5. The defining relations verify that $\mathcal{U}_q(\mathfrak{n}_J)$ is generated as an algebra by its degree one elements. By using (4.3), we have $\mathcal{C}(\mathcal{U}_q(\mathfrak{n}_J))$ is generated by x_1, x_3, and x_5. Thus, $\mathcal{C}(\mathcal{U}_q(\mathfrak{n}_J)) = \mathcal{U}_q(\mathfrak{n}_J)$ and Theorem 4.1 be can applied to conclude that every automorphism of $\mathcal{U}_q(\mathfrak{n}_J)$ preserves the $\mathbb{N}$-grading. The elements Θ_1 and Θ_2 of the normal subalgebra have degrees 4 and 6, respectively. Hence, by Theorem 4.3, every automorphism sends Θ_1 and Θ_2 to nonzero multiples of themselves. Finally, one can verify that Theorem 4.5 can be applied to conclude that every automorphism sends x_1 and x_5 to nonzero multiples of themselves. Therefore, every automorphism of $\mathcal{U}_q(\mathfrak{n}_J)$ is a diagonal automorphism. As a CGL extension, the algebra $\mathcal{U}_q(\mathfrak{n}_J)$ has rank 2. Thus, by [**15**, Theorems 5.3 and 5.5],

$$\operatorname{Aut}(\mathcal{U}_q(\mathfrak{n}_J)) \cong \left(\mathbb{K}^\times\right)^2.$$

The case when $J = \{\alpha_2\}$ is treated similarly, except now we choose the coweight $\varpi_2 \in P^\vee$ to equip the corresponding quantized nilradical with a $\mathbb{N}$-gradation. Here, we identify $\mathcal{U}_q(\mathfrak{n}_J)$ with the subalgebra of $\mathcal{U}_q^+(\mathfrak{g})$ generated by $x_2, \dots, x_6$. With this identification, the degree one root vectors of $\mathcal{U}_q(\mathfrak{n}_J)$ are x_2, x_3, x_5, and x_6. Again, we see that $\mathcal{U}_q(\mathfrak{n}_J)$ is generated as an algebra by its elements of degree one, and the core $\mathcal{C}(\mathcal{U}_q(\mathfrak{n}_J))$ coincides with $\mathcal{U}_q(\mathfrak{n}_J)$. All of the hypotheses of Theorem 4.1 apply. Hence, every automorphism preserves the $\mathbb{N}$-grading. The elements Θ_1 and Θ_2 of the normal subalgebra have degrees 2 and 4, respectively, in this setting. Thus, by Theorem 4.3, every automorphism of $\mathcal{U}_q(\mathfrak{n}_J)$ sends Θ_1 to a nonzero multiple of itself. Finally, by applying Theorem 4.4, we can conclude that every automorphism sends each degree one root vector to a nonzero multiple of itself. Thus, every automorphism is a diagonal automorphism. Finally, [**15**, Theorems 5.3 and 5.5] imply that

$$\operatorname{Aut}(\mathcal{U}_q(\mathfrak{n}_J)) \cong \left(\mathbb{K}^\times\right)^2.$$

The case when $J = \{\alpha_1, \alpha_2\}$ has already been established in [**34**]. □

Theorems 4.4 and 4.5 can be applied in several other cases to conclude that every graded automorphism of a quantized nilradical sends each degree one generator to a multiple of itself. If it can also be established that some particular quantized nilradical $\mathcal{U}_q(\mathfrak{n}_J)$ satisfies all of the hypotheses of Theorem 4.1, then the results of Goodearl and Yakimov in [**15**, Theorems 5.3 and 5.5] can be applied to conclude that

$$\operatorname{Aut}(\mathcal{U}_q(\mathfrak{n}_J)) \cong \left(\mathbb{K}^\times\right)^{\operatorname{rank}(\mathfrak{g})}.$$

PROPOSITION 4.7. *As above, let $\mathfrak{g}$ be a finite dimensional complex simple Lie algebra, and let J be a nonempty set of simple roots. Choose the coweight*

$$\lambda := \sum_{j \in J} \varpi_j \in P^\vee$$

to equip the quantized nilradical $\mathcal{U}_q(\mathfrak{n}_J)$ with a $\mathbb{N}$-gradation. The following list is an exhaustive list of all cases with $\operatorname{rank}(\mathfrak{g}) \leq 9$ *such that Theorems* 4.4 *and* 4.5 *can be applied to conclude that every graded automorphism of $\mathcal{U}_q(\mathfrak{n}_J)$ sends each degree one Lusztig root vector X_β to a multiple of itself.*

(1) $\mathfrak{g}$ *is of type* B_6 *and* J *is* $\{\alpha_2, \alpha_5\}$ *or* $\{\alpha_2, \alpha_4, \alpha_5\}$.
(2) $\mathfrak{g}$ *is of type* B_7 *or* B_8 *and* $J = \{\alpha_2, \alpha_6, \alpha_7\}$.
(3) $\mathfrak{g}$ *is of type* C_5 *and* J *is one of* $\{\alpha_3, \alpha_5\}$, $\{\alpha_1, \alpha_3, \alpha_5\}$, $\{\alpha_2, \alpha_3, \alpha_5\}$, $\{\alpha_2, \alpha_4, \alpha_5\}$.
(4) $\mathfrak{g}$ *is of type* C_6 *and* J *is* $\{\alpha_2, \alpha_4, \alpha_6\}$ *or* $\{\alpha_1, \alpha_3, \alpha_5, \alpha_6\}$.
(5) $\mathfrak{g}$ *is of type* D_7 *and* J *is* $\{\alpha_2, \alpha_4, \alpha_6\}$ *or* $\{\alpha_2, \alpha_4, \alpha_7\}$.
(6) $\mathfrak{g}$ *is of type* D_8 *and* J *is* $\{\alpha_2, \alpha_4, \alpha_6, \alpha_7\}$ *or* $\{\alpha_2, \alpha_4, \alpha_6, \alpha_8\}$.
(7) $\mathfrak{g}$ *is of type* E_7 *and* J *is one of* $\{\alpha_3, \alpha_5\}$, $\{\alpha_4, \alpha_5\}$, $\{\alpha_4, \alpha_7\}$, $\{\alpha_1, \alpha_4, \alpha_5\}$, $\{\alpha_1, \alpha_4, \alpha_7\}$, $\{\alpha_2, \alpha_3, \alpha_6\}$, $\{\alpha_2, \alpha_4, \alpha_6\}$, $\{\alpha_3, \alpha_4, \alpha_5\}$, $\{\alpha_3, \alpha_4, \alpha_7\}$, $\{\alpha_3, \alpha_5, \alpha_6\}$, $\{\alpha_4, \alpha_5, \alpha_6\}$, $\{\alpha_4, \alpha_6, \alpha_7\}$, $\{\alpha_1, \alpha_2, \alpha_4, \alpha_6\}$, $\{\alpha_1, \alpha_4, \alpha_5, \alpha_6\}$, $\{\alpha_1, \alpha_4, \alpha_6, \alpha_7\}$, $\{\alpha_2, \alpha_3, \alpha_4, \alpha_6\}$, $\{\alpha_2, \alpha_3, \alpha_5, \alpha_7\}$, $\{\alpha_3, \alpha_4, \alpha_5, \alpha_6\}$, $\{\alpha_3, \alpha_4, \alpha_6, \alpha_7\}$.
(8) $\mathfrak{g}$ *is of type* E_8 *and* J *is one of* $\{\alpha_4, \alpha_6\}$, $\{\alpha_4, \alpha_7\}$, $\{\alpha_4, \alpha_8\}$, $\{\alpha_1, \alpha_4, \alpha_6\}$, $\{\alpha_1, \alpha_4, \alpha_7\}$, $\{\alpha_1, \alpha_4, \alpha_8\}$, $\{\alpha_2, \alpha_3, \alpha_7\}$, $\{\alpha_3, \alpha_4, \alpha_6\}$, $\{\alpha_3, \alpha_4, \alpha_7\}$, $\{\alpha_3, \alpha_4, \alpha_8\}$, $\{\alpha_3, \alpha_5, \alpha_7\}$, $\{\alpha_4, \alpha_5, \alpha_7\}$, $\{\alpha_4, \alpha_6, \alpha_7\}$, $\{\alpha_4, \alpha_6, \alpha_8\}$, $\{\alpha_4, \alpha_7, \alpha_8\}$, $\{\alpha_1, \alpha_2, \alpha_4, \alpha_6\}$, $\{\alpha_1, \alpha_2, \alpha_4, \alpha_8\}$, $\{\alpha_1, \alpha_4, \alpha_5, \alpha_7\}$, $\{\alpha_1, \alpha_4, \alpha_6, \alpha_7\}$, $\{\alpha_1, \alpha_4, \alpha_6, \alpha_8\}$, $\{\alpha_2, \alpha_3, \alpha_5, \alpha_7\}$, $\{\alpha_2, \alpha_4, \alpha_6, \alpha_8\}$, $\{\alpha_3, \alpha_4, \alpha_5, \alpha_7\}$, $\{\alpha_3, \alpha_4, \alpha_6, \alpha_7\}$, $\{\alpha_3, \alpha_4, \alpha_6, \alpha_8\}$, $\{\alpha_1, \alpha_2, \alpha_4, \alpha_6, \alpha_7\}$, $\{\alpha_1, \alpha_2, \alpha_4, \alpha_6, \alpha_8\}$, $\{\alpha_2, \alpha_3, \alpha_4, \alpha_6, \alpha_8\}$.
(9) $\mathfrak{g}$ *is of type* F_4 *and* J *is one of* $\{\alpha_3\}$, $\{\alpha_1, \alpha_3\}$, $\{\alpha_2, \alpha_3\}$, $\{\alpha_2, \alpha_4\}$.
(10) $\mathfrak{g}$ *is of type* G_2 *and* $J = \{\alpha_1\}$ *or* $J = \{\alpha_2\}$

We choose the first case listed above, namely when $\mathfrak{g}$ is of type B_6 and $J = \{\alpha_2, \alpha_5\}$, and completely determine its automorphism group. It remains to show that every automorphism preserves the $\mathbb{N}$-grading in this case.

THEOREM 4.8. *If $\mathfrak{g}$ is of type B_6 and $J = \{\alpha_2, \alpha_5\}$, then* $\operatorname{Aut}(\mathcal{U}_q(\mathfrak{n}_J)) \cong (\mathbb{K}^\times)^6$.

PROOF. Consider the reduced expression

$$(s_5s_6)(s_4s_5s_6)(s_3s_4s_5s_6)(s_2s_3s_4s_5s_6)(s_1s_2s_3s_4s_5s_6)(s_1s_2s_3s_4s_5)(s_3s_4)(s_2s_3)(s_1s_2)$$

for the parabolic element $w_o^J w_o$. The length of this Weyl group element is 31. With this, let $\beta_1, \dots, \beta_{31}$ and $x_1, \dots, x_{31}$ denote the corresponding radical roots and root vectors, respectively. We will choose the coweight $\varpi_2 + \varpi_5$ to equip the quantized nilradical $\mathcal{U}_q(\mathfrak{n}_J)$ with a $\mathbb{N}$-gradation. By applying (4.3), we obtain that the core $\mathcal{C}(\mathcal{U}_q(\mathfrak{n}_J))$ coincides with $\mathcal{U}_q(\mathfrak{n}_J)$. Secondly, for every root vector x_i ($i = 1, \dots, 31$), there is another root vector x_j ($j \neq i$) such that $x_i x_j = \kappa x_j x_i$ for some scalar $\kappa \neq 1$. One can always simply let $j = i+1$ or $j = i-1$. Observe that in the reduced word above, every letter s_k is always adjacent to either s_{k-1} or s_{k+1}. If the i-th letter is s_k and the $(i+1)$-th letter is $s_{k\pm 1}$, then $x_i x_{i+1} = q^{\langle \alpha_k, s_k(\alpha_{k\pm 1})\rangle} x_{i+1} x_i$. On the other hand, if the i-th letter is s_k and the $(i-1)$-th letter is $s_{k\pm 1}$, then $x_{i-1} x_i = q^{\langle \alpha_{k\pm 1}, s_{k\pm 1}(\alpha_k)\rangle} x_i x_{i-1}$. In either case, the relevant hypothesis of Theorem 4.1 is satisfied.

Among the 31 Lusztig root vectors exactly 15 of them are of degree one. We leave it to the reader to verify that each remaining root vector x_i can be written, up to a scalar multiple, as a q-commutator $x_j x_k - q^{\langle \beta_j, \beta_k\rangle} x_k x_j$ for some j, k. Thus

$\mathcal{U}_q(\mathfrak{n})$ is generated by its elements of degree one. Hence all of the hypotheses of Theorem 4.1 are satisfied. Therefore every automorphism of $\mathcal{U}_q(\mathfrak{n}_J)$ preserves the $\mathbb{N}$-grading.

The degrees of the normal elements $\Theta_1, \ldots, \Theta_6$ are $4, 8, 10, 12, 14$, and 7, respectively. Hence, Theorem 4.3 tells us that every automorphism of $\mathcal{U}_q(\mathfrak{n}_J)$ sends Θ_1, Θ_3, and Θ_6 to multiples of themselves. Next, Theorem 4.5 can be applied to conclude that every automorphism sends each degree one generator x_i to a multiple of itself. Hence, every automorphism is a diagonal automorphism. Finally, [**15**, Theorems 5.3 and 5.5] imply that $\mathrm{Aut}(\mathcal{U}_q(\mathfrak{n}_J)) \cong (\mathbb{K}^{\times})^6$. □

5. The automorphism group of $\mathcal{U}_q(\mathfrak{n}_J)$ for $\mathfrak{g} = F_4$

We now consider the case when $\mathfrak{g}$ is the Lie algebra of type F_4. In this section we prove that, for every nonempty subset J of simple roots of $\mathfrak{g}$, the automorphism group of the quantized nilradical $\mathcal{U}_q(\mathfrak{n}_J)$ is isomorphic to the algebraic torus $(\mathbb{K}^{\times})^4$.

We consider the following reduced expressions for the longest element w_o of the Weyl group of $\mathfrak{g}$:

$$\mathbf{R}[w_o] = (1,2,1,3,2,3,1,2,4,3,2,1,3,2,3,4,3,2,3,1,2,3,4,3), \tag{5.1}$$

$$\mathbf{R}'[w_o] = (4,1,2,3,4,2,1,3,2,3,1,2,4,3,2,1,3,2,3,4,3,2,3,2). \tag{5.2}$$

The main reason we consider these two particular reduced expressions for w_o is that for every nonempty subset $J \subseteq \Pi$, a reduced expression for the corresponding parabolic element w_J appears either as a substring of $\mathbf{R}[w_o]$ or as a substring of $\mathbf{R}'[w_o]$. Thus, if we treat $\mathcal{U}_q(\mathfrak{n}^+)$ as an iterated Ore extension over $\mathbb{K}$, then it follows that each quantized nilradical $\mathcal{U}_q(\mathfrak{n}_J)$ can be viewed as an *interval subalgebra* (defined in (3.2)) of $\mathcal{U}_q(\mathfrak{n}^+)$. This is advantageous for finding explicit presentations of $\mathcal{U}_q(\mathfrak{n}_J)$ more efficiently. To make this more explicit, we will denote the Lusztig root vectors corresponding to the reduced expressions, $\mathbf{R}[w_o]$ and $\mathbf{R}'[w_o]$, by

$$x_i := X_{\beta_i}, \qquad y_i := X_{\beta'_i}, \qquad (1 \le i \le \ell(w_o) = 24), \tag{5.3}$$

respectively. Since every parabolic element w_J appears as a substring of $\mathbf{R}[w_o]$ or $\mathbf{R}'[w_o]$, then every quantized nilradical $\mathcal{U}_q(\mathfrak{n}_J)$ is isomorphic, as an algebra, to a subalgebra of $\mathcal{U}_q(\mathfrak{n}^+)$ generated by either a contiguous sequence of x_i's or y_i's. The following example illustrates this when $J = \{\alpha_2, \alpha_4\}$.

EXAMPLE 5.1. *Suppose $J = \{\alpha_2, \alpha_4\}$. The parabolic element $w_J \in W$ has a reduced expression*

$$w_J = s_2 s_1 s_3 s_2 s_3 s_1 s_2 s_4 s_3 s_2 s_1 s_3 s_2 s_3 s_4 s_3 s_2 s_3 s_1 s_2 s_3 s_4.$$

This expression is obtained by removing the first and last letters from $\mathbf{R}[w_o]$. Thus $\mathcal{U}_q(\mathfrak{n}_J)$ is isomorphic to the interval subalgebra $\mathbf{U}_{[2,23]} \subseteq \mathcal{U}_q(\mathfrak{n}^+)$,

$$\mathcal{U}_q(\mathfrak{n}_J) \cong \mathbb{K}\langle x_2, \ldots, x_{23} \rangle \subseteq \mathcal{U}_q(\mathfrak{n}^+).$$

With this identification, $x_2, x_3, x_4, x_6, x_7, x_8, x_{22}, x_{23}$ is a list of Lusztig root vectors of degree 1.

THEOREM 5.2. *If $\mathfrak{g}$ is the Lie algebra of type F_4 and J is a nonempty subset of simple roots of $\mathfrak{g}$, then every automorphism of the quantized nilradical $\mathcal{U}_q(\mathfrak{n}_J)$ is a diagonal automorphism.*

PROOF. The case when J is the full set of simple roots was handled in [**34**, Theorem 5.1]. Thus, we suppose J is a nonempty proper subset of the set of simple roots. Throughout this proof we will let ϕ be an arbitrary algebra automorphism of $\mathcal{U}_q(\mathfrak{n}_J)$. The algebra $\mathcal{U}_q(\mathfrak{n}_J)$ satisfies the hypotheses of Theorem 4.1. Hence, ϕ is a graded automorphism. Thus, the sets C_d^m and $\gamma_{d,\ell}^m$ are ϕ-invariant.

Our objective is to prove that ϕ is a diagonal automorphism. The commutation relations given in Lemmas 6.1 and 6.2 show that $\mathcal{U}_q(\mathfrak{n}_J)$ is generated by the degree one component $\mathcal{U}_q(\mathfrak{n}_J)_1$. Hence it suffices to show that each of the degree one generators of $\mathcal{U}_q(\mathfrak{n}_J)$ gets sent to a scalar multiple of itself under the map ϕ. We handle this on a case by case basis for each subset J of simple roots. The strategy is the same in each case. One key step is to observe that the sets C_d^m and $\gamma_{d,\ell}^m$ (defined in (4.5) and (4.6)) are ϕ-invariant. In several instances we will be able to characterize the C_d^m's and $\gamma_{d,\ell}^m$'s, or intersections of them, as either the set of all scalar multiples or nonzero scalar multiples of a generator of $\mathcal{U}_q(\mathfrak{n}_J)$. In these cases, we can immediately conclude that ϕ sends that particular generator to a multiple of itself. In other cases, we can show that the C_d^m's (or intersections of C_d^m's) are vector subspaces of $\mathcal{U}_q(\mathfrak{n}_J)_1$ spanned by either two or three generators of $\mathcal{U}_q(\mathfrak{n}_J)$. In these cases we will need to appeal to the defining relations of Lemmas 6.1 and 6.2 as well as Corollary 3.1 in order to conclude that ϕ indeed sends every degree one generator of $\mathcal{U}_q(\mathfrak{n}_J)$ to itself. Recall that Corollary 3.1 gives us sufficient conditions to conclude that certain q-commutators, say $[x_i, x_j]$ or $[y_i, y_j]$, equal 0. Throughout this proof, we are tacitly applying this result whenever we state that a q-commutator equals 0.

$\underline{J = \{\alpha_1\}}$ In this case, $\mathcal{U}_q(\mathfrak{n}_J) \cong \mathbf{U}'_{[2,16]}$. The degree one generators are y_2, y_3, y_4, y_5, y_6, y_7, y_8, y_{10}, y_{11}, y_{12}, y_{13}, y_{14}, y_{15}, and y_{16}. We have $\mathbb{K}^\times y_2 = \gamma_{6,3}^6$, $\mathbb{K}^\times y_3 = \gamma_{6,2}^6$, $\mathbb{K}^\times y_4 = \gamma_{4,1}^4 \cap \gamma_{4,1}^2$, $\mathbb{K}^\times y_5 = \gamma_{4,1}^2 \cap \gamma_{4,2}^0$, $\mathbb{K}^\times y_6 = \gamma_{6,3}^2$, $\mathbb{K}^\times y_7 = \gamma_{6,2}^2 \cap \gamma_{6,1}^{-2}$, $\mathbb{K}^\times y_8 = \gamma_{6,3}^0 \cap \gamma_{6,1}^2$, $\mathbb{K}^\times y_{10} = \gamma_{6,3}^0 \cap \gamma_{6,1}^{-2}$, $\mathbb{K}^\times y_{11} = \gamma_{6,2}^{-2} \cap \gamma_{6,1}^2$, $\mathbb{K}^\times y_{12} = \gamma_{6,3}^{-2}$, $\mathbb{K}^\times y_{13} = \gamma_{4,1}^{-2} \cap \gamma_{4,2}^0$, $\mathbb{K}^\times y_{14} = \gamma_{4,1}^{-2} \cap \gamma_{4,1}^{-4}$, $\mathbb{K}^\times y_{15} = \gamma_{6,2}^{-6}$, $\mathbb{K}^\times y_{16} = \gamma_{6,3}^{-6}$. Hence ϕ is a diagonal automorphism.

$\underline{J = \{\alpha_2\}}$ In this case, $\mathcal{U}_q(\mathfrak{n}_J) \cong \mathbf{U}_{[2,21]}$. The degree one generators are x_2, x_3, x_4, x_6, x_7, x_8, x_{13}, x_{15}, x_{17}, x_{19}, x_{20}, and x_{21}. We have $\mathbb{K}x_2 = C_4^2 \cap C_6^2$, $\mathbb{K}x_3 = C_4^2 \cap C_6^{-2}$, $\mathbb{K}x_4 = C_4^1 \cap C_6^2$, $\mathbb{K}x_6 = C_4^1 \cap C_6^{-2}$, $\mathbb{K}x_{17} = C_4^{-1} \cap C_6^2$, $\mathbb{K}x_{19} = C_4^{-1} \cap C_6^{-2}$, $\mathbb{K}x_{20} = C_4^{-2} \cap C_6^2$, $\mathbb{K}x_{21} = C_4^{-2} \cap C_6^{-2}$, $\mathbb{K}x_7 \oplus \mathbb{K}x_{13} = C_4^0 \cap C_6^2$, and $\mathbb{K}x_8 \oplus \mathbb{K}x_{15} = C_4^0 \cap C_6^{-2}$.

Thus, $\phi(x_7) = ax_7 + bx_{13}$ for some $a, b \in \mathbb{K}$. By applying ϕ to the relation $[x_7, x_{17}] = 0$ and using $[x_{13}, x_{17}] = 0$, we conclude $b = 0$. Hence, $\phi(x_7) \in \mathbb{K}x_7$. Similarly, by applying ϕ to the relation $[x_{13}, x_{17}] = 0$ we conclude $\phi(x_{13}) \in \mathbb{K}x_{13}$. Analogously, we can conclude that $\phi(x_8) \in \mathbb{K}x_8$ and $\phi(x_{15}) \in \mathbb{K}x_{15}$ by applying ϕ to the relations $[x_8, x_{19}] = 0$ and $[x_{15}, x_{19}] = 0$. Hence ϕ is a diagonal automorphism.

$\underline{J = \{\alpha_3\}}$ Here, $\mathcal{U}_q(\mathfrak{n}_J) \cong \mathbf{U}'_{[4,23]}$. The degree one generators are y_4, y_5, y_{17}, y_{19}, y_{21}, and y_{23}. We have $\mathbb{K}y_4 = C_6^1 \cap C_8^2$, $\mathbb{K}y_5 = C_6^{-1} \cap C_8^2$, $\mathbb{K}y_{17} = C_6^1 \cap C_8^0$, $\mathbb{K}y_{19} = C_6^1 \cap C_8^{-2}$, $\mathbb{K}y_{21} = C_6^{-1} \cap C_8^0$, and $\mathbb{K}y_{23} = C_6^{-1} \cap C_8^{-2}$. Therefore, ϕ is a diagonal automorphism.

$J=\{\alpha_4\}$ In this case, $\mathcal{U}_q(\mathfrak{n}_J)\cong \mathbf{U}_{[9,23]}$. The degree one generators are x_9, x_{10}, x_{13}, x_{15}, x_{17}, x_{19}, x_{22}, and x_{23}. We have $\mathbb{K}x_9=C_6^3$, $\mathbb{K}^\times x_{10}=C_6^1\cap\gamma_{8,2}^4$, $\mathbb{K}^\times x_{13}=C_6^1\cap\gamma_{8,1}^4$, $\mathbb{K}^\times x_{15}=C_6^1\cap\gamma_{8,0}^4$, $\mathbb{K}^\times x_{17}=C_6^{-1}\cap\gamma_{8,0}^{-4}$, $\mathbb{K}^\times x_{19}=C_6^{-1}\cap\gamma_{8,1}^{-4}$, $\mathbb{K}^\times x_{22}=C_6^{-1}\cap\gamma_{8,2}^{-4}$, and $\mathbb{K}x_{23}=C_6^{-3}$. Therefore, ϕ is a diagonal automorphism.

$J=\{\alpha_1,\alpha_2\}$ Here, $\mathcal{U}_q(\mathfrak{n}_J)\cong \mathbf{U}_{[1,21]}$. The degree one generators are x_1, x_3, x_6, x_8, x_{15}, x_{19}, and x_{21}. We have $\mathbb{K}x_3=C_6^2$, $\mathbb{K}x_6=C_6^1$, $\mathbb{K}x_{19}=C_6^{-1}$, $\mathbb{K}x_{21}=C_6^{-2}$, and $\mathbb{K}x_1\oplus\mathbb{K}x_8\oplus\mathbb{K}x_{15}=C_6^0$. Hence $\phi(x_3)\in\mathbb{K}x_3$, $\phi(x_6)\in\mathbb{K}x_6$, $\phi(x_{19})\in\mathbb{K}x_{19}$, $\phi(x_{21})\in\mathbb{K}x_{21}$, and there exist scalars $a_{ij}\in\mathbb{K}$, $(1\le i,j\le 3)$ such that

$$\phi(x_1)=a_{11}x_1+a_{12}x_8+a_{13}x_{15}$$
$$\phi(x_8)=a_{21}x_1+a_{22}x_8+a_{23}x_{15}$$
$$\phi(x_{15})=a_{31}x_1+a_{32}x_8+a_{33}x_{15}$$

By applying ϕ to the relation $[x_6,x_{15}]=0$ and using the relations $[x_6,x_8]=0$ and $[x_1,x_6]=x_4$ to straighten unordered monomials, we obtain

$$a_{31}\left(\left(q^2-q\right)x_1x_6-q^2x_4\right)+a_{32}\left(1-q^{-1}\right)x_6x_8=0.$$

Hence $a_{31}=a_{32}=0$. Therefore $\phi(x_{15})\in\mathbb{K}x_{15}$.

Next we apply ϕ to the relation $[x_6,x_8]=0$ and use the relations $[x_1,x_6]=x_4$ and $[x_6,x_{15}]=0$ to straighten unordered monomials to conclude that $-q^2a_{21}x_4+a_{23}(1-q)x_6x_{15}=0$. Hence $a_{21}=a_{23}=0$. Thus, $\phi(x_8)\in\mathbb{K}x_8$.

The relations $[x_1,x_{19}]=x_{17}$ and $[x_8,x_{17}]=0$ imply $[x_8,[x_1,x_{19}]]=0$. By applying ϕ to this relation and using the relations $[x_8,x_{15}]=[x_8,x_{19}]=[x_{15},x_{19}]=0$ to straighten any unordered monomials, we get

$$a_{12}\left(1-q^{-4}\right)x_8^2x_{19}+a_{13}\left(1-q^{-3}\right)x_8x_{15}x_{19}=0.$$

Hence $a_{12}=a_{13}=0$. Thus, $\phi(x_1)\in\mathbb{K}x_1$ and we conclude that ϕ is a diagonal automorphism.

$J=\{\alpha_1,\alpha_3\}$ In this situation, $\mathcal{U}_q(\mathfrak{n}_J)\cong \mathbf{U}'_{[2,23]}$. The degree one generators are y_2, y_3, y_{17}, y_{19}, y_{21}, and y_{23}. We have $\mathbb{K}y_2=C_8^0\cap C_{22}^2$, $\mathbb{K}y_3=C_8^0\cap C_{22}^{-2}$, $\mathbb{K}y_{17}=C_8^1\cap C_{22}^2$, $\mathbb{K}y_{19}=C_8^1\cap C_{22}^{-2}$, $\mathbb{K}y_{21}=C_8^{-1}\cap C_{22}^2$, and $\mathbb{K}y_{23}=C_8^{-1}\cap C_{22}^{-2}$. Hence ϕ is a diagonal automorphism.

$J=\{\alpha_1,\alpha_4\}$ In this case, $\mathcal{U}_q(\mathfrak{n}_J)\cong \mathbf{U}'_{[1,20]}$. The degree one generators are y_1, y_2, y_3, y_5, y_{11}, y_{12}, y_{17}, y_{19}, and y_{20}. We have $\mathbb{K}^\times y_1=C_{10}^2\cap\gamma_{14,1}^2$, $\mathbb{K}^\times y_2=C_{10}^2\cap\gamma_{14,1}^4$, $\mathbb{K}y_3=C_{10}^2\cap C_{14}^0$, $\mathbb{K}y_5=C_{10}^0\cap C_{14}^0$, $\mathbb{K}y_{11}=C_{10}^{-2}\cap C_{14}^0$, $\mathbb{K}^\times y_{12}=C_{10}^{-2}\cap\gamma_{14,1}^{-4}$, $\mathbb{K}^\times y_{17}=C_{10}^0\cap\gamma_{14,1}^2$, $\mathbb{K}^\times y_{19}=C_{10}^0\cap\gamma_{14,1}^{-2}$, and $\mathbb{K}^\times y_{20}=C_{10}^{-2}\cap\gamma_{14,1}^{-2}$. Hence ϕ is a diagonal automorphism.

$J=\{\alpha_2,\alpha_3\}$ Here, $\mathcal{U}_q(\mathfrak{n}_J)\cong \mathbf{U}'_{[3,24]}$. The degree one generators are y_3, y_{17}, y_{21}, and y_{24}. We have $\mathbb{K}y_3=C_{14}^2$, $\mathbb{K}y_{17}=C_{10}^1$, $\mathbb{K}y_{21}=C_{10}^{-1}$, and $\mathbb{K}y_{24}=C_{14}^{-2}$. Hence, ϕ is a diagonal automorphism.

$J=\{\alpha_2,\alpha_4\}$ In this case, $\mathcal{U}_q(\mathfrak{n}_J)\cong \mathbf{U}_{[2,23]}$. The degree one generators are x_2, x_3, x_4, x_6, x_7, x_8, x_{22}, and x_{23}. We have $\mathbb{K}x_2=C_{10}^2\cap C_{14}^2$, $\mathbb{K}x_3=C_{10}^{-2}\cap C_{14}^2$,

$\mathbb{K}x_4 = C_{10}^2 \cap C_{14}^0$, $\mathbb{K}x_6 = C_{10}^{-2} \cap C_{14}^0$, $\mathbb{K}x_7 = C_{10}^2 \cap C_{14}^{-2}$, $\mathbb{K}x_8 = C_{10}^{-2} \cap C_{14}^{-2}$, $\mathbb{K}x_{22} = C_{14}^1$, and $\mathbb{K}x_{23} = C_{14}^{-1}$. Hence, ϕ is a diagonal automorphism.

$\underline{J = \{\alpha_3, \alpha_4\}}$ We have $\mathcal{U}_q(\mathfrak{n}_J) \cong \mathbf{U}_{[4,24]}$. The degree one generators are x_4, x_6, x_{22}, and x_{24}. We have $\mathbb{K}x_4 = C_{12}^2$, $\mathbb{K}x_{24} = C_{12}^{-2}$, and $\mathbb{K}x_6 \oplus \mathbb{K}x_{22} = C_{12}^0$. Thus, there exist scalars $a_{11}, a_{12}, a_{21}, a_{22}, \gamma, \delta \in \mathbb{K}$ such that $\phi(x_6) = a_{11}x_6 + a_{12}x_{22}$, $\phi(x_{22}) = a_{21}x_6 + a_{22}x_{22}$, $\phi(x_4) = \gamma x_4$, and $\phi(x_{24}) = \delta x_{24}$.

Observe first that the relation $[x_4, x_{24}] = (q+q^{-1})x_7$ implies $\phi(x_7) = \gamma\delta x_7$. Next apply ϕ to the relation $[x_6, x_7] = 0$ and use the relation $[x_7, x_{22}] = x_{17}$ to straighten unordered monomials to obtain

$$a_{12}\left((q^2-1)\,x_7x_{22} - q^2x_{17}\right) = 0.$$

Hence $a_{12} = 0$. Thus $\phi(x_6) \in \mathbb{K}x_6$.

The relations $[x_4, x_{22}] = x_{13}$ and $[x_4, x_{13}] = 0$ imply $[x_4, [x_4, x_{22}]] = 0$. Applying ϕ to this relation and using the relations $[x_4, x_6] = (q+q^{-1})\,x_5$ and $[x_4, x_5] = 0$ to straighten unordered monomials we get

$$a_{21}\left(1-q^{-1}\right)\left((1-q)\,x_4^2x_6 + \left(q+q^{-1}\right)^2 x_4x_5\right) = 0.$$

Hence $a_{21} = 0$. Therefore $\phi(x_{22}) \in \mathbb{K}x_{22}$. Thus ϕ is a diagonal automorphism.

$\underline{J = \{\alpha_1, \alpha_2, \alpha_3\}}$ In this case, $\mathcal{U}_q(\mathfrak{n}_J) \cong \mathbf{U}'_{[2,24]}$. The degree one generators are y_2, y_{17}, y_{21}, and y_{24}. We have $\mathbb{K}y_{17} = C_{12}^1$, $\mathbb{K}y_{21} = C_{12}^{-1}$, and $\mathbb{K}y_2 \oplus \mathbb{K}y_{24} = C_{12}^0$. Thus, there exist $b_2, b_{24}, c_2, c_{24} \in \mathbb{K}$ such that $\phi(y_2) = b_2y_2 + b_{24}y_{24}$ and $\phi(y_{24}) = c_2y_2 + c_{24}y_{24}$. Applying ϕ to the relation $[y_2, y_{17}] = 0$ and using the identity $[y_{17}, y_{24}] = y_{19}$ gives us

$$0 = (b_2y_2 + b_{24}y_{24})y_{17} - y_{17}(b_2y_2 + b_{24}y_{24}) = b_{24}\left(q\widehat{q}y_{17}y_{24} - q^2y_{19}\right).$$

Hence $b_{24} = 0$. Next we observe that $[y_2, y_{24}] = y_3$ and $[y_2, y_3] = 0$. Hence we have the relation $[y_2, [y_2, y_{24}]] = 0$. Applying ϕ to this relation gives us $c_2\widehat{q}^2y_2^3 = 0$. Therefore $c_2 = 0$ and we conclude that ϕ is a diagonal automorphism.

$\underline{J = \{\alpha_1, \alpha_2, \alpha_4\}}$ In this case, $\mathcal{U}_q(\mathfrak{n}_J) \cong \mathbf{U}_{[1,23]}$. The degree one generators are x_1, x_3, x_6, x_8, x_{22}, and x_{23}. We have $\mathbb{K}x_3, = C_{18}^2$, $\mathbb{K}x_8 = C_{18}^{-2}$, $\mathbb{K}x_{22} = C_{18}^1$, $\mathbb{K}x_{23} = C_{18}^{-1}$, and $\mathbb{K}x_1 \oplus \mathbb{K}x_6 = C_{18}^0$. Hence, there exist $b_1, b_6, c_1, c_6 \in \mathbb{K}$ such that $\phi(x_1) = b_1x_1 + b_6x_6$ and $\phi(x_6) = c_1x_1 + c_6x_6$. Applying ϕ to the relation $[x_1, x_{22}] = 0$ gives us

$$0 = (b_1x_1 + b_6x_6)x_{22} - x_{22}(b_1x_1 + b_6x_6) = b_6\left((1-q)x_6x_{22} + qx_{15}\right).$$

because $[x_6, x_{22}] = x_{15}$. Hence $b_6 = 0$. Similarly, by applying ϕ to the relation $[x_3, x_6] = 0$ we obtain

$$0 = x_3(c_1x_1 + c_6x_6) - q^2(c_1x_1 + c_6x_6)x_3 = -c_1q^2x_2$$

because $[x_1, x_3] = x_2$. Hence $c_1 = 0$. Therefore ϕ is a diagonal automorphism.

$\underline{J = \{\alpha_1, \alpha_3, \alpha_4\}}$ In this situation, $\mathcal{U}_q(\mathfrak{n}_J) \cong \mathbf{U}'_{[1,23]}$. The degree one generators are y_1, y_2, y_3, y_{21}, and y_{23}. We have $\mathbb{K}y_1 = C_{30}^0$, $\mathbb{K}y_2 \oplus \mathbb{K}y_{21} = C_{30}^2$, and $\mathbb{K}y_3 \oplus \mathbb{K}y_{23} = C_{30}^{-2}$. Hence there exist $a_2, a_{21}, b_2, b_{21}, c_3, c_{23}, d_3, d_{23} \in \mathbb{K}$ such that $\phi(y_2) = a_2y_2 + a_{21}y_{21}$, $\phi(y_{21}) = b_2y_2 + b_{21}y_{21}$, $\phi(y_3) = c_3y_3 + c_{23}y_{23}$, and $\phi(y_{23}) = d_3y_3 + d_{23}y_{23}$.

The relations $[y_1, y_{17}] = 0$ and $[y_1, y_{21}] = y_{17}$ give us $[y_1, [y_1, y_{21}]] = 0$. Applying ϕ to this relation and using the commutation relation $y_1y_2 = y_2y_1$ gives us $b_2(2 - q - q^{-1})y_1^2y_2 = 0$. Hence $b_2 = 0$. Next, since $[y_1, y_{21}] = y_{17}$ and $[y_2, y_{17}] = 0$, we have the relation $[y_2, [y_1, y_{21}]] = 0$. Applying ϕ to this relation and using the identity $[y_{17}, y_{21}] = 0$ gives us $a_{21}\left(q^{-1} - 1\right) y_{17}y_{21} = 0$. Therefore $a_{21} = 0$. Observe next that $[y_{21}, y_{23}] = [2]_q y_{22}$ and $[y_{21}, y_{22}] = 0$. Therefore $[y_{21}, [y_{21}, y_{23}]] = 0$. Applying ϕ to this relation and using the identities $[y_3, y_{21}] = y_5$ and $[y_5, y_{21}] = [2]_q y_{11}$ give us $d_3[2]_q q^2 y_{11} = 0$. Thus $d_3 = 0$. Finally, applying ϕ to the relation $[y_1, y_3] = 0$ and using the identity $[y_1, y_{23}] = y_{19}$ gives us $c_{23}\left((1 - q)y_1y_{23} + qy_{19}\right) = 0$. Hence $c_{23} = 0$. Therefore ϕ is a diagonal automorphism.

$J = \{\alpha_2, \alpha_3, \alpha_4\}$ In this case, $\mathcal{U}_q(\mathfrak{n}_J) \cong \mathbf{U}_{[2,24]}$. The degree one generators are x_2, x_3, x_{22}, and x_{24}. We have $\mathbb{K}x_2 = C_{18}^{2}$, $\mathbb{K}x_3 = C_{18}^{-2}$, and $\mathbb{K}x_{22} \oplus \mathbb{K}x_{24} = C_{18}^{0}$. By applying ϕ to the relation $[x_2, x_{22}] = 0$ and using the relation $[x_2, x_{24}] = x_4$ to straighten unordered monomials, we can conclude that $\phi(x_{22}) \in \mathbb{K}x_{22}$.

Since $\mathbb{K}x_{22} \oplus \mathbb{K}x_{24}$ is a ϕ-invariant subspace, there exist scalars $\gamma, \delta \in \mathbb{K}$ such that $\phi(x_{24}) = \gamma x_{22} + \delta x_{24}$. The relations $[x_2, x_{24}] = x_4$ and $[x_3, x_4] = 0$ give us the relation $[x_3, [x_2, x_{24}]] = 0$. Applying ϕ to this relation and using the relations $[x_2, x_{22}] = [x_2, x_3] = [x_3, x_{22}] = 0$ to straighten any unordered monomials gives us $\gamma\left(q^{-2} - 1\right) x_2x_3x_{22} = 0$. Hence $\gamma = 0$. Therefore $\phi(x_{24}) \in \mathbb{K}x_{24}$ and ϕ is a diagonal automorphism. □

We are now able to prove the main result of this section. The following theorem proves Conjecture 1.1 when the underlying Lie algebra $\mathfrak{g}$ is of type F_4.

THEOREM 5.3. *If $\mathfrak{g}$ is the Lie algebra of type F_4 and J is a nonempty subset of simple roots of $\mathfrak{g}$, then* $\mathrm{Aut}(\mathcal{U}_q(\mathfrak{n}_J)) \cong (\mathbb{K}^\times)^4$.

PROOF. By Theorem 5.2, every automorphism of $\mathcal{U}_q(\mathfrak{n}_J)$ is a diagonal automorphism. As a CGL extension, the algebra $\mathcal{U}_q(\mathfrak{n}_J)$ has rank 4. Thus, by [**15**, Theorems 5.3 and 5.5], $\mathrm{Aut}(\mathcal{U}_q(\mathfrak{n}_J)) \cong (\mathbb{K}^\times)^4$. □

6. Two lemmas regarding $\mathcal{U}_q(\mathfrak{n}_J)$ when $\mathfrak{g} = F_4$

In this section we prove two lemmas regarding the quantized nilradicals $\mathcal{U}_q(\mathfrak{n}_J)$ for the case when the underlying Lie algebra $\mathfrak{g}$ is of type F_4 and J is any nonempty subset of simple roots.

Recall from Section 2.1 that for each coweight $\lambda \in P^\vee$, there is an induced $\mathbb{Z}$-grading on $\mathcal{U}_q(\mathfrak{n}_J)$. We will use the coweight $\lambda = \sum_{i \in J} \varpi_i^\vee \in P^\vee$. The two lemmas in this section explicitly show how every Lusztig root vector X_β in $\mathcal{U}_q(\mathfrak{n}_J)$ with $\mathrm{height}(\beta) > 1$ can be written, up to a scalar multiple, as a q-commutator of other Lusztig root vectors. As a direct consequence of these lemmas, one can readily verify that each quantized nilradical $\mathcal{U}_q(\mathfrak{n}_J)$ is generated, as an algebra, by the Lusztig root vectors of degree 1. Hence, $\mathcal{U}_q(\mathfrak{n}_J)$ is a locally finite, connected, $\mathbb{N}$-graded algebra generated by its graded component of degree 1.

LEMMA 6.1. *Let $\mathfrak{g}$ be the Lie algebra of type F_4, and let $x_1, \ldots, x_{24}$ be the Lusztig root vectors (recall* (5.3)*) corresponding to the reduced expression* $\mathbf{R}[w_o]$ *(see* (5.1)*) of the longest element of the Weyl group of $\mathfrak{g}$. Then*

$$
\begin{aligned}
x_2 &= [x_1, x_3], & x_{10} &= [x_8, x_{13}], & x_{17} &= [x_7, x_{22}],\\
x_4 &= [x_1, x_6], & x_{11} &= \tfrac{1}{[2]_q}[x_{10}, x_{13}], & x_{18} &= \tfrac{1}{[2]_q}[x_{17}, x_{19}],\\
x_4 &= [x_2, x_{24}], & x_{12} &= \tfrac{1}{[2]_q}[x_{10}, x_{15}], & x_{19} &= [x_8, x_{22}],\\
x_5 &= \tfrac{1}{[2]_q}[x_4, x_6], & x_{13} &= [x_1, x_{15}], & x_{20} &= [x_1, x_{21}],\\
x_6 &= [x_3, x_{24}], & x_{13} &= [x_4, x_{22}], & x_{20} &= \tfrac{1}{[2]_q}[x_{17}, x_{22}],\\
x_7 &= [x_1, x_8], & x_{14} &= \tfrac{1}{[2]_q}[x_{13}, x_{15}], & x_{21} &= \tfrac{1}{[2]_q}[x_{19}, x_{22}],\\
x_7 &= \tfrac{1}{[2]_q}[x_4, x_{24}], & x_{15} &= [x_6, x_{22}], & x_{23} &= [x_{22}, x_{24}].\\
x_8 &= \tfrac{1}{[2]_q}[x_6, x_{24}], & x_{16} &= [x_{15}, x_{17}], & &\\
x_9 &= [x_6, x_{13}], & x_{17} &= [x_1, x_{19}], & &
\end{aligned}
$$

PROOF. Throughout the proof of this lemma, we adopt the abbreviation

$$E_{ij} := [E_i, E_j], \qquad i, j \in \mathbf{I},$$

for a q-commutator and inductively define the nested q-commutator

$$E_{i_1 i_2 \cdots i_n} := [E_{i_1 i_2 \cdots i_{n-1}}, E_{i_n}]$$

for $i_1, i_2, \ldots, i_n \in \mathbf{I}$.

Since s_1 is the first simple reflection appearing the reduced expression $R[w_o]$, it is clear that $x_1 = E_1$. As a direct consequence of Proposition 2.1, $x_3 = E_2$, $x_{22} = E_4$, and $x_{24} = E_3$. At this point we have identified each Chevalley generator E_i with a corresponding Lusztig root vector. Next we will show how the remaining Lusztig root vectors x_i can be written, up to a scalar multiple, as q-commutators of other Lusztig root vectors. The next step will be to focus on those x_i with $\langle \deg_Q(x_i), \varpi_1^\vee + \varpi_2^\vee + \varpi_3^\vee + \varpi_4^\vee \rangle = 2$. In other words, we focus on the x_i's such that $\deg_Q(x_i)$ has height 2. For short, we will say that the $\mathbb{N}$-degree of x_i is d whenever $\deg_Q(x_i)$ has height d. We compute

$$
\begin{aligned}
x_2 &= \mathbf{T}_1(E_2) = E_{12} = [x_1, x_3],\\
x_6 &= \mathbf{T}_{\underline{121}32}(E_3) = \mathbf{T}_{2\cdot 1232}(E_3) = \mathbf{T}_2(E_3) = E_{23} = [x_3, x_{24}],\\
x_{23} &= \mathbf{T}_{\underline{121323124321323432312 3}}(E_4) = \mathbf{T}_{4\cdot 121321324321323432132}(E_4) = \mathbf{T}_4(E_3)\\
&= E_{43} = [x_{22}, x_{24}],
\end{aligned}
$$

where, in the above computations, we adopt the underlining notation, as in $\mathbf{T}_{\underline{121}32}$ above, to highlight that braid relations in the Weyl group are being applied to the underlined part in moving from one step in the calculations to the next. We use the dot notation, as in $\mathbf{T}_{2\cdot 1232}$ above, to split a reduced word into two parts in order to indicate which Lusztig symmetries are being applied at that particular step. We continue to compute

$$
\begin{aligned}
x_4 &= \mathbf{T}_{12\cdot 1}(E_3) = \mathbf{T}_{1\cdot 2}(E_3) = T_1(E_{23}) = E_{123} = [E_{12}, E_3] = [x_2, x_{24}],\\
[2]_q x_8 &= [2]_q \mathbf{T}_{\underline{121323}1}(E_2) = [2]_q \mathbf{T}_{23\cdot 12321}(E_2) = [2]_q \mathbf{T}_{23}(E_2) = E_{233} = [E_{23}, E_3]\\
&= [x_6, x_{24}],\\
x_{15} &= \mathbf{T}_{\underline{121323124321}32}(E_3) = \mathbf{T}_{23\cdot 123412321232}(E_3) = \mathbf{T}_{23}(E_4) = E_{234}\\
&= [E_{23}, E_4] = [x_6, x_{22}].
\end{aligned}
$$

So far we have identified how each Lusztig root vector x_i with $\mathbb{N}$-degree at most 3 can be written as a q-commutator of Lusztig root vectors of smaller $\mathbb{N}$-degree.

We continue in this manner focusing next on the Lusztig root vectors x_i having $\mathbb{N}$-degree equal to 4. We compute

$$\begin{aligned}
[2]_q x_7 &= [2]_q \mathbf{T}_{12\underline{13}23}(E_1) = [2]_q \mathbf{T}_{123\cdot 123}(E_1) = [2]_q \mathbf{T}_{1\cdot 23}(E_2) = \mathbf{T}_1(E_{233}) \\
&= E_{1233} = [E_{123}, E_3] = [x_4, x_{24}], \\
x_{13} &= \mathbf{T}_{12\underline{13}23124321}(E_3) = \mathbf{T}_{123\cdot 123124321}(E_3) = \mathbf{T}_{123}(E_4) = E_{1234} = [E_{123}, E_4] \\
&= [x_4, x_{22}], \\
[2]_q x_{19} &= [2]_q \mathbf{T}_{\underline{121323124321323}432}(E_3) = [2]_q \mathbf{T}_{2342\cdot 12321432132432}(E_3) \\
&= [2]_q \mathbf{T}_{234\cdot 2}(E_3) = [2]_q \mathbf{T}_{234}(E_{23}) = [2]_q [\mathbf{T}_{23\cdot 4}(E_2), \mathbf{T}_{234}(E_3)] \\
&= [2]_q [\mathbf{T}_{23}(E_2), E_4] = [E_{233}, x_{22}] = [2]_q [x_8, x_{22}], \\
x_{19} &= \mathbf{T}_{\underline{121323124321323}432}(E_3) = \mathbf{T}_{2324\cdot 12321432132432}(E_3) \\
&= \mathbf{T}_{232\cdot 4}(E_3) = \mathbf{T}_{232}(E_{43}) = [\mathbf{T}_{232}(E_4), \mathbf{T}_{232}(E_3)] = [E_{234}, E_3] = [x_{15}, x_{24}].
\end{aligned}$$

Next we show how each Lusztig root vector x_i of $\mathbb{N}$-degree equal to 5 (i.e. x_5, x_{17}, and x_{21}) can be written as a q-commutator of Lusztig root vectors of smaller $\mathbb{N}$-degree. We have

$$\begin{aligned}
[2]_q x_5 &= [2]_q \mathbf{T}_{121\cdot 3}(E_2) = \mathbf{T}_{121}([E_3, E_{32}]) = \mathbf{T}_{121}([E_3, \mathbf{T}_{32}(E_3)]) \\
&= [\mathbf{T}_{121}(E_3), \mathbf{T}_{12132}(E_3)] = [x_4, x_6], \\
x_{17} &= \mathbf{T}_{12\underline{13231243213}234}(E_3) = \mathbf{T}_{12324\cdot 12324321234}(E_3) = \mathbf{T}_{1232\cdot 4}(E_3) \\
&= \mathbf{T}_{1232}(E_{43}) = [\mathbf{T}_{123\cdot 2}(E_4), \mathbf{T}_{1232}(E_3)] = [\mathbf{T}_{123}(E_4), E_3] = [E_{1234}, E_3] \\
&= [x_{13}, x_{24}], \\
[2]_q x_{17} &= [2]_q \mathbf{T}_{12\underline{13231243213}234}(E_3) = [2]_q \mathbf{T}_{12324\cdot 12324321234}(E_3) = [2]_q \mathbf{T}_{123\underline{24}}(E_3) \\
&= [2]_q \mathbf{T}_{1234\cdot 2}(E_3) = [2]_q \mathbf{T}_{1234}([E_2, E_3]) = [2]_q [\mathbf{T}_{123\cdot 4}(E_2), \mathbf{T}_{1234}(E_3)] \\
&= [2]_q [\mathbf{T}_{1\cdot 23}(E_2), E_4] = [\mathbf{T}_1(E_{233}), E_4] = [E_{1233}, E_4] = [2]_q [x_7, x_{22}], \\
[2]_q x_{21} &= [2]_q \mathbf{T}_{\underline{1213231243213234323}1}(E_2) = [2]_q \mathbf{T}_{23243\cdot 123241321324321}(E_2) \\
&= [2]_q \mathbf{T}_{2324\cdot 3}(E_2) = \mathbf{T}_{2324}([E_3, E_{32}]) = [\mathbf{T}_{232\cdot 4}(E_3), \mathbf{T}_{23\underline{24}}(E_{32})] \\
&= [\mathbf{T}_{232}(E_{43}), \mathbf{T}_{234\cdot 2}(E_{32})] = [[\mathbf{T}_{23\cdot 2}(E_4), \mathbf{T}_{232}(E_3)], \mathbf{T}_{234}(E_3)] \\
&= [[\mathbf{T}_{23}(E_4), E_3], E_4] = [E_{234}, E_3], E_4] = [E_{2343}, E_4] = [x_{19}, x_{22}].
\end{aligned}$$

Continuing in this manner, we get

$$\begin{aligned}
x_9 &= \mathbf{T}_{121323\cdot 12}(E_4) = \mathbf{T}_{12132\cdot 3}(E_4) = \mathbf{T}_{12132}(E_{34}) = [\mathbf{T}_{12132}(E_3), \mathbf{T}_{12\underline{13}2}(E_4)] \\
&= [x_6, \mathbf{T}_{123\cdot 12}(E_4)] = [x_6, \mathbf{T}_{123}(E_4)] = [x_6, x_{13}], \\
x_{10} &= \mathbf{T}_{1213231\underline{24}}(E_3) = \mathbf{T}_{12132314\cdot 2}(E_3) = \mathbf{T}_{12132314}(E_{23}) \\
&= [\mathbf{T}_{1213231\cdot 4}(E_2), \mathbf{T}_{1213\cdot 2314}(E_3)] = [\mathbf{T}_{1213231}(E_2), \mathbf{T}_{12\underline{13}}(E_4)] \\
&= [x_8, \mathbf{T}_{123\cdot 1}(E_4)] = [x_8, \mathbf{T}_{123}(E_4)] = [x_8, x_{13}], \\
[2]_q x_{11} &= [2]_q \mathbf{T}_{121323124\cdot 3}(E_2) = \mathbf{T}_{121323124}([E_3, E_{32}]) \\
&= [\mathbf{T}_{121323124}(E_3), \mathbf{T}_{12\underline{1323124}}(E_{32})] = [x_{10}, \mathbf{T}_{12312314\cdot 2}(E_{32})] \\
&= [x_{10}, \mathbf{T}_{123\cdot 12314}(E_3)] = [x_{10}, \mathbf{T}_{123}(E_4)] = [x_{10}, x_{13}], \\
[2]_q x_{14} &= [2]_q \mathbf{T}_{121323124321\cdot 3}(E_2) = \mathbf{T}_{121323124321}([E_3, E_{32}]) \\
&= [\mathbf{T}_{121323124321}(E_3), \mathbf{T}_{121323124321}(E_{32})] = [x_{13}, \mathbf{T}_{121323124321}\mathbf{T}_{32}(E_3)] \\
&= [x_{13}, x_{15}],
\end{aligned}$$

$$\begin{aligned}
[2]_q x_{12} &= [2]_q \mathbf{T}_{\underline{12132312432}}(E_1) = [2]_q \mathbf{T}_{231213423\cdot 12}(E_1) = [2]_q \mathbf{T}_{2312134\cdot 23}(E_2) \\
&= \mathbf{T}_{2312134}(E_{233}) = [\mathbf{T}_{2312134}(E_{23}), \mathbf{T}_{23\cdot 12134}(E_3)] \\
&= [\mathbf{T}_{23121342}(E_3), \mathbf{T}_{23}(E_4)] = [\mathbf{T}_{23121342\cdot 1}(E_3), E_{234}] \\
&= [\mathbf{T}_{\underline{121323124}}(E_3), E_{234}] = [x_{10}, x_{15}], \\
x_{16} &= \mathbf{T}_{12132312432132\cdot 3}(E_4) = \mathbf{T}_{12132312432132}(E_{34}) \\
&= [\mathbf{T}_{12132312432132}(E_3), \mathbf{T}_{12132312432132}(E_4)] \\
&= [\mathbf{T}_{12132312432132}(E_3), \mathbf{T}_{12132312432132}\mathbf{T}_{34}(E_3)] = [x_{15}, x_{17}], \\
[2]_q x_{18} &= [2]_q \mathbf{T}_{1213231243213234\cdot 3}(E_2) = \mathbf{T}_{1213231243213234}([E_3, E_{32}]) \\
&= [\mathbf{T}_{1213231243213234}(E_3), \mathbf{T}_{1213231243213234}(E_{32})] \\
&= [x_{17}, \mathbf{T}_{1213231243213234}\mathbf{T}_{32}(E_3)] = [x_{17}, x_{19}], \\
[2]_q x_{20} &= [2]_q \mathbf{T}_{12\underline{1323124321323432}3}(E_1) = [2]_q \mathbf{T}_{123243\cdot 1232143231423}(E_1) \\
&= [2]_q \mathbf{T}_{12324\cdot 3}(E_2) = \mathbf{T}_{12324}([E_3, E_{32}]) = [\mathbf{T}_{1232\cdot 4}(E_3), \mathbf{T}_{123\underline{24}}(E_{32})] \\
&= [\mathbf{T}_{1232}(E_{43}), \mathbf{T}_{1234\cdot 2}(E_{32})] = [[\mathbf{T}_{123\cdot 2}(E_4), \mathbf{T}_{1232}(E_3)], \mathbf{T}_{1234}(E_3)] \\
&= [[\mathbf{T}_{123}(E_4), E_3], E_4] = [E_{1234}, E_3], E_4] = [E_{12343}, E_4] = [x_{17}, x_{22}].
\end{aligned}$$

Finally, we can use q-associativity to prove the remaining identities,

$$\begin{aligned}
&x_4 = [x_2, x_{24}] = [[x_1, x_3], x_{24}] = [x_1, [x_3, x_{24}]] = [x_1, x_6], \\
&[2]_q x_7 = [x_4, x_{24}] = [[x_1, x_6], x_{24}] = [x_1, [x_6, x_{24}]] = [2]_q [x_1, x_8], \\
&x_{13} = [x_4, x_{22}] = [[x_1, x_6], x_{22}] = [x_1, [x_6, x_{22}]] = [x_1, x_{15}], \\
&x_{17} = [x_7, x_{22}] = [[x_1, x_8], x_{22}] = [x_1, [x_8, x_{22}]] = [x_1, x_{19}], \\
&[2]_q x_{20} = [x_{17}, x_{22}] = [[x_1, x_{19}], x_{22}] = [x_1, [x_{19}, x_{22}]] = [2]_q [x_1, x_{21}].
\end{aligned}$$

□

LEMMA 6.2. *Let $\mathfrak{g}$ be the Lie algebra of type F_4, and let $y_1, \ldots, y_{24}$ be the Lusztig root vectors (recall* (5.3)*) corresponding to the reduced expression $\mathbf{R}'[w_o]$ (see* (5.1)*) of the longest element of the Weyl group of $\mathfrak{g}$. Then*

$$\begin{aligned}
y_3 &= [y_2, y_{24}], & y_{10} &= [y_5, y_{19}], & y_{17} &= [y_1, y_{21}], \\
y_4 &= [y_3, y_{17}], & y_{11} &= \tfrac{1}{[2]_q}[y_5, y_{21}], & y_{18} &= \tfrac{1}{[2]_q}[y_{17}, y_{19}], \\
y_5 &= [y_3, y_{21}], & y_{12} &= \tfrac{1}{[2]_q}[y_5, y_{23}], & y_{19} &= [y_1, y_{23}], \\
y_6 &= \tfrac{1}{[2]_q}[y_4, y_{17}], & y_{13} &= [y_{10}, y_{17}], & y_{19} &= [y_{17}, y_{24}], \\
y_7 &= \tfrac{1}{[2]_q}[y_4, y_{19}], & y_{14} &= [y_{12}, y_{17}], & y_{20} &= [y_{19}, y_{21}], \\
y_8 &= [y_5, y_{17}], & y_{15} &= \tfrac{1}{[2]_q}[y_{14}, y_{17}], & y_{22} &= \tfrac{1}{[2]_q}[y_{21}, y_{23}], \\
y_9 &= \tfrac{1}{[2]_q}[y_8, y_{10}], & y_{16} &= \tfrac{1}{[2]_q}[y_{14}, y_{19}], & y_{23} &= [y_{21}, y_{24}].
\end{aligned}$$

PROOF. In the proof of this lemma we adopt the same abbreviation for q-commutators as used in the proof of Lemma 6.1.

Observe first that the two reduced expressions for w_o in (5.1) share a common substring, namely $\mathbf{R}'[6, 23] = \mathbf{R}[2, 19]$. Hence, there is an algebra isomorphism $\Phi : \mathbf{U}'_{[6,23]} \to \mathbf{U}_{[2,19]}$ such that $\Phi(y_i) = x_{i-4}$ for all $i \in [6, 23]$. Thus, the commutation relations among the x_i's given in Lemma 6.1 translate into commutation relations among the y_i's. In particular, we have $y_9 = \frac{1}{[2]_q}[y_8, y_{10}]$, $y_{13} = [y_{10}, y_{17}]$, $y_{14} = [y_{12}, y_{17}]$, $y_{15} = \frac{1}{[2]_q}[y_{14}, y_{17}]$, $y_{16} = \frac{1}{[2]_q}[y_{14}, y_{19}]$, $y_{18} = \frac{1}{[2]_q}[y_{17}, y_{19}]$, $y_{20} = [y_{19}, y_{21}]$, and $y_{22} = \frac{1}{[2]_q}[y_{21}, y_{23}]$.

Next we apply Proposition 2.1 to identify which of the y_i's correspond to the standard Chevalley generators E_i. We get $y_1 = E_4$, $y_2 = E_1$, $y_{21} = E_3$, and $y_{24} = E_2$. We next identify the y_i's that can be written as q-commutators of these E_i's. For example, we have $y_3 = \mathbf{T}_{\underline{41}}(E_2) = \mathbf{T}_{1\cdot 4}(E_2) = \mathbf{T}_1(E_2) = E_{12} = [y_2, y_{24}]$. We note here we have adopted the same underlining notation, as in $\mathbf{T}_{\underline{41}}$ above, as well as the dot notation, as in $\mathbf{T}_{1\cdot 4}$ above, used in the proof of Lemma 6.1. With this, we also have $y_{17} = \mathbf{T}_{4\cdot 123421323124321}(E_3) = \mathbf{T}_4(E_3) = E_{43} = [y_1, y_{21}]$ and $y_{23} = \mathbf{T}_{\underline{41234213231243213234 32}}(E_3) = \mathbf{T}_{32\cdot 12321432341232143234}(E_3) = \mathbf{T}_{32}(E_3) = E_{32} = [y_{21}, y_{24}]$.

Now that we have established some of the identities of this lemma, we can continue with this same strategy to establish further identities. We have

$$\begin{aligned}
y_5 &= \mathbf{T}_{\underline{412}3}(E_4) = \mathbf{T}_{12\cdot 43}(E_4) = \mathbf{T}_{12}(E_3) = E_{123} = [y_3, y_{21}],\\
y_{19} &= \mathbf{T}_{4\underline{12342132312432132}}(E_3) = \mathbf{T}_{432\cdot 123214323412321}(E_3) = \mathbf{T}_{4\cdot 32}(E_3) = \mathbf{T}_4(E_{32})\\
&= E_{432} = [y_{17}, y_{24}],\\
y_4 &= \mathbf{T}_{41\cdot 2}(E_3) = \mathbf{T}_{41}([E_2, E_3]) = [\mathbf{T}_{4\cdot 1}(E_2), \mathbf{T}_{4\cdot 1}(E_3)] = [\mathbf{T}_4(E_{12}), \mathbf{T}_4(E_3)]\\
&= [E_{12}, E_{43}] = [y_3, y_{17}].
\end{aligned}$$

Next we compute a few more identities that build off of the identities already established. We have

$$\begin{aligned}
[2]_q y_6 &= [2]_q \mathbf{T}_{4123\cdot 4}(E_2) = [2]_q \mathbf{T}_{41\cdot 23}(E_2) = \mathbf{T}_{41}(E_{233})\\
&= [\mathbf{T}_{41}(E_{23}), \mathbf{T}_{4\cdot 1}(E_3)] = [y_4, \mathbf{T}_4(E_3)] = [y_4, y_{17}],\\
y_7 &= \mathbf{T}_{41234\cdot 2}(E_1) = \mathbf{T}_{41234}([E_2, E_1]) = [\mathbf{T}_{41234}(E_2), \mathbf{T}_{41234}(E_1)] = [y_6, E_2]\\
&= [y_6, y_{24}],\\
y_8 &= \mathbf{T}_{412342\cdot 1}(E_3) = \mathbf{T}_{4123\underline{42}}(E_3) = \mathbf{T}_{41232\cdot 4}(E_3) = \mathbf{T}_{41232}([E_4, E_3])\\
&= [\mathbf{T}_{4123\cdot 2}(E_4), \mathbf{T}_{4\cdot 1232}(E_3)] = [\mathbf{T}_{4123\cdot 2}(E_4), \mathbf{T}_{4\cdot 1232}(E_3)]\\
&= [\mathbf{T}_{4123}(E_4), \mathbf{T}_4(E_3)] = [y_5, y_{17}],\\
y_{10} &= \mathbf{T}_{4123421\cdot 32}(E_3) = \mathbf{T}_{4123421}([E_3, E_2]) = [\mathbf{T}_{412342\cdot 1}(E_3), \mathbf{T}_{4123421}(E_2)]\\
&= [\mathbf{T}_{412342}(E_3), E_2] = [y_8, y_{24}],\\
[2]_q y_{11} &= [2]_q \mathbf{T}_{\underline{412342}1323}(E_1) = [2]_q \mathbf{T}_{123\cdot 4321323}(E_1) = [2]_q \mathbf{T}_{1\cdot 23}(E_2)\\
&= \mathbf{T}_1([E_{23}, E_3]) = [\mathbf{T}_1(E_{23}), \mathbf{T}_1(E_3)] = [E_{123}, E_3] = [y_5, y_{21}],\\
y_{12} &= \mathbf{T}_{\underline{41234213231}}(E_2) = \mathbf{T}_{1232\cdot 4321323}(E_2) = \mathbf{T}_{123\cdot 2}(E_1) = \mathbf{T}_{123}([E_2, E_1])\\
&= [\mathbf{T}_{123}(E_2), \mathbf{T}_{123}(E_1)] = [\mathbf{T}_{123}(E_2), E_2] = [y_{11}, y_{24}].
\end{aligned}$$

Finally, we can use q-associativity to establish the remaining identities,

$$\begin{aligned}
&[2]_q y_7 = [2]_q [y_6, y_{24}] = [[y_4, y_{17}], y_{24}] = [y_4, [y_{17}, y_{24}]] = [y_4, y_{19}],\\
&y_{19} = [[E_4, E_3], E_2] = [E_4, [E_3, E_2]] = [y_1, y_{23}],\\
&y_{10} = [y_8, y_{24}] = [[y_5, y_{17}], y_{24}] = [y_5, [y_{17}, y_{24}] = [y_5, y_{19}],\\
&[2]_q y_{12} = [2]_q [y_{11}, y_{24}] = [[y_5, y_{21}], y_{24}] = [y_5, [y_{21}, y_{24}]] = [y_5, y_{23}],
\end{aligned}$$

□

7. Quantum Symmetric Matrices

The algebra of $n \times n$ quantum symmetric matrices [**21, 30**] is a quantized nilradical $\mathcal{U}_q(\mathfrak{n}_J)$ for the case when the underlying Lie algebra $\mathfrak{g}$ is of type C_n and $J = \{\alpha_n\}$. In this section, we prove that Conjecture 1.1 holds in this case.

We let x_{ij} $(1 \le i \le j \le n)$ denote the standard generators of the algebra of quantum symmetric matrices. The defining relations are given in [**21**, Proposition 5.2]. Let $\mathcal{N}_J$ be the normal subalgebra of $\mathcal{U}_q(\mathfrak{n}_J)$. It is generated by the normal elements $\Theta_1, \dots, \Theta_n$. Here, the quantized nilradical $\mathcal{U}_q(\mathfrak{n}_J)$ is $\mathbb{N}$-graded with $\deg(x_{ij}) = 1$ for all $1 \le i \le j \le n$. In view of this, $\deg(\Theta_i) = i$ (for $1 \le i \le n$). The simple roots are $\alpha_i = e_i - e_{i+1}$ (for $1 \le i < n$) and $\alpha_n = 2e_n$. The fundamental weights are $\varpi_i = e_1 + \cdots + e_i$ (for $1 \le i \le n$). Let $Q = \mathbb{Z}\alpha_1 + \cdots + \mathbb{Z}\alpha_n$ and $P = \mathbb{Z}\varpi_1 + \cdots + \mathbb{Z}\varpi_n$ be the root lattice and weight lattice respectively, and let $\langle -, - \rangle$ be the symmetric bilinear form on P defined by the rule $\langle e_i, e_j \rangle = \delta_{ij}$. The algebra $\mathcal{U}_q(\mathfrak{n}_J)$ is Q-graded with

$$\deg_Q(x_{ij}) = e_i + e_j = (\alpha_i + \alpha_{i+1} + \cdots \alpha_n) + (\alpha_j + \alpha_{j+1} + \cdots + \alpha_{n-1}).$$

Consider the parabolic element $w_J := w_o^J w_o \in W$. We have the reduced expression

$$w_J = (s_n s_{n-1} \cdots s_1)(s_n s_{n-1} \cdots s_2) \cdots (s_n s_{n-1}) s_n. \tag{7.1}$$

We recall (4.4) the commutation relations,

$$x_{ij}\Theta_k = q^{-\langle \deg_Q(x_{ij}), (1+w_J)\varpi_k \rangle} \Theta_k x_{ij}$$

for $1 \le i \le j \le n$ and $1 \le k \le n$.

PROPOSITION 7.1. *Suppose $\mathfrak{g}$ is the Lie algebra of type C_n with $n > 1$, and suppose $J = \{\alpha_n\}$. If ψ is an automorphism of the quantized nilradical $\mathcal{U}_q(\mathfrak{n}_J)$, then $\psi(\Theta_i) \in \mathbb{K}^\times \Theta_i$ for every $i \in \{1, \dots, n\}$.*

PROOF. As in the proof of Theorem 4.8, the hypotheses in Theorem 4.1 involving the core and the existence of relations of the form $xy = \kappa yx$ can be seen to be satisfied by observing the relevant properties of the reduced expression (7.1). Hence, every automorphism of $\mathcal{U}_q(\mathfrak{n}_J)$ preserves the $\mathbb{N}$-grading.

The normal subalgebra $\mathcal{N}_J$ is invariant under any algebra automorphism of $\mathcal{U}_q(\mathfrak{n}_J)$. Since $\dim_{\mathbb{K}}(\mathcal{N}_J)_1 = 1$ (in fact, $(\mathcal{N}_J)_1 = \mathbb{K}\Theta_1$), then $\psi(\Theta_1) \in \mathbb{K}^\times \Theta_1$. Furthermore, since $\mathbb{K}x_{nn} = \{x \in (\mathcal{U}_q(\mathfrak{n}_J))_1 : x\Theta_1 = q^2\Theta_1 x\}$, $\psi(x_{nn}) = \mathbb{K}^\times x_{nn}$.

For $k \in \{1, \dots, n\}$, $\dim_{\mathbb{K}}((\mathcal{N}_J)_k) = \mathcal{P}(k)$, where $\mathcal{P}$ is the partition function. We also have $x_{nn}\Theta_k = q^2\Theta_k x_{nn}$ for every $k \in \{1, \dots, n-1\}$.

For a fixed natural number $k \in \mathbb{N}$, we represent a partition ν of $k \in \mathbb{N}$ (and write $\nu \vdash k$) by a weakly increasing sequence of natural numbers $\nu_1 \le \nu_2 \le \cdots$ such that $\sum_i \nu_i = k$. For a partition $\nu \vdash k$, we let $\mathrm{parts}(\nu)$ be the number of parts of ν and write $\nu = (\nu_1, \nu_2, \dots, \nu_{\mathrm{parts}(\nu)})$, and we define the monomial

$$\Theta^\nu := \Theta_{\nu_1}\Theta_{\nu_2} \cdots \Theta_{\nu_{\mathrm{parts}(\nu)}} \in (\mathcal{N}_J)_k.$$

Suppose

$$\psi(\Theta_k) = \sum_{\nu \vdash k} c_{\nu,k}\Theta^\nu, \qquad (1 \le k \le n).$$

for scalars $c_{\nu,k} \in \mathbb{K}$. For $1 \le k < n$, we apply the automorphism ψ to the relation $x_{nn}\Theta_k = q^2\Theta_k x_{nn}$ to conclude that

$$\sum_{\nu \vdash k} c_{\nu,k} x_{nn}\Theta^\nu = \sum_{\nu \vdash k} q^2 c_{\nu,k}\Theta^\nu x_{nn}.$$

However since $x_{nn}\Theta_k = q^2\Theta_k x_{nn}$ for every $k \in \{1,\ldots,n-1\}$, then in the above sum we can replace $x_{nn}\Theta^\nu$ with $q^{2\cdot\text{parts}(\nu)}\Theta^\nu x_{nn}$. Thus,

$$\sum_{\nu\vdash k}\left(q^{2\cdot\text{parts}(\nu)} - q^2\right) c_{\nu,k}\Theta^\nu x_{nn} = 0.$$

Since q is not a root of unity, the only nonzero coefficients $c_{\nu,k}$ appearing in the above sum are those such that $\text{parts}(\nu) = 1$. In other words, there is at most one monomial Θ^ν in the sum $\sum_{\nu\vdash k} c_{\nu,k}\Theta^\nu$ with a nonzero coefficient, namely Θ_k. Hence $\psi(\Theta_k) \in \mathbb{K}^\times\Theta_k$ for $k < n$.

Finally, consider Θ_n. Since Θ_n generates the center of $\mathcal{U}_q(\mathfrak{n}_J)$ (see e.g. [**18**]), then $\psi(\Theta_n) \in \mathbb{K}^\times\Theta_n$. □

THEOREM 7.2. *If $\mathfrak{g}$ is the Lie algebra of type C_n with $n > 1$ and $J = \{\alpha_n\}$ (i.e. $\mathcal{U}_q(\mathfrak{n}_J)$ is the algebra of $n \times n$ quantum symmetric matrices), then* $\text{Aut}(\mathcal{U}_q(\mathfrak{n}_J)) \cong (\mathbb{K}^\times)^n$.

PROOF. Suppose ϕ is an automorphism of $\mathcal{U}_q(\mathfrak{n}_J)$. As established in Proposition 7.1, all hypotheses of Theorem 4.1 are satisfied. Hence, ϕ is a graded automorphism.

Next we will apply Theorem 4.5 to show that each x_{ij} with $i+j \neq n+1$ gets sent to a multiple of itself by ϕ. First, suppose i and j are chosen such that $1 \le i \le j \le n$ and $i+j < n+1$, and consider the corresponding element x_{ij}. We show that for every other Lusztig root vector $x_{k\ell}$, there is a normal element Θ_p such that x_{ij} and $x_{k\ell}$ commute differently with Θ_p. Equivalently, this means $\langle (e_i+e_j) - (e_k+e_\ell), (1+w_J)\varpi_p\rangle \neq 0$. With this, Theorem 4.5 implies x_{ij} gets sent to a multiple of itself by the automorphism ϕ. There are two cases to consider. If $i \neq k$, let $p = \min(i,k)$. On the other hand, if $i = k$, let $p = \min(j,\ell)$. Now suppose i and j are chosen so that $1 \le i \le j \le n$ and $i+j > n+1$, and consider the corresponding Lusztig root vector x_{ij}. As before, we will show that for every other Lusztig root vector $x_{k\ell}$, there is a normal element Θ_p such that x_{ij} and $x_{k\ell}$ commute differently with Θ_p. Here, if $j \neq \ell$, put $p = \max(j-1, \ell-1)$. However, if $j = \ell$, put $p = \max(i-1, k-1)$.

Now we will show that every Lusztig root vector x_{ij} with $i+j = n+1$ gets sent to a multiple of itself by ϕ. Theorem 4.5 does not apply here because these x_{ij} all commute the same way with each normal element Θ_p. In fact, each of these x_{ij} commute with all of the elements in the normal subalgebra. None of the other x_{ij}'s behave this way. This means $S := \text{span}_{\mathbb{K}}\{x_{ij} \mid i+j = n+1\}$ is a ϕ-invariant vector subspace of $\mathcal{U}_q(\mathfrak{n}_J)_1$ For every $y \in S$, define

$$C(y) := \left\{x \in (\mathcal{U}_q(\mathfrak{n}_J))_1 \mid yx = qxy\right\}.$$

Each $C(y)$ is a vector space. Observe $\dim(C(y)) = \dim(C(\phi(y))$. From the defining relations of $\mathcal{U}_q(\mathfrak{n}_J)$, we obtain that $C(x_{ij})$ is spanned by

$$\{x_{ik} : k < j\} \cup \{x_{kj} : k < i\}.$$

Hence $\dim(C(x_{ij})) = n - i$. The only elements $y \in S$ with $\dim(C(y)) = n - i$ are the nonzero multiples of x_{ij}. Hence $\phi(x_{ij}) \in \mathbb{K}^\times x_{ij}$.

We have shown now that ϕ is a diagonal automorphism. Since $\mathcal{U}_q(\mathfrak{n}_J)$ has rank n as a CGL extension, [**15**, Theorems 5.3 and 5.5] imply that $\text{Aut}\left(\mathcal{U}_q\left(\mathfrak{n}_J\right)\right) \cong (\mathbb{K}^\times)^n$. □

References

[1] J. Alev and M. Chamarie, *Dérivations et automorphismes de quelques algèbres quantiques* (French), Comm. Algebra **20** (1992), no. 6, 1787–1802, DOI 10.1080/00927879208824431. MR1162608

[2] N. Andruskiewitsch and F. Dumas, *On the automorphisms of $U_q^+(\mathfrak{g})$*, Quantum groups, IRMA Lect. Math. Theor. Phys., vol. 12, Eur. Math. Soc., Zürich, 2008, pp. 107–133, DOI 10.4171/047-1/4. MR2432991

[3] G. Cauchon, *Effacement des dérivations et spectres premiers des algèbres quantiques* (French, with English summary), J. Algebra **260** (2003), no. 2, 476–518, DOI 10.1016/S0021-8693(02)00542-2. MR1967309

[4] S. Ceken, J. H. Palmieri, Y.-H. Wang, and J. J. Zhang, *The discriminant controls automorphism groups of noncommutative algebras*, Adv. Math. **269** (2015), 551–584, DOI 10.1016/j.aim.2014.10.018. MR3281142

[5] S. Ceken, J. H. Palmieri, Y.-H. Wang, and J. J. Zhang, *The discriminant criterion and automorphism groups of quantized algebras*, Adv. Math. **286** (2016), 754–801, DOI 10.1016/j.aim.2015.09.024. MR3415697

[6] K. Chan, J. Gaddis, R. Won, and J. J. Zhang, *Reflexive hull discriminants and applications*, Selecta Math. (N.S.) **28** (2022), no. 2, Paper No. 40, 35 pp., DOI 10.1007/s00029-021-00755-x. MR4370373

[7] C. De Concini, V. G. Kac, and C. Procesi, *Some quantum analogues of solvable Lie groups*, Geometry and analysis (Bombay, 1992), Tata Inst. Fund. Res., Bombay, 1995, pp. 41–65. MR1351503

[8] N. Yu. Reshetikhin, L. A. Takhtadzhyan, and L. D. Faddeev, *Quantization of Lie groups and Lie algebras* (Russian), Algebra i Analiz **1** (1989), no. 1, 178–206; English transl., Leningrad Math. J. **1** (1990), no. 1, 193–225. MR1015339

[9] J. Gaddis, E. Kirkman, and W. F. Moore, *On the discriminant of twisted tensor products*, J. Algebra **477** (2017), 29–55, DOI 10.1016/j.jalgebra.2016.12.019. MR3614142

[10] J. Gaddis and T. Lamkin, *Centers and automorphisms of PI quantum matrix algebras*, `arXiv:2207.11956`, to appear in *Contemporary Mathematics*.

[11] C. Geiß, B. Leclerc, and J. Schröer, *Cluster structures on quantum coordinate rings*, Selecta Math. (N.S.) **19** (2013), no. 2, 337–397, DOI 10.1007/s00029-012-0099-x. MR3090232

[12] K. R. Goodearl and E. S. Letzter, *The Dixmier-Moeglin equivalence in quantum coordinate rings and quantized Weyl algebras*, Trans. Amer. Math. Soc. **352** (2000), no. 3, 1381–1403, DOI 10.1090/S0002-9947-99-02345-4. MR1615971

[13] K. R. Goodearl and M. T. Yakimov, *Quantum cluster algebra structures on quantum nilpotent algebras*, Mem. Amer. Math. Soc. **247** (2017), no. 1169, vii+119, DOI 10.1090/memo/1169. MR3633289

[14] K. R. Goodearl and M. T. Yakimov, *Unipotent and Nakayama automorphisms of quantum nilpotent algebras*, Commutative algebra and noncommutative algebraic geometry. Vol. II, Math. Sci. Res. Inst. Publ., vol. 68, Cambridge Univ. Press, New York, 2015, pp. 181–212. MR3496865

[15] K. R. Goodearl and M. T. Yakimov, *From quantum Ore extensions to quantum tori via noncommutative UFDs*, Adv. Math. **300** (2016), 672–716, DOI 10.1016/j.aim.2016.03.029. MR3534843

[16] N. Hu and X. Wang, *Convex PBW-type Lyndon basis and restricted two-parameter quantum groups of type G_2*, Pacific J. Math. **241** (2009), no. 2, 243–273, DOI 10.2140/pjm.2009.241.243. MR2507577

[17] J. E. Humphreys, *Introduction to Lie algebras and representation theory*, Graduate Texts in Mathematics, Vol. 9, Springer-Verlag, New York-Berlin, 1972. MR0323842

[18] H. P. Jakobsen, *The center of $\mathcal{U}_q(\mathfrak{n}_\omega)$*, Comm. Algebra **46** (2018), no. 1, 262–282, DOI 10.1080/00927872.2017.1332197. MR3764862

[19] J. C. Jantzen, *Lectures on quantum groups*, Graduate Studies in Mathematics, vol. 6, American Mathematical Society, Providence, RI, 1996, DOI 10.1090/gsm/006. MR1359532

[20] A. Jaramillo and G. Johnson, *Quantized nilradicals of parabolic subalgebras of $\mathfrak{sl}(n)$ and algebras of coinvariants*, Comm. Algebra **50** (2022), no. 11, 4997–5015, DOI 10.1080/00927872.2022.2080216. MR4469940

[21] A. Kamita, *Quantum deformations of certain prehomogeneous vector spaces. III*, Hiroshima Math. J. **30** (2000), no. 1, 79–115. MR1753385

[22] M. Kashiwara, *Crystalizing the q-analogue of universal enveloping algebras*, Comm. Math. Phys. **133** (1990), no. 2, 249–260. MR1090425

[23] S. Launois and T. H. Lenagan, *Primitive ideals and automorphisms of quantum matrices*, Algebr. Represent. Theory **10** (2007), no. 4, 339–365, DOI 10.1007/s10468-007-9059-0. MR2333441

[24] S. Launois and T. H. Lenagan, *Automorphisms of quantum matrices*, Glasg. Math. J. **55** (2013), no. A, 89–100, DOI 10.1017/S0017089513000529. MR3110806

[25] S. Levendorskiĭ and Y. Soibelman, *Algebras of functions on compact quantum groups, Schubert cells and quantum tori*, Comm. Math. Phys. **139** (1991), no. 1, 141–170. MR1116413

[26] G. Lusztig, *Introduction to quantum groups*, Progress in Mathematics, vol. 110, Birkhäuser Boston, Inc., Boston, MA, 1993. MR1227098

[27] G. Lusztig, *Canonical bases arising from quantized enveloping algebras*, J. Amer. Math. Soc. **3** (1990), no. 2, 447–498, DOI 10.2307/1990961. MR1035415

[28] A. Mériaux and G. Cauchon, *Admissible diagrams in quantum nilpotent algebras and combinatoric properties of Weyl groups*, Represent. Theory **14** (2010), 645–687, DOI 10.1090/S1088-4165-2010-00382-9. MR2736313

[29] I. M. Musson, *Ring-theoretic properties of the coordinate rings of quantum symplectic and Euclidean space*, Ring theory (Granville, OH, 1992), World Sci. Publ., River Edge, NJ, 1993, pp. 248–258. MR1344235

[30] M. Noumi, *Macdonald's symmetric polynomials as zonal spherical functions on some quantum homogeneous spaces*, Adv. Math. **123** (1996), no. 1, 16–77, DOI 10.1006/aima.1996.0066. MR1413836

[31] E. Strickland, *Classical invariant theory for the quantum symplectic group*, Adv. Math. **123** (1996), no. 1, 78–90, DOI 10.1006/aima.1996.0067. MR1413837

[32] M. Yakimov, *Invariant prime ideals in quantizations of nilpotent Lie algebras*, Proc. Lond. Math. Soc. (3) **101** (2010), no. 2, 454–476, DOI 10.1112/plms/pdq006. MR2679698

[33] M. Yakimov, *On the spectra of quantum groups*, Mem. Amer. Math. Soc. **229** (2014), no. 1078, vi+91. MR3185525

[34] M. Yakimov, *Rigidity of quantum tori and the Andruskiewitsch-Dumas conjecture*, Selecta Math. (N.S.) **20** (2014), no. 2, 421–464, DOI 10.1007/s00029-013-0145-3. MR3177924

[35] M. Yakimov, *The Launois-Lenagan conjecture*, J. Algebra **392** (2013), 1–9, DOI 10.1016/j.jalgebra.2013.07.001. MR3085018

Department of Mathematics and Physics, North Carolina Central University, Durham, North Carolina 27707

Email address: gjohns62@nccu.edu

Department of Mathematics and Physics, North Carolina Central University, Durham, North Carolina 27707

Email address: gmelikian@nccu.edu

Contemporary Mathematics
Volume **791**, 2024
https://doi.org/10.1090/conm/791/15873

On Wronskians and qq-systems

Anton M. Zeitlin

ABSTRACT. We discuss the qq-systems, the functional form of the Bethe ansatz equations for the twisted Gaudin model from a new geometric point of view. We use a concept of G-Wronskians, which are certain meromorphic sections of principal G-bundles on the projective line. In this context, the qq-system, similar to its difference analog, is realized as the relation between generalized minors of the G-Wronskian. We explain the link between G-Wronskians and twisted G-oper connections, which are the traditional source for the qq-systems.

CONTENTS

1. Introduction

The impact of quantum integrable models on modern mathematics is enormous. The important examples of this kind are the so-called spin chain models [**B1**, **KBI**, **R**]. While many of the algebraic structures observed there found themselves in pure mathematics, in particular in the modern theory of quantum groups, the original method of solution of these models, known as algebraic Bethe ansatz [**TF**, **R**] remained popular mainly in the framework of mathematical physics. The centerpiece of this method, which on the folklore level is a method of diagonalization of a mutually commuting set of operators (transfer-matrices) in finite-dimensional vector space [**KBI**, **R**] is the resulting algebraic equations, known as *Bethe equations* [**B2**, **KBI**] which at first sight have no particular mathematical meaning. In recent years a lot of activity has been devoted to finding the geometric context in which these equations appear naturally. In a particular case of XXX/XXZ spin chains [**TF**, **OW**], the integrable models, based on Yangians $(Y_\hbar(\mathfrak{g}))$/quantum

2020 *Mathematics Subject Classification.* Primary 82B23, 14D24, 13F60.
The author is indebted to E. Frenkel, P. Koroteev, and D. Sage for fruitful discussions. The author is partially supported by Simons Collaboration Grant 578501 and NSF grant DMS-2203823.

affine algebras ($U_\hbar(\mathfrak{g})$), the Bethe equations emerge as the relations for the quantum equivariant cohomology/K-theory of a certain variety [**PSZ,KPSZ,AO,O**] as conjectured in the theoretical physics context [**NS1,NS2**]. At the same time, these relations may straightforwardly be written in terms of a system of difference equations, known as QQ-systems. Incidentally for $U_\hbar(\mathfrak{g})$ the QQ-systems emerge as the relations in the extended Grothendieck ring of finite-dimensional representations of quantum affine algebras [**BLZ,HJ,FH1,FH2**].

Much earlier, another geometric realization due to B. Feigin, E. Frenkel and their collaborators [**FFR1,F1,F2,FFTL,FFR2**] was achieved for the semiclassical version of the aforementioned integrable models, the so-called Gaudin model. It turned out the Bethe equations, in this case, describe certain principal ^LG-bundle connections (group ^LG corresponds to the Langlands dual $^L\mathfrak{g}$) on the projective line, called opers, with a prescribed singularity structure. This geometrization of Gaudin Bethe equations was a part of a far bigger story: this correspondence is an example of a geometric Langlands correspondence.

A naturally arising question is whether there exists a deformation of this example if a similar correspondence holds for Bethe equations of XXX/XXZ models, and what is a proper generalization of the principal bundle connection. In [**KSZ, FKSZ**] we introduced the deformed version of the connection, which we called $(^LG,\hbar)$-opers for Bethe equations of XXZ type [1]. While [**FKSZ**] we treated $(^LG,\hbar)$-opers on Lie-theoretic level, in [**KSZ, KZ2**] for $(SL(N),\hbar)$-opers we exploited a different approach, which used interpretations of QQ-systems as minors in the deformed and twisted version of the Wronskian matrix. At the time, it seemed like a construction specific for defining the representation of $SL(N)$. Still, later, it was observed in [**KZ1**] that a QQ-system emerges from a new object, $(^LG,\hbar)$-Wronskian: a meromorphic section of a principal ^LG-bundle satisfying a certain difference equation. We established the explicit correspondence between $(^LG,\hbar)$-opers and $(^LG,\hbar)$-Wronskians in [**KZ1**], so that the QQ-system emerges as relations between generalized minors [**FZ1,FZ2**] of $(G,\hbar)$-Wronskian.

In [**BSZ**] we described the classical limit of the QQ-system: we called it qq-system, which is a system of differential equations representing the original Gaudin model/oper context. On the level of equations one obtains QQ-system from qq-system via proper application of $\hbar \to 1$ limit and rescaling. In this note we explain how to obtain the differential G-Wronskian and generalized minor interpretation of the qq-system.

The exposition is as follows. First, Section 2 reviews the concept of generalized minors following Fomin and Zelevinsky. Then, in Section 3, we introduce the idea of G-Wronskian for simply connected simple group G and its relation to the qq-system. Next, in Section 4, we discuss the class of G-opers, which correspond to the qq-system, following [**BSZ**]. Finally, in Section 5, we establish the relation between two objects: G-opers and G-Wronskians and discuss the differences between G-Wronskians and their deformed analogs.

2. Generalized minors

2.1. Group-theoretic data. Let G be a connected, simply connected, simple algebraic group of rank r over $\mathbb{C}$. We fix a Borel subgroup B_- with unipotent radical $N_- = [B_-, B_-]$ and a maximal torus $H \subset B_-$. Let B_+ be the opposite

[1] In the non-simply laced case of $\mathfrak{g}$, the situation is more involved, see recent paper [**FHR**].

Borel subgroup containing H. Let $\{\alpha_1, \dots, \alpha_r\}$ be the set of positive simple roots for the pair $H \subset B_+$. Let $\{\check{\alpha}_1, \dots, \check{\alpha}_r\}$ be the corresponding coroots; the elements of the Cartan matrix of the Lie algebra $\mathfrak{g}$ of G are given by $a_{ij} = \langle \alpha_j, \check{\alpha}_i \rangle$.

The Lie algebra $\mathfrak{g}$ has Chevalley generators $\{e_i, f_i, \check{\alpha}_i\}_{i=1,\dots,r}$, so that $\mathfrak{b}_- = \mathrm{Lie}(B_-)$ is generated by the f_i's and the $\check{\alpha}_i$'s and $\mathfrak{b}_+ = \mathrm{Lie}(B_+)$ is generated by the e_i's and the $\check{\alpha}_i$'s, while $\mathfrak{h} = \mathfrak{b}_\pm / [\mathfrak{b}_\pm, \mathfrak{b}_\pm]$ is generated by $\check{\alpha}_i$'s. Let $\{\omega_i\}_{i=1,\dots,r}$ and $\{\check{\omega}_i\}_{i=1,\dots,r}$ be the fundamental weights and coweights correspondingly, defined by $\langle \omega_i, \check{\alpha}_j \rangle = \langle \check{\omega}_i, \alpha_j \rangle \delta_{ij}$. The element $\mathrm{ad}_{\check{\rho}}$, where $\check{\rho} = \sum_{i=1}^r \check{\omega}_i$ defines a principal gradation on $\mathfrak{b}_+ = \oplus_{i \geq 0} \mathfrak{b}_{+,i}$.

Let $W_G = N(H)/H$ be the Weyl group of G. Let $w_i \in W$, $(i = 1, \dots, r)$ denote the simple reflection corresponding to α_i. We also denote by w_0 be the longest element of W, so that $B_+ = w_0(B_-)$. We also fix representatives $s_i \in N(H)$ of w_i, and in general will denote $\tilde{w}$ the representative of w in $N(H)$.

2.2. Generalized Minors and their properties. Consider the big cell $G_0 \in G$ in Bruhat decomposition: $G_0 = N_- H N_+$. Any element $g \in G_0$ can be represented as follows:

$$g = n_- \, h \, n_+. \tag{2.1}$$

for some $n_+ \in N_+$, $h \in H$, and $n_- \in N_-$. Let V_i be the irreducible representation of G with highest weight ω_i and highest weight vector ν_i which is the eigenvector for any element $h \in H$, i.e. $h\nu_i = [h]^{\omega_i} \nu_i$, where $[h]^{\omega_i} \in \mathbb{C}^\times$ denotes an eigenvalue of h.

We formulate the following definition.

Definition 2.1. [**FZ1**] Regular functions $\{\Delta^{\omega_i}\}_{i=1,\dots,r}$ on G, whose values on a dense set G_0 are given by

$$\Delta^{\omega_i}(g) = [h]^{\omega_i}, \quad i = 1, \dots, r \tag{2.2}$$

will be referred to as *principal minors* of a group element g.

In case of $G = SL(r+1)$, these functions coincide with the principal minors of the standard matrix realization of $SL(r+1)$. We define other generalized minors using the action of the lifts of the Weyl group elements on the right and the left and then applying principal minors to the result. In other words, we have the following definition.

Definition 2.2. [**FZ1**] For $u, v \in W_G$, we define a regular function $\Delta_{u \cdot \omega_i, v \cdot \omega_i}$ on G by setting

$$\Delta_{u \cdot \omega_i, v \cdot \omega_i}(g) = \Delta^{\omega_i}(\tilde{u}^{-1} g \, \tilde{v}), \tag{2.3}$$

where $\tilde{u}, \tilde{v}$ are lifts of Weyl group elements u, v to G.

Notice that in this notation $\Delta_{\omega_i, \omega_i}(g) = \Delta^{\omega_i}(g)$. Consider the orbit $\mathcal{O}_{W_G} = W_G \cdot \mathbb{C}\nu_i$. Then we have the following Proposition.

Proposition 2.3. Action of the group element on the highest weight vector $\nu_i \in V_i$ is as follows:

$$g \cdot \nu_i = \sum_{w \in W} \Delta_{w \cdot \omega_i, \omega_i}(g) \tilde{w} \cdot \nu_i + \dots, \tag{2.4}$$

where dots stand for the vectors, which do not belong to the orbit $\mathcal{O}_W$.

The set of generalized minors $\{\Delta_{w\cdot\omega_i,\omega_i}\}_{w\in W; i=1,\dots,r}$ creates a set of coordinates on G/B_+, known as *generalized Plücker coordinates.* In particular, the set of zeroes of each of $\Delta_{w\cdot\omega_i,\omega_i}$ is a uniquely and unambiguously defined hypersurface in G/B. This feature is important for characterizing Schubert cells as quasi-projective subvarieties of a generalized flag variety, see [**FZ2**] for details.

We started this section from the explicit definition of the principal minors by means of Gaussian decomposition. The following proposition (see Corollary 2.5 in [**FZ1**]) provides a necessary and sufficient condition of its existence for a given group element.

PROPOSITION 2.4. An element $g \in G$ admits the Gaussian decomposition if and only if $\Delta^{\omega_i}(g) \neq 0$ for any $i = 1, \dots, r$.

Finally, we introduce the fundamental relation ([**FZ1**], Theorem 1.17) between generalized minors, which will be crucial in the following.

PROPOSITION 2.5. Let, $u, v \in W$, such that for $i \in \{1, \dots, r\}$, $\ell(uw_i) = \ell(u)+1$, $\ell(vw_i) = \ell(v) + 1$. Then

$$\Delta_{u\cdot\omega_i,v\cdot\omega_i}(g)\Delta_{uw_i\cdot\omega_i,vw_i\cdot\omega_i}(g) - \Delta_{uw_i\cdot\omega_i,v\cdot\omega_i}(g)\Delta_{u\cdot\omega_i,vw_i\cdot\omega_i}(g) = \prod_{j\neq i}\left[\Delta_{u\cdot\omega_j,v\cdot\omega_j}(g)\right]^{-a_{ji}}. \tag{2.5}$$

3. G-Wronskians

3.1. Differential equations and G-Wronskian. Consider the irreducible representation V_i of G with highest weight ω_i. It comes equipped with a 2-dimensional subspace $W_i = L_i^+ \oplus L_i^-$, $L_i^+ = \mathbb{C}\nu_i$, $L_i^- = \mathbb{C}f_i\nu_i$, which is invariant under the action of B_+.

Suppose we have a principal G-bundle $\mathcal{F}_G$ and its B_+-reduction $\mathcal{F}_{B_+}$ and thus an H-reduction $\mathcal{F}_H$, where $H = B_+/[B_+, B_+]$ as well. Therefore for each $i = 1, \dots, r$, the vector bundle

$$\mathcal{V}_i = \mathcal{F}_{B_+} \underset{B_+}{\times} V_i = \mathcal{F}_G \underset{G}{\times} V_i$$

associated to V_i contains an H-line subbundles

$$\mathcal{L}_i^+ = \mathcal{F}_H \underset{H}{\times} L_i^+, \quad \mathcal{L}_i^- = \mathcal{F}_H \underset{H}{\times} L_i^-$$

associated to $L_i^\pm \subset V_i$.

DEFINITION 3.1. G-Wronskian on $\mathbb{P}^1$ is the quadruple $(\mathcal{F}_G, \mathcal{F}_{B_+}, \mathcal{G}, \nabla^Z)$, where $\mathcal{G}$ is a meromorphic section of a principle bundle $\mathcal{F}_G$, $\mathcal{F}_{B_+}$ is a reduction of $\mathcal{F}_G$ to B_+, ∇^Z is an H-connection, so that $H = B_+/[B_+, B_+]$, satisfying the following condition. There exist a Zariski open dense subset $U \subset \mathbb{P}^1$ together with the trivialization $\imath_{B_+}$ of $\mathcal{F}_{B_+}$ so that $i_{\frac{d}{dz}}\nabla^Z = \nabla^Z_z = \partial_z - Z$, where $Z \in \mathfrak{h} = \mathfrak{b}_+/[\mathfrak{b}_+, \mathfrak{b}_+]$, so that for certain sections $\{v_i^\pm\}_{i=1,\dots,r}$ of line bundles $\{\mathcal{L}_i^\pm\}_{i=1,\dots,r}$ on U we have $\mathcal{G}$ as an element of $G(z)$ satisfy the following conditions:

$$\nabla^Z_z(\mathcal{G} \cdot v_i^+) = \mathcal{G} \cdot v_i^-, \quad i = 1, \dots, r \tag{3.1}$$

In local terms we have the following. Representing the corresponding section $\mathcal{G} = \mathcal{G}(z) \in G(z)$, we have:

$$(\partial_z - Z)\mathcal{G}(z)\nu_i = \mathcal{G}(z)p^\phi_{-1}(z)\nu_i, \tag{3.2}$$

where $p^{\phi}_{-1}(z) = \sum_{i=1}^{r} \phi_i(z) f_i$, and $\phi_i(z) \in \mathbb{C}(z)$. We are interested in the case when $\{\phi_i(z)\}_{i=1,\dots,r}$ are polynomials.

DEFINITION 3.2. G-Wronskian has regular singularities if $p^{\phi}_{-1}(z) = p^{\Lambda}_{-1}(z) = \sum_{i=1}^{r} \Lambda_i(z) f_i$, where $\Lambda_i(z) \in \mathbb{C}[z]$.

We also are interested to impose some non-degeneracy conditions on $\mathscr{G}(z)$, which has to do with their generalized minor structure.

DEFINITION 3.3. We say that G-Wronskian with regular singularities is *nondegenerate* if $\Delta_{w\cdot\omega_i,\omega_i}(\mathscr{G}(z))$ are nonzero polynomials for all $w \in W$ and $i = 1, \dots, r$.

Now let us look at the equations, satisfied by non-degenerate genralized G-Wronskians:

$$(\partial_z - Z)\mathscr{G}(z)\nu_i = \mathscr{G}(z)p^{\Lambda}_{-1}(z)\nu_i \tag{3.3}$$

Restricting this equation to $W_i \subset V_i$, we have the following:

$$\begin{aligned}(\partial_z - \langle Z, \omega_i\rangle)\Delta_{\omega_i,\omega_i}(\mathscr{G}(z)) &= \Lambda_i(z)\Delta_{\omega_i,w_i\cdot\omega_i}(\mathscr{G}(z))\\ (\partial_z - \langle Z, \omega_i - \alpha_i\rangle)\Delta_{w_i\cdot\omega_i,\omega_i}(\mathscr{G}(z)) &= \Lambda_i(z)\Delta_{w_i\cdot\omega_i,w_i\cdot\omega_i}(\mathscr{G}(z))\end{aligned} \tag{3.4}$$

Let us make use of the relation (2.5), when $u, v = 1$, applying it to $\mathscr{G}(z)$:

$$\begin{aligned}\Delta_{\omega_i,\omega_i}(\mathscr{G}(z))\Delta_{w_i\cdot\omega_i,w_i\cdot\omega_i}(\mathscr{G}(z)) - \Delta_{w_i\cdot\omega_i,\omega_i}(\mathscr{G}(z))\Delta_{\omega_i,w_i\cdot\omega_i}(\mathscr{G}(z))\\ = \prod_{j\neq i}\left[\Delta_{\omega_j,\omega_j}(\mathscr{G}(z))\right]^{-a_{ji}}.\end{aligned} \tag{3.5}$$

With the help of (3.4) we obtain:

$$\begin{aligned}&\Lambda_i^{-1}(z)\Delta_{\omega_i,\omega_i}(\mathscr{G}(z))(\partial_z - \langle Z, \omega_i - \alpha\rangle)\Delta_{w_i\omega_i,\omega_i}(\mathscr{G}(z)) -\\ &\Lambda_i^{-1}(z)\Delta_{w_i\cdot\omega_i,\omega_i}(\mathscr{G}(z))(\partial_z - \langle Z, \omega_i\rangle)\Delta_{\omega_i,\omega_i}(\mathscr{G}(z)) = \prod_{j\neq i}\Delta^{-a_{ji}}_{\omega_j,\omega_j}(\mathscr{G}(z)).\end{aligned} \tag{3.6}$$

Denoting $\Delta_{\omega_i,\omega_i}(\mathscr{G}(z)) - q^i_+(z)$, $\Delta_{w_i\cdot\omega_i,\omega_i}(\mathscr{G}(z)) = q^i_-(z)$, then simplifying, and collecting such equations for all fundamental weights, we arrive to the following system:

$$\begin{aligned}&q^i_+(z)\partial_z q^i_-(z) - q^i_-(z)\partial_z q^i_+(z) + \langle Z, \alpha_i\rangle q^i_+(z)q^i_-(z) =\\ &\Lambda_i(z)\prod_{j\neq i}\left[q^j_+(z)\right]^{-a_{ji}}, \quad i = 1, \dots, r.\end{aligned} \tag{3.7}$$

Applying $\tilde{w} \in G$ which is a lift of the element of $w \in W$ to (3.3), we have:

$$(\partial_z - Z^w)\mathscr{G}^w(z)\nu_i = \mathscr{G}^w(z)p^{\Lambda}_{-1}(z)\nu_i\ , \tag{3.8}$$

where $Z^w \in \tilde{w}Z\tilde{w}^{-1}$ and $\mathscr{G}^w(z) = \tilde{w}\ \mathscr{G}(z)$. Let us denote $\Delta_{w^{-1}\cdot\omega_i,\omega_i}(\mathscr{G}(z)) = q^{i,w}_+(z)$. Then, if $\ell(w^{-1}s_i) = \ell(w^{-1}) + 1$, we have $\Delta_{w^{-1}\cdot\omega_i,\omega_i}(\mathscr{G}(z)) = q^{i,w}_+(z)$, so that $\Delta_{s_i\cdot\omega_i,\omega_i}(\mathscr{G}(z)) = q^i_-(z)$ and the following relations hold:

$$\begin{aligned}&q^{i,w}_+(z)\partial_z q^{i,w}_-(z) - q^{i,w}_-(z)\partial_z q^{i,w}_+(z) + \langle Z^w, \alpha_i\rangle q^{i,w}_+(z)q^{i,w}_-(z) =\\ &\Lambda_i(z)\prod_{j\neq i}\left[q^{j,w}_+(z)\right]^{-a_{ji}}, \quad i = 1, \dots, r; \quad w \in W.\end{aligned} \tag{3.9}$$

DEFINITION 3.4. The collection of equations (3.7) on polynomials $\{q^i_+\}_{i=1,\dots,r}(z)$ is called the *qq-system*. The collection of equations (3.9) is called *the full qq-system*.

Using this definition we can formulate the following Theorem.

THEOREM 3.5. *i) The element $\mathscr{G}(z) \in G(z)$, which defines the nondegenerate G-Wronskian has a Gaussian decomposition $\mathscr{G}(z) = \mathscr{B}_-(z)\mathscr{N}_+(z)$.*
ii) Generalized minors, which determine $\mathscr{B}_-(z)$ are the solutions of the qq-system, according to the formula $q^{i,w}_+(z) = \Delta_{w^{-1}\cdot\omega_i,\omega_i}(\mathscr{G}(z))$.
iii) Representing $\mathscr{N}_+(z) = e^{n_+(z)}$, the restriction $n_+(z)|_{\mathfrak{b}_{+,1}(z)}$ is determined by the solution of the qq-system.
iv) Given a solution $\mathscr{G}(z)$ of (3.8), we obtain that $\mathscr{G}(z)U_+(z)$, where $U_+(z) \in [N_+, N_+](z)$, i.e. $U_+(z) = e^{u_+(z)}$, so that $u_+(z) \in \oplus_{k\geq 2}\mathfrak{n}_k$.

PROOF. To prove i), it is enough to refer to nondegeneracy condition of $\mathscr{G}(z)$ and use 2.4. We derived ii) above. Part iii) follows from the fact that all components of $n_+(z)|_{\mathfrak{b}_{+,1}(z)}$ which is a linear combination of e_i's appear in the right hand side of (3.3), since they act on $f_i\nu_i$. For the same reason iv) is true since $e^{u_+(z)}|_{W_i} = \mathrm{id}_{W_i}$, if $u_+(z) \in \oplus_{k\geq 2}\mathfrak{n}_k$. $\square$

Part iv) of the above Theorem motivates the following definition.

DEFINITION 3.6. Two G-Wronskians $\mathscr{G}_1(z), \mathscr{G}_2(z)$ are called *equivalent* if $\mathscr{G}_1(z) = \mathscr{G}_2(z)U_+(z)$, where $U_+(z) \in [N_+, N_+](z)$.

Theorem (3.5) implies that there is a map from equivalence classes of nondegenerate G-Wronskians to the solution of the full qq-systems, where $\{q^{w,i}_+(z)\}_{w\in W, i=1,\dots,r}$ are nonzero. As one could guess, we would like an inverse map. We will construct it in the last section (we also refer to the last section for the explicit $SL(2)$ example). An important tool for that is the concept of Z-twisted Miura G-oper with regular singularities, which we review in the next section.

4. Z-twisted G-opers on $\mathbb{P}^1$ and the qq-systems

4.1. Z-twisted G-opers on $\mathbb{P}^1$. We now define meromorphic G-oper conenctions, or, simply, G-opers, on $\mathbb{P}^1$. Let us consider the pair $(\mathcal{F}_G, \nabla)$ of a principal G-bundle on $\mathbb{P}^1$ and a connection, which is automatically flat. Let $\mathcal{F}_{B_+}$ be a reduction of $\mathcal{F}_G$ to the Borel subgroup B_+. If ∇' is any connection which preserves $\mathcal{F}_{B_+}$, then $\nabla - \nabla'$ induces a well-defined one-form on $\mathbb{P}^1$ with values in the associated bundle $(\mathfrak{g}/\mathfrak{b}_+)_{\mathcal{F}_{B_+}}$. We denote this 1-form by $\nabla/\mathcal{F}_{B_-}$.

Following [**BD**] (see also [**F3**]) we will define a G-oper as a G-connection $(\mathcal{F}_G, \nabla)$ together with a reduction $\mathcal{F}_{B_+}$, such that this reduction is not preserved by the connection but instead satisfies a certain condition on the 1-form $\nabla/\mathcal{F}_{B_+}$.

Let $\mathbf{O} \in [\mathfrak{n}_+, \mathfrak{n}_+]^\perp/\mathfrak{b}_+ \in \mathfrak{g}/\mathfrak{b}_+$ be the open B_+-orbit consisting of vectors stabilized by N_+ and such that all of the simple root components with respect to the adjoint action of B_+/N_+, are non-zero, where the orthogonal complement is taken with respect to the Killing form.

DEFINITION 4.1. A meromorphic *G-oper* on $\mathbb{P}^1$ is a triple $(\mathcal{F}_G, \nabla, \mathcal{F}_{B_+})$, where pair $(\mathcal{F}_G, \Delta)$ is a principal G-bundle on $\mathbb{P}^1$ with a meromorphic connection and $\mathcal{F}_{B_+}$ is a reduction of $\mathcal{F}_G$ to B_+ satisfying the following condition: there exists a Zariski open dense subset $U \subset \mathbb{P}^1$ together with a trivialization of $\mathcal{F}_{B_+}$ such that the restriction of the 1-form $\nabla/\mathcal{F}_{B_+}$ to U, written as an element of $\mathfrak{g}/\mathfrak{b}_+(z)$, belongs to $\mathbf{O}(z)$.

Using the trivialization $\imath_{B_-}$, the G-oper can be written as a differential operator:

$$\nabla = \partial_z + \sum_{i=1}^{r} \phi_i(z) f_i + b(z) \tag{4.1}$$

where $\phi_i(z) \in \mathbb{C}(z)$ and $b(z) \in \mathfrak{b}_+(z)$ are regular on U and moreover $\phi_i(z)$ has no zeros in U.

One can impose the following restrictions on ϕ_i:

DEFINITION 4.2. We say that a meromorphic G-oper has regular singularities if $\phi_i(z) = \Lambda_i(z)$, where $\Lambda_i(z) \in \mathbb{C}[z]$ for all $i = 1, \ldots, r$.

4.2. Z-twisted Miura G-opers with regular singularities.

DEFINITION 4.3. A *Miura G-oper* on $\mathbb{P}^1$ is a quadruple $(\mathcal{F}_G, \nabla, \mathcal{F}_{B_+}, \mathcal{F}_{B_-})$, where $(\mathcal{F}_G, \nabla, \mathcal{F}_{B_+})$ is a meromorphic G-oper on $\mathbb{P}^1$ and $\mathcal{F}_{B_-}$ is a reduction of the G-bundle $\mathcal{F}_G$ to B_- that is preserved by the connection ∇.

Let us discuss the relative position of the two reductions over any $x \in \mathbb{P}^1$. This relative position will be an element of the Weyl group. To define this, first note that the fiber $\mathcal{F}_{G,x}$ of $\mathcal{F}_G$ at x is a G-torsor with reductions $\mathcal{F}_{B_-,x}$ and $\mathcal{F}_{B_+,x}$ to B_- and B_+ respectively. Under this isomorphism, $\mathcal{F}_{B_-,x}$ gets identified with $gB_- \subset G$ and $\mathcal{F}_{B_+,x}$ with hB_+ for some $g, h \in G$. The quotient $g^{-1}h$ specifies an element of the double coset space $B_-\backslash G/B_+$. The Bruhat decomposition gives a bijection between this spaces and the Weyl group, so we obtain a well-defined element of G. We say that $\mathcal{F}_{B_-}$ and $\mathcal{F}_{B_+}$ have *generic relative position* at $x \in \mathbb{P}^1$ if the relative position is the identity element of W. More concretely, this mean that the quotient $g^{-1}h$ belongs to the open dense Bruhat cell $B_-B_+ \subset G$. It turns out the following theorem holds.

THEOREM 4.4. *i) For any Miura G-oper on $\mathbb{P}^1$, there exists an open dense subset $V \subset \mathbb{P}^1$ such that the reductions $\mathcal{F}_{B_-}$ and $\mathcal{F}_{B_+}$ are in generic relative position for all $x \in V$.*
ii) For any Miura G-oper with regular singularities on $\mathbb{P}^1$, there exists a trivialization of the underlying G-bundle $\mathcal{F}_G$ on an open dense subset of $\mathbb{P}^1$ for which the oper connection has the form

$$\nabla = \partial_z - \sum_{i=1}^{r} g_i(z)\check{\alpha}_i + \sum_{i=1}^{r} \Lambda_i(z) f_i, \tag{4.2}$$

where $g_i(z), \phi_i(z) \in \mathbb{C}(z)$.

Let us impose a strong condition, which picks up a subset of opers we are interested in, namely the Miura G-opers, which are gauge equivalent to a constant connection.

DEFINITION 4.5. A *Z-twisted Miura G-oper* on $\mathbb{P}^1$ is a Miura G-oper that is equivalent to the constant element $Z \in \mathfrak{b}_- \subset \mathfrak{g}(z)$ under the gauge action of $G(z)$.

Here we immediately can decompose the twist Z into

$$Z = Z^H + Z^{N_-}, \quad Z^H = \sum_{i=1}^{r} \zeta_i \check{\alpha}_i \ , \tag{4.3}$$

breaking it into Cartan and nilpotent part.

For untwisted opers, there is a full flag variety G/B_- of associated Miura opers. If $Z = Z^H$, this space is discrete and is one-to-one correspondence with Weyl group W. In, general, in the twisted case, we must introduce certain closed subvarieties of the flag manifold of the form $(G/B_-)_Z = \{gB_- \mid g^{-1}Zg \in \mathfrak{b}_-\}$, known as *Springer fibers* (see, for example, Chapter 3 of [**CG**]). For $SL(n)$ (or $GL(n)$), a Springer fiber may be viewed as the space of complete flags in $\mathbb{C}^n$ preserved by a fixed endomorphism.

PROPOSITION 4.6. [**BSZ**] The map from Miura Z-twisted opers to Z-twisted opers is a fiber bundle with fiber $(G/B_-)_Z$.

Taking the quotient of $\mathcal{F}_{B_-}$ by $N_- = [B_-, B_-]$, we obtain an H-bundle $\mathcal{F}_{B_-}/N_-$ endowed with an H-connection, which we will refer to as *associated Cartan connection*: $\nabla^H = \partial_z + A^H(z)$, so that

$$A^H(z) = \sum_{i=1}^{r} g_i(z)\check{\alpha}_i. \tag{4.4}$$

For Z-twisted Miura G-opers, we immediately obtain that

$$g_i(z) = \zeta_i - y_i(z)^{-1}\partial_z y_i(z), \tag{4.5}$$

where $y_i(z) \in \mathbb{C}(z)$. We refer to Section 5 for the explicit $SL(2)$ example.

4.3. Nondegenerate Z-twisted Miura G-opers, qq-systems and Bäcklund transformations. Now we will impose nondegeneracy conditions on $y_i(z)$, which will lead us to the relation between Miura G-opers and the qq-systems. We will formulate it in the algebraic manner and refer to [**BSZ**] for more geometric formulation.

DEFINITION 4.7. A Z-twisted Miura G-oper is called nondegenerate, if:
i) it has the form (4.2) with $g_i(z)$ satisfying (4.5), where:

(1) $y_i(z)$ are polynomials with no multiple zeros;
(2) if $a_{ik} \neq 0$, then the roots of $\Lambda_k(z)$ are distinct from the the zeros and poles of $y_i(z)$; and
(3) if $i \neq j$ and there exists k for which $a_{ik}, a_{jk} \neq 0$, then the zeros and poles of $y_i(z)$ and $y_j(z)$ are distinct from each other.

A related definition, which imposes a similar type of conditions on the solution of the qq-system is as follows:

DEFINITION 4.8. A polynomial solution $\{q^i_+(z), q^i_-(z)\}_{i=1,\dots,r}$ of (3.7) is called *nondegenerate* if each $q^i_+(z)$ is relatively prime to $q^i_-(z)$, and the $q^i_+(z)$'s satisfy the conditions in Definition 4.7.

Then the following statement is true.

THEOREM 4.9. [**BSZ**] *There is one-to-one correspondence between nondegenerate Z-twisted Miura G-opers and nondegenerate solutions of the qq-system, which allows polynomial solutions to the full qq-system, so that*

$$y_i(z) = q^i_+(z), \quad i = 1, \dots, r. \tag{4.6}$$

Moreover, any Z-twisted Miura G-oper is Z^H-twisted.

One can write explicit algebraic equations on zeroes of $q_+^i(z) = \prod_{\ell=1}^{\deg(q_+^i(z))}(z - w_\ell^i)$ polynomials (without the loss of generality we can assume $q_+^i(z)$ to be monic). These algebraic equations are known as *Bethe equations* for Z-twisted ${}^L\mathfrak{g}$-Gaudin model [**FFTL**], [**FFR2**]:

$$
\begin{gathered}
\langle \alpha_i, Z^H\rangle + \sum_{j=1}^{N} \frac{\langle \alpha_i, \check{\lambda}_j\rangle}{w_\ell^i - z_j} - \sum_{(j,s)\neq(i,\ell)} \frac{a_{ji}}{w_\ell^i - w_s^j} = 0, \\
i = 1, \dots, r, \quad \ell = 1, \dots, \deg(q_+^i(z)).
\end{gathered} \tag{4.7}
$$

The correspondence between qq-systems and Bethe ansatz equations is summarized in the following Theorem (see [**MV2, MV3**], [**BSZ**].)

THEOREM 4.10. *i) If Z^H is regular, there is a bijection between the solutions of the Bethe ansatz equations* (4.7) *and the nondegenerate polynomial solutions of the qq-system* (3.7).
ii) If $\langle \alpha_l, Z^H\rangle = 0$, for $l = i_1, \dots, i_k$ and is nonzero otherwise, then $\{q_+^i(z)\}_{i=1,\dots,r}$ and $\{q_-^i(z)\}_{i\neq i_1,\dots,i_k}$, are uniquely determined by the Bethe ansatz equations, but each $\{q_-^{i_j}(z)\}$ for $j = 1, \dots k$ is only determined up to an arbitrary transformation $q_-^{i_j}(z) \to q_-^{i_j}(z) + c_j q_+^{i_j}(z)$, where $c_j \in \mathbb{C}$.

From now on we will drop the superscript H over Z and consider Z to be an element of Cartan subalgebra. We also remark, that in the case of Gaudin model, one puts the restrictions on degrees of $\{q_+^i(z)\}_{i=1,\dots,r}$, so that they determine a certain weight:

$$
\check{\Lambda} = \sum_{i=1}^{N} \check{\lambda}_i - \sum_i \deg(q_+^i(z))\check{\alpha}_i \tag{4.8}
$$

in the representation of ${}^L\mathfrak{g}$, and the degrees of $\{q_+^{i,w}\}_{w\in W,\ i=1,\dots,r}$ in the full qq-system determine $w \cdot \check{\Lambda} = \sum_{i=1}^N \check{\lambda}_i - \sum_i \deg(q_+^{w,i}(z))\check{\alpha}_i$. This means that with this restriction on the degrees, the qq-system can be extended to the full qq-system, allowing polynomial solutions. From now on we will assume that qq-systems allow such extension to the full qq-system.

The Miura G-oper connection operator, which correspond to the qq-system can be explicitly written as follows:

$$
\nabla = \partial_z - Z + \sum_{i=1}^{r} \partial_z \log(q_+^i(z))\check{\alpha}_i + \sum_{i=1}^{r} \Lambda_i(z) f_i\ . \tag{4.9}
$$

Let us explain what role the full qq-system will play in this context. To do that, we use the following Proposition to introduced *Bäcklund transformations*, aligned with the action of the generators of the Weyl group [2].

PROPOSITION 4.11. [**BSZ**] Let $\{q_+^j, q_-^j\}_{j=1,\dots,r}$ be a polynomial solution of the qq-system (3.7), and let ∇ be the connection in the form (4.9). Let $\nabla^{(i)}$ be the

[2]We note here that the degenerate version of the qq-system (Z=0), as well as degenerate version of the Proposition below were introduced by Mukhin and Varchenko [**MV1**].

connection obtained from ∇ via the gauge transformation by $e^{\mu_i(z)e_i}$, where

$$\mu_i(z) = \Lambda_i(z)^{-1}\left[\partial_z \log\left(\frac{q^i_-(z)}{q^i_+(z)}\right) + \langle \alpha_i, Z^H\rangle\right]. \tag{4.10}$$

Then $\nabla^{(i)}$ is obtained by making the following substitutions in (4.9):

$$\begin{aligned} &q^j_+(z) \mapsto q^j_+(z), && j \neq i,\\ &q^i_+(z) \mapsto q^i_-(z), && Z \mapsto s_i(Z^H) = Z^H - \langle \alpha_i, Z^H\rangle\, \check{\alpha}_i\,. \end{aligned} \tag{4.11}$$

Thus the sequence of Bäcklund transformations produces Z^w-twisted Miura G-oper, where

$$Z^w = \tilde{w} Z \tilde{w}^{-1}, \quad w \in W, \tag{4.12}$$

corresponding to a given G-oper, each of which is constructed from $q^{i,w}_\pm(z)$.

5. From Z-twisted Miura G-opers to G-Wronskians

We see that there is a crtain relation between nondegenerate G-Wronskians and nondegenerate G-opers. Let us make this correspondence explicit. The Z-twisted condition implies that that there exist an element $\mathscr{B}_-(z)$, such that

$$\nabla = \mathcal{B}_-^{-1}(z)(\partial_z - Z)\mathcal{B}_-(z), \tag{5.1}$$

where

$$\mathcal{B}_-(z) = \prod_{i=1}^r e^{\frac{q^i_-(z)}{q^i_+(z)} f_i} \prod_{i=1}^r \left[q^i_+(z)\right]^{\check{\alpha}_i} \dots, \tag{5.2}$$

where dots stand for the terms from $[N_-, N_-](z)$. Let us use the following upper-triangular transformation, choosing a certain order in the product below:

$$\mathcal{N}_+(z) = \prod_{i=1}^r e^{\frac{(\partial_z \log(q^i_+(z)) - \zeta_i)e_i}{\Lambda_i(z)}} \tag{5.3}$$

such that

$$\mathcal{N}_+^{-1}(z)\nabla\mathcal{N}_+(z) = \partial_z + p^\Lambda_{-1} + n_+(z) \tag{5.4}$$

where $n_+(z) \in \mathfrak{n}_+(z)$. Then applying this operator to the highest weight vector ν in some representation V the product $\mathcal{G}(z) = \mathcal{B}_-(z)\mathcal{N}_+(z)$ satisfies the equation

$$\mathcal{G}(z)^{-1}(\partial_z - Z)\mathcal{G}(z)\nu = p^\Lambda_{-1}\nu, \tag{5.5}$$

which coincides with equations 3.3 when ν varies over highest weights corresponding to fundamental representations. Turns out this correspondence works in a different direction as well, namely we want to show the following.

THEOREM 5.1. *There is a one-to-one correspondence between nondegnerate Z-twisted Miura G-opers, and equivalence classes of G-Wronskians, corresponding to the solution of the nondegenerate full qq-system. The element $\mathcal{G}(z) = \mathcal{B}_-(z)\mathcal{N}_+(z) \in G(z)$, where $\mathscr{B}_-(z)$, $\mathscr{N}_+(z)$ are defined in (5.2), (5.3) correspondingly, is a representative in the class of G-Wronskians corresponding to a given solution of the full qq-system.*

PROOF. Given $\mathscr{G}(z)=\mathscr{B}_-(z)\mathscr{N}_+(z)\in G(z)$ corresponding to a G-Wronskian, we have the following equations:

$$\mathscr{N}_+(z)^{-1}\mathscr{B}_-(z)^{-1}(\partial_z-Z)\mathscr{B}_-(z)\mathscr{N}_+(z)\nu_i=p_{-1}^{\Lambda}\nu_i, \quad ,i=1,\dots,r. \tag{5.6}$$

This means $\mathscr{N}_+(z)^{-1}A(z)\mathscr{N}_+(z)\nu_i=p_{-1}^{\Lambda}\nu_i$, where

$$\partial_z+A(z)=\mathscr{B}_-(z)^{-1}(\partial_z-Z)\mathscr{B}_-(z), \tag{5.7}$$

so that $A(z)\in\mathfrak{b}_-(z)$. Then $A(z)\in p_{-1}^{\Lambda}+\mathfrak{h}_+(z)$, namely a Miura oper connection. That, however, implies that $A(z)$ defines a Miura oper connection. The diagonal part of $\mathscr{B}_-(z)$ is given by principal minors, which are exactly the $q_+^i(z)$, thus reproducing the diagonal part of Z-twisted Miura-Plucker oper. □

At the same time, the nondegeneracy conditions on qq-systems is an open condition, which was only needed for Miura G-oper connection itself and the corresponding Z-twisted condition, which is not the case for G-Wronskians, which only care about the operator (4.9), which satisfies Z-twisted condition as long as $\{q_+^i(z)\}_{i=1,\dots,r}$ satisfy the qq-system. Therefore we have the following theorem.

THEOREM 5.2. *There is a one-to-one correspondence between solutions of the full qq-system, such that $q_+^{i,w}(z)\neq 0$ and equivalence classes of nondegenerate G-Wronskians.*

Example: $SL(2)$-opers and $SL(2)$-Wronskians. The Z-twisted Miura $SL(2)$-oper connection is given by the following matrix operator in defining representation of $\mathfrak{sl}(2)$:

$$\partial_z+\begin{pmatrix}\partial_z\log(q_+)-\zeta & \Lambda(z)\\ 0 & \zeta-\partial_z\log(q_+)\end{pmatrix} \tag{5.8}$$

where $\Lambda(z)$ is a polynomial, defining regular singularities $q_+(z)$ is a polynomial, satisfying qq-system:

$$q_+(z)\partial_z q_-(z)-q_-(z)\partial_z q_+(z)+2\zeta q_+(z)q_-(z)=\Lambda(z) \tag{5.9}$$

The matrices corresponding to $\mathcal{B}_+(z)$ and $\mathcal{N}_+(z)$ are represented by the following matrices:

$$\mathcal{B}_-(z)=\begin{pmatrix}q_+(z) & 0\\ q_-(z) & q_+(z)^{-1}\end{pmatrix}\quad \mathcal{N}_+(z)=\begin{pmatrix}1 & \Lambda(z)^{-1}(\partial_z\log[q_+(z)] & \zeta)\\ 0 & 1\end{pmatrix}, \tag{5.10}$$

so that their product give a version of a Wronskian (one needs to use the qq-system (5.9) to obtain the answer):

$$\mathcal{G}(z)=\mathcal{B}_-(z)\mathcal{N}_+(z)=\begin{pmatrix}q_+(z) & \Lambda^{-1}(\partial_z-\zeta)q_+(z)\\ q_-(z) & \Lambda^{-1}(\partial_z+\zeta)q_-(z)\end{pmatrix}. \tag{5.11}$$

Comparison to $(G,\hbar)$-oper case. In [**KZ1**] we considered a similar construction related to $(G,\hbar)$-opers. The $(G,\hbar)$-Wronskian equations, which is the $\hbar$-difference version of (3.3) is as follows:

$$Z^{-1}\mathscr{G}(\hbar z)\nu_i=\mathscr{G}(z)s_{\Lambda}^{-1}(z)\nu_i, \quad i=1,\dots r, \tag{5.12}$$

where $\mathscr{G}(z)\in G(z)$, $Z\in H$, $s_\Lambda(z)=\prod_{i=1}^r\Lambda_i^{\check{\alpha}_i}(z)s_i$ is a lift of a chosen Coxeter element to $G(z)$ and $\Lambda_i(z)$ are polynomials, corresponding to regular singularities. As in the differential case, one can define a class of those: namely, if $\mathscr{G}(z)$ is a solution of (5.12), then so is $\mathscr{G}(z)n_+(z)$, as long as $s_\Lambda n_+(z)s_\Lambda^{-1}\in N_+(z)$. Choosing Coxeter element in such a way that $s_\Lambda^{h/2}=w_0^{\Lambda}$, where h is the Coxeter number and

w_0^Λ is a lift of the Weyl reflection, corresponding to the longest root to $G(z)$, we can iterate the equations 5.12:

$$Z^{-k}\mathscr{G}(\hbar^k z)\nu_i = \mathscr{G}(z)s_\Lambda^{-1}(z)s_\Lambda^{-1}(\hbar z)\dots s_\Lambda^{-1}(\hbar^{k-1}z)\nu_i,$$
$$i = 1,\dots r, \quad k = 1,\dots h/2. \tag{5.13}$$

The solution of this kind picks up a unique element in the equivalence class of $(G,\hbar)$-Wronskians, moreover, we were able to write down a universal formula for it. In the $SL(N)$ case that gives a suitably twisted $(G,\hbar)$-Wronskian matrix. Unfortunately it is non-obvious whether if is possible to write a similar universal formula in differential case: the blunt approach using differential analogue of equations (5.13) do not seem to work beyond defining representation of $SL(N)$.

References

[AO] M. Aganagic and A. Okounkov, *Quasimap counts and Bethe eigenfunctions*, Mosc. Math. J. **17** (2017), no. 4, 565–600, DOI 10.17323/1609-4514-2016-16-4-565-600. MR3734654

[B1] R. J. Baxter, *Exactly solved models in statistical mechanics*, Academic Press, 1982, `http://www.amazon.com/dp/0486462714`.

[B2] H. Bethe, *Zur Theorie der Metalle I. Eigenwerte und Eigenfunktionen der linearen Atomkette*, Z. Phys. **71** (1931), 205–226.

[BD] A. Beilinson and V. Drinfeld, *Opers* (2005), `math/0501398`, available at math/0501398.

[BLZ] V. V. Bazhanov, S. L. Lukyanov, and A. B. Zamolodchikov, *Integrable structure of conformal field theory. III. The Yang-Baxter relation*, Comm. Math. Phys. **200** (1999), no. 2, 297–324, DOI 10.1007/s002200050531. MR1673992

[BSZ] T. J. Brinson, D. S. Sage, and A. M. Zeitlin, *Opers on the projective line, Wronskian relations, and the Bethe Ansatz*, `https://arxiv.org/abs/2112.02711`, available at 2112.02711.

[CG] N. Chriss and V. Ginzburg, *Representation theory and complex geometry*, Birkhäuser Boston, Inc., Boston, MA, 1997. MR1433132

[F1] E. Frenkel, *Opers on the projective line, flag manifolds and Bethe ansatz* (English, with English and Russian summaries), Mosc. Math. J. **4** (2004), no. 3, 655–705, 783, DOI 10.17323/1609-4514-2004-4-3-655-705. MR2119144

[F2] E. Frenkel, *Gaudin model and opers*, Infinite dimensional algebras and quantum integrable systems, Progr. Math., vol. 237, Birkhäuser, Basel, 2005, pp. 1–58, DOI 10.1007/3-7643-7341-5_1. MR2160841

[F3] E. Frenkel, *Langlands correspondence for loop groups*, Cambridge Studies in Advanced Mathematics, vol. 103, Cambridge University Press, Cambridge, 2007. MR2332156

[FFR1] B. Feigin, E. Frenkel, and N. Reshetikhin, *Gaudin model, Bethe ansatz and critical level*, Comm. Math. Phys. **166** (1994), no. 1, 27–62. MR1309540

[FFR2] B. Feigin, E. Frenkel, and L. Rybnikov, *Opers with irregular singularity and spectra of the shift of argument subalgebra*, Duke Math. J. **155** (2010), no. 2, 337–363, DOI 10.1215/00127094-2010-057. MR2736168

[FFTL] B. Feigin, E. Frenkel, and V. Toledano Laredo, *Gaudin models with irregular singularities*, Adv. Math. **223** (2010), no. 3, 873–948, DOI 10.1016/j.aim.2009.09.007. MR2565552

[FH1] E. Frenkel and D. Hernandez, *Baxter's relations and spectra of quantum integrable models*, Duke Math. J. **164** (2015), no. 12, 2407–2460, DOI 10.1215/00127094-3146282. MR3397389

[FH2] E. Frenkel and D. Hernandez, *Spectra of Quantum KdV Hamiltonians, Langlands Duality, and Affine Opers*, Commun. Math. Phys. **362** (2018), 361–414, `https://arxiv.org/abs/1606.05301`, available at 1606.05301.

[FHR] E. Frenkel, D. Hernandez, and N. Reshetikhin, *Folded quantum integrable models and deformed W-algebras*, Lett. Math. Phys. **112**, 80, `https://arxiv.org/abs/2110.14600`, available at 2110.14600.

[FKSZ] E. Frenkel, P. Koroteev, D. S. Sage, and A. M. Zeitlin, *q-Opers, QQ-Systems, and Bethe Ansatz*, J. Europ. Math. Soc. (2020), to appear, `https://arxiv.org/pdf/2002.07344.pdf`, available at 2002.07344.

[FZ1] S. Fomin and A. Zelevinsky, *Double Bruhat cells and total positivity*, J. Amer. Math. Soc. **12** (1999), no. 2, 335–380, DOI 10.1090/S0894-0347-99-00295-7. MR1652878

[FZ2] S. Fomin and A. Zelevinsky, *Recognizing Schubert Cells*, J. Algeb. Comb. **12** (2000), 37–57, `https://arxiv.org/pdf/math/9807079.pdf`, available at math/9807079.

[HJ] D. Hernandez and M. Jimbo, *Asymptotic representations and Drinfeld rational fractions*, Compos. Math. **148** (2012), no. 5, 1593–1623, DOI 10.1112/S0010437X12000267. MR2982441

[KBI] V. Korepin, N. Bogoliubov, and A. Izergin, *Quantum Inverse Scattering Method and Correlation Functions*, Cambridge University Press (1993), `http://dx.doi.org/10.1017/CBO9780511628832`.

[KPSZ] P. Koroteev, P. Pushkar, A. Smirnov, and A. Zeitlin, *Quantum K-theory of Quiver Varieties and Many-Body Systems*, Selecta Math. in press (2017), `https://arxiv.org/pdf/1705.10419`, available at 1705.10419.

[KSZ] P. Koroteev, D. Sage, and A. Zeitlin, *(SL(N),q)-Opers, the q-Langlands Correspondence, and Quantum/Classical Duality*, Commun. Math. Phys. **381** (2021), no. 2, 641–672, `https://arxiv.org/abs/1508.02690`, available at 1811.09937.

[KZ1] P. Koroteev and A. M. Zeitlin, *q-Opers, QQ-systems, and Bethe Ansatz II: Generalized Minors*, Journal fur die Reine und angewandte Matematik (Crelles Journal) **2023** (2023), no. 795, 271–296, DOI https://doi.org/10.1515/crelle-2022-0084.

[KZ2] P. Koroteev and A. M. Zeitlin, *Toroidal q-Opers*, Journal of the Institute of Mathematics of Jussieu **22** (2023), no. 2, 581–642, DOI https://doi:10.1017/S1474748021000220.

[MV1] E. Mukhin and A. Varchenko, *Miura Opers and Critical Points of Master Functions*, Cent. Eur. J. Math. **3**, 155–182, `https://arxiv.org/abs/math/0312406`, available at math/0312406.

[MV2] E. Mukhin and A. Varchenko, *Critical points of master functions and flag varieties*, Commun. Contemp. Math. **06** (2004), no. 01, 113–163, `https://arxiv.org/abs/math/0209017`, available at math/0209017.

[MV3] E. Mukhin and A. Varchenko, *Quasi-polynomials and the Bethe Ansatz*, Geom. Top. Mon. **13** (2008), 385–420, available at arXiv:math/0604048.

[NS1] N. Nekrasov and S. Shatashvili, *Quantum integrability and supersymmetric vacua*, Prog. Theor. Phys. Suppl. **177** (2009), 105–119, available at 0901.4748.

[NS2] N. A. Nekrasov and S. L. Shatashvili, *Supersymmetric vacua and Bethe ansatz*, Nuclear Phys B Proc. Suppl. **192/193** (2009), 91–112, DOI 10.1016/j.nuclphysbps.2009.07.047. MR2570974

[O] A. Okounkov, *Lectures on K-theoretic computations in enumerative geometry* (2015), available at 1512.07363.

[OW] E. Ogievetsky and P. Wiegmann, *Factorized S-matrix and the Bethe ansatz for simple Lie groups*, Phys. Lett. B **168** (1986), no. 4, 360–366, DOI 10.1016/0370-2693(86)91644-8. MR831897

[PSZ] P. P. Pushkar, A. V. Smirnov, and A. M. Zeitlin, *Baxter Q-operator from quantum K-theory*, Adv. Math. **360** (2020), 106919, 63, DOI 10.1016/j.aim.2019.106919. MR4035952

[R] N. Reshetikhin, *Lectures on the integrability of the 6-vertex model* (2010), available at 1010.5031.

[TF] L. A. Tahtadžjan and L. D. Faddeev, *The quantum method for the inverse problem and the XYZ Heisenberg model* (Russian), Uspekhi Mat. Nauk **34** (1979), no. 5(209), 13–63, 256. MR562799

DEPARTMENT OF MATHEMATICS, LOUISIANA STATE UNIVERSITY, BATON ROUGE, LOUISIANA 70803

Email address: `zeitlin@lsu.edu`

Contemporary Mathematics
Volume **791**, 2024
https://doi.org/10.1090/conm/791/15874

One-dimensional topological theories with defects: the linear case

Mee Seong Im and Mikhail Khovanov

ABSTRACT. The paper studies the Karoubi envelope of a one-dimensional topological theory with defects and inner endpoints, defined over a field. It turns out that the Karoubi envelope is determined by a symmetric Frobenius algebra $\mathcal{K}$ associated to the theory. The Karoubi envelope is then equivalent to the quotient of the Frobenius–Brauer category of $\mathcal{K}$ modulo the ideal of negligible morphisms. Symmetric Frobenius algebras, such as $\mathcal{K}$, describe two-dimensional TQFTs for the category of thin flat surfaces, and elements of the algebra can be turned into defects on the side boundaries of these surfaces. We also explain how to couple $\mathcal{K}$ to the universal construction restricted to closed surfaces to define a topological theory of open-closed two-dimensional cobordisms which is usually not an open-closed 2D TQFT.

CONTENTS

1. Introduction

Universal construction [**BHMV95, Kho20a**] starts with an evaluation function for closed n-manifolds to produce state space for closed $(n-1)$-manifolds and maps between these spaces associated to n-cobordisms. This results in a functor from the category of n-dimensional cobordisms to the category of vector spaces (if

2020 *Mathematics Subject Classification.* Primary 18M05, 18M30, 57K16, 16W60, 15A63.

Key words and phrases. Rational noncommutative power series, topological theory, universal construction, topological quantum field theory (TQFT), defects in TQFT, Brauer categories, symmetric Frobenius algebras.

the evaluation function takes values in a field) which usually fails to be a TQFT, with the tensor product of state spaces for two $(n-1)$-manifolds N_1, N_2 properly embedded into the state space for their union:

$$A(N_1) \otimes A(N_2) \hookrightarrow A(N_1 \sqcup N_2). \tag{1}$$

Universal construction for foams in $\mathbb{R}^3$ in place of n-cobordisms is used as an intermediate step in constructing link homology theories [**Kho04, MV07, RW20**], see also a review in [**KK20**] and papers [**Kho20b, Mei21**] for other uses and references for the universal construction.

The universal construction turns out to be interesting already in low dimensions, including in dimensions two [**Kho20b, KS21, KOK22, KQR21**] and one [**Kho20a, IK22, IZ22, IKO23, GIK⁺23**]. In the latter case, one needs to add zero-dimensional defects (0-submanifolds) with labels in a set Σ. An oriented interval with a collection of Σ-labelled defects encodes a word ω, that is, an element of the free monoid Σ^* on the set Σ. An oriented circle with labels in Σ encodes a word up to cyclic equivalence. Given an evaluation of each word and a separate evaluation of words up to cyclic equivalence, there is an associated rigid linear monoidal category, defined in studied in [**Kho20a**]. It is straightforward to see [**Kho20a**] that the hom spaces in the resulting categories are finite-dimensional if and only if the evaluations are given by *rational* noncommutative power series [**BR90, RRV99**].

In the present paper we study this category for a rational evaluation α. The Karoubi closure of the resulting category can be reduced to the Karoubi closure of a category built from a symmetric Frobenius algebra $\mathcal{K}$ that can be extracted from α, as explained in Section 2.4. Sections 2.1-2.3 are devoted to the setup, basic theory and various examples. In Section 3 we review thin flat surface 2D TQFTs associated to symmetric Frobenius algebras and explain how to enhance these TQFTs by 0-dimensional defects floating along the boundary that carry elements of the algebra. Throughout the paper we run comparisons between one-dimensional theories with defects and two-dimensional theories without defects and discuss nonsemisimple versus semisimple TQFTs in two dimensions.

The Boolean analogues of these categories and their relation to automata and regular languages are investigated in [**IK22**], where the absence of linear structure creates additional complexities.

2. One-dimensional topological theories with defects over a field

We fix a ground field $\mathbf{k}$.

2.1. A one-dimensional defect TQFT from a noncommutative power series. Start with a finite set (or alphabet) Σ. Let Σ^* be the set of finite words in letters of Σ (elements of Σ), and $\Sigma^*_\circ$ be the set of circular words, *i.e.*, elements of Σ^* up to the equivalence relation $\omega_1\omega_2 \sim \omega_2\omega_1$. The empty word $\varnothing$ is included in both Σ^* and $\Sigma^*_\circ$. Suppose we are given an evaluation

$$\alpha = (\alpha_{\mathsf{I}}, \alpha_\circ), \tag{2}$$

where

$$\alpha_{\mathsf{I}} : \Sigma^* \longrightarrow \mathbf{k}, \quad \alpha_\circ : \Sigma^*_\circ \longrightarrow \mathbf{k} \tag{3}$$

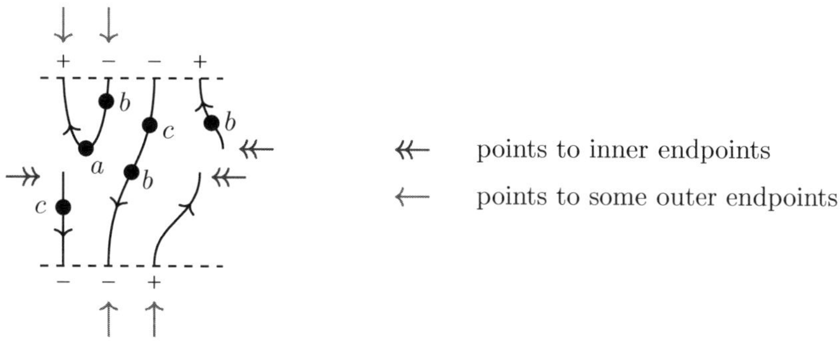

FIGURE 2.1.1. A morphism from $--+$ to $+--+$ in $\mathcal{C}_\alpha$ represented by a diagram with 3 inner endpoints. This diagram has 7 outer endpoints, 3 at the bottom and 4 at the top. Some of these endpoints are indicated by smaller (red and blue) arrows.

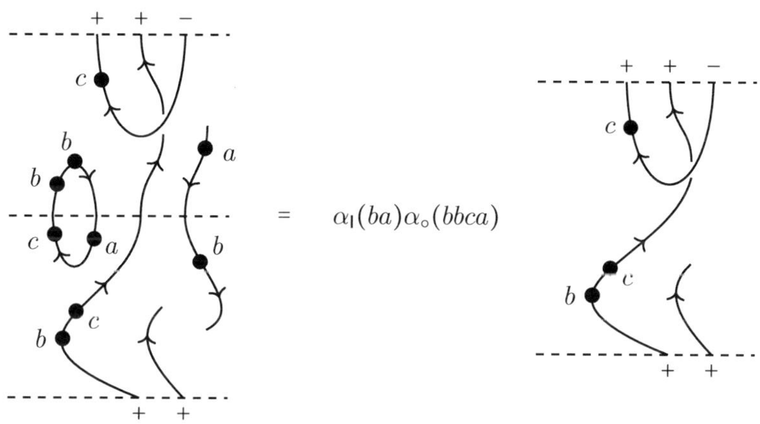

FIGURE 2.1.2. An example of computing the composition of morphisms in $\mathcal{C}_\alpha$.

are two functions on words and circular words in Σ, respectively, with values in a field $\mathbf{k}$. To α, following [**Kho20a**] (also see earlier work [**KS21, KOK22**] for a similar framework), there is assigned a symmetric $\mathbf{k}$-linear monoidal category $\mathcal{C}_\alpha$. Its objects are finite sign sequences ε, thought of as oriented 0-manifolds, and morphisms are $\mathbf{k}$-linear combinations of oriented 1-cobordisms with 0-dimensional defects, the latter decorated by elements of Σ. See 2.1.1. One-cobordisms can have "inner" boundary points, in addition to "outer" boundary points that define the objects for the morphism. Forming the composition of two such cobordisms may result in components without any "outer" boundary points, called *floating* intervals and circles. See 2.1.2. These floating components are evaluated via the interval and circle evaluation functions $\alpha_{|}$ and $\alpha_{\circ}$, respectively, and the composition is then reduced to a diagram without floating connected components.

Two linear combinations of such morphisms between sequences ε and ε' are equal if any closures of these two linear combinations evaluate to the same element of $\mathbf{k}$ via α, see [**KS21, Kho20a**].

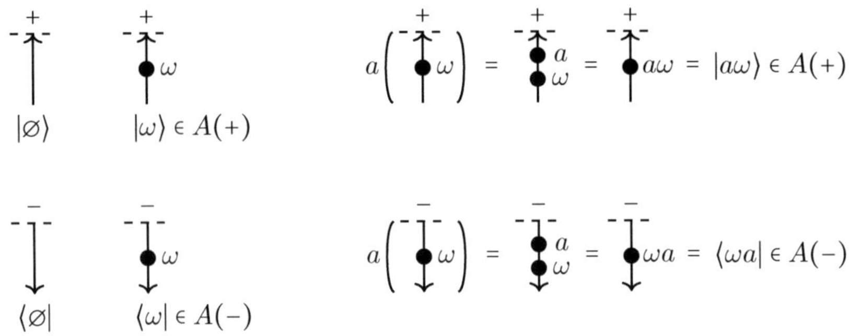

FIGURE 2.1.3. Top row: diagrams of $|\varnothing\rangle, |\omega\rangle \in A(+)$ and the action of a on $|\omega\rangle$. Bottom row: similar diagrams for $A(-)$.

The state space $A_\alpha(\varepsilon) := \mathsf{Hom}_{\mathcal{C}_\alpha}(\varnothing_0, \varepsilon)$ of a sequence ε is defined as the space of homs from the empty sign sequence $\varnothing_0$ to ε in $\mathcal{C}_\alpha$ (we use different notations $\varnothing \in \mathbf{k}\Sigma^*$ for the empty word and $\varnothing_0$ for the empty oriented 0-manifold and the corresponding object of $\mathcal{C}_\alpha$).

We say that α is *rational* if the following equivalent conditions hold:

- State spaces $A_\alpha(\varepsilon)$ are finite-dimensional for all ε.
- Hom spaces in $\mathcal{C}_\alpha$ are finite-dimensional.
- Spaces $A(+)$ and $A(+-)$ are finite-dimensional.
- $A(+-)$ is finite-dimensional.

Each of these conditions is equivalent to both of the noncommutative power series $\alpha_{\mathsf{I}}, \alpha_\circ$ in (2), (3) being rational in the sense of [**BR90**] (having a finite-dimensional state space, equivalently, a finite-dimensional syntactic algebra, see also [**Kho20a**]).

Category $\mathcal{C}_\alpha$ is $\mathbf{k}$-linear and preadditive. For rational α, it is convenient to consider the Karoubi closure $\mathsf{Kar}(\mathcal{C}_\alpha)$ of $\mathcal{C}_\alpha$ (also denoted $\underline{\mathcal{DC}}_\alpha$ in [**KS21, Kho20a**] in this and related cases) given by forming finite direct sums of objects and then adding objects for idempotent endomorphisms of these direct sums. The category $\mathsf{Kar}(\mathcal{C}_\alpha)$ is $\mathbf{k}$-linear, additive, idempotent-closed, with finite-dimensional hom spaces over $\mathbf{k}$. There is a fully-faithful functor

$$\mathcal{C}_\alpha \longrightarrow \mathsf{Kar}(\mathcal{C}_\alpha). \tag{4}$$

Assume from now on that evaluation α is rational. $\mathbf{k}$-vector space $A(+)$ is a left $\mathbf{k}\Sigma^*$-module, that is, a module over the ring of noncommutative polynomials in letters in Σ, equivalently the monoid algebra of the free monoid Σ^*. Module $A(+)$ has a distinguished element $|\varnothing\rangle$ corresponding to the diagram with an empty word, and action of $\omega \in \Sigma^*$ that takes it to $|\omega\rangle$. Element $|\varnothing\rangle$ is a cyclic vector in $A(+)$, so $A(+) = \mathbf{k}\Sigma^*|\varnothing\rangle$; see Figure 2.1.3, top row.

$A(-)$ is the dual vector space of $A(+)$, also carrying a distinguished vector $\langle\varnothing|$, with a right action of $\mathbf{k}\Sigma^*$; see Figure 2.1.3, bottom row.

The space $A(+)$ comes with the trace map

$$\mathsf{tr} : A(+) \longrightarrow \mathbf{k}, \quad \mathsf{tr}(|\omega\rangle) = \alpha_{\mathsf{I}}(\omega). \tag{5}$$

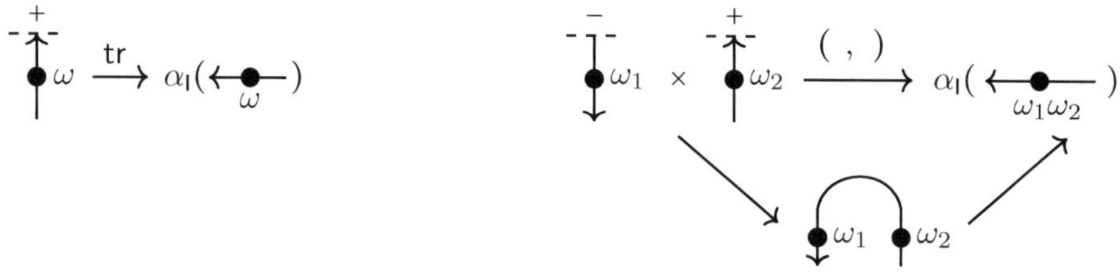

FIGURE 2.1.4. Figure of the trace map on $A(+)$ and of the pairing (6).

Diagrammatically, we evaluate an oriented interval with the word ω written on it using α_{I}. The trace map is nondegenerate: if $x \in A(+)$ with $x \neq 0$, then there exists $\omega \in \Sigma^*$ such that $\mathsf{tr}(\omega x) \neq 0$.

The trace map is part of the perfect pairing

$$(6) \qquad (\, , \,) \; : \; A(-) \otimes A(+) \longrightarrow \mathbf{k}, \text{ where } \langle \omega_1 | \otimes | \omega_2 \rangle \; \mapsto \; \alpha_{\mathsf{I}}(\omega_1 \omega_2),$$

given by concatenating words ω_1, ω_2 written on "half-intervals" into the word $\omega_1\omega_2$ on an interval and evaluating it, see Figure 2.1.4.

The pairing makes the left action of $\mathbf{k}\Sigma^*$ on $A(+)$ and the right action of $\mathbf{k}\Sigma^*$ on $A(-)$ adjoint:

$$(x\omega, y) = (x, \omega y) = \alpha_{\mathsf{I}}(x\omega y), \quad x \in A(-), \quad y \in A(+), \quad \omega \in \Sigma^*.$$

Action of Σ^* on $A(+)$ induces a $\mathbf{k}$-algebra homomorphism

$$(7) \qquad \mathbf{k}\Sigma^* \xrightarrow{\phi_\alpha} \mathsf{End}_{\mathbf{k}}(A(+))$$

of the algebra of noncommutative polynomials into a finite-dimensional matrix algebra. Denote by

$$(8) \qquad B_0 \; := \; \mathsf{im}(\phi_\alpha)$$

the image of $\mathbf{k}\Sigma^*$ in $\mathsf{End}_{\mathbf{k}}(A(+))$. It is a unital subalgebra of the matrix algebra. The algebra B_0 is the entire $\mathsf{End}_{\mathbf{k}}(A(+))$ if and only if the representation $A(+)$ of $\mathbf{k}\Sigma^*$ is absolutely irreducible. Note that $A(+), A(-), B_0$ and the actions above depend only on α_{I}, not on $\alpha_\circ$.

Vice versa, suppose we are given a finite-dimensional representation V of $\mathbf{k}\Sigma^*$ with a cyclic vector v^0 and a nondegenerate trace $\mathsf{tr} : V \longrightarrow \mathbf{k}$. Here v^0 corresponds to the undecorated upward-oriented half-interval (half-interval with the empty word $\varnothing$ on it). The trace is nondegenerate in the sense that for any $v \in V, v \neq 0$ there exists a word ω such that $\mathsf{tr}(\omega v) \neq 0$.

To this data one assigns rational noncommutative series via the evaluation $\alpha_{\mathsf{I}}(\omega) = \mathsf{tr}(\omega v^0)$ for $\omega \in \Sigma^*$ so that $A(+) = V = \mathbf{k}\Sigma^* v^0$ and $A(-) = V^*$, where v^0 is a cyclic vector. This gives a bijection between isomorphism classes of rational evaluations (rational noncommutative power series in Σ) and nondegenerate triples (V, v^0, tr) with an action of Σ^*. The action of infinite-dimensional algebra $\mathbf{k}\Sigma^*$ on V factors through a faithful action of the finite-dimensional algebra $B_0 \subset \mathsf{End}_{\mathbf{k}}(V)$.

Also, given a finite-dimensional algebra B_0 with a set of generators Σ, its faithful action on a finite-dimensional vector space V with a cyclic vector v^0 and a nondegenerate trace form on V, this recovers the noncommutative series α_{I} via the above recipe, with $A(+) \cong V$.

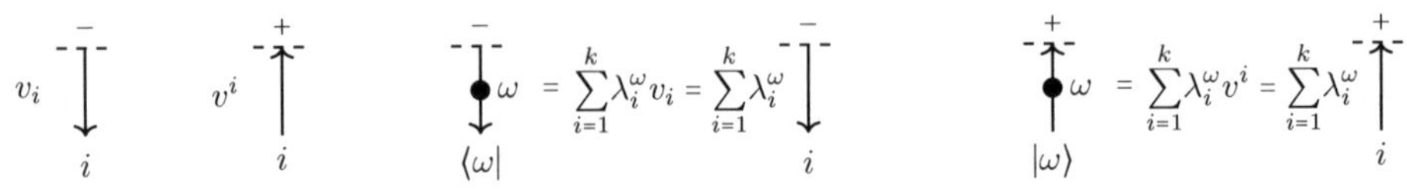

FIGURE 2.1.5. Left: basis vectors v_i and v^i. Middle and right: writing $\langle\omega| \in A(-)$ and $|\omega\rangle \in A(+)$ in these bases. In the special case $\omega = \varnothing$, the half-intervals are undecorated and the corresponding equations are shown in (9).

There is a minimalist way to extend the above structure of $A(+)$ with an action of Σ^*, a cyclic vector and a trace to a symmetric monoidal category $\mathcal{C}'_{\alpha_{\mathsf{I}}}$, which turns out to be a TQFT with defects. In this construction evaluation of decorated circles is derived from that for decorated intervals. To define $\mathcal{C}'_{\alpha_{\mathsf{I}}}$, first enhance the graphical calculus for decorated half-intervals by picking a basis $\{v^1, \dots, v^k\}$ of $A(+)$ and the dual basis $\{v_1, \dots, v_k\}$ of $A(-)$, with $k = \dim A(-) = \dim A(+)$. Denote vectors v^i, v_i by placing the label i at the end of the suitably oriented half-interval, see Figure 2.1.5, left. A half-interval decorated by $\omega \in \Sigma^*$ can be written as a linear combination of the basis vectors, by writing $\langle\omega|$ and $|\omega\rangle$ in these bases, see Figure 2.1.5, middle and right. In the special case of undecorated half-intervals, $\omega = \varnothing$, and

$$\langle\varnothing| = \sum_{i=1}^{k} \lambda_i^{\varnothing} v_i, \quad |\varnothing\rangle = \sum_{i=1}^{k} \lambda_i^{\varnothing} v^i, \quad \lambda_i^{\varnothing} \in \mathbf{k}. \tag{9}$$

There is then the surgery formula (the dual basis relation), shown in Figure 2.1.6 left, for cutting any half-interval in the middle, where any floating intervals that appear are evaluated via α_{I}. This formula also tells us that, for consistency, a circle carrying word ω should be evaluated to the trace of ω acting on $A(+)$ (equivalently, on $A(-)$),

$$\alpha_{\circ}(\omega) := \mathsf{tr}_{A(+)}(\omega) = \mathsf{tr}_{A(-)}(\omega), \tag{10}$$

see Figure 2.1.7. Choose an interval on the circle, replace it by the sum of i,i-colored half-intervals, $1 \le i \le k$, and evaluate via α_{I}. We denote the circular series associated to a rational series α_{I} in this way by $\alpha_{\mathsf{I}}^{\mathsf{tr}}$, so that

$$\alpha_{\mathsf{I}}^{\mathsf{tr}}(\omega) = \alpha_{\circ}(\omega) = \mathsf{tr}_{A(+)}(\omega). \tag{11}$$

Necessarily, $\alpha_{\mathsf{I}}^{\mathsf{tr}}$ is a rational circular series in the same set Σ of variables as α_{I}. The coefficient of of the empty word $\varnothing$ in this series equals $\dim A(+)$.

To a rational series α_{I} we assign a symmetric monoidal category $\mathcal{C}'_{\alpha_{\mathsf{I}}}$ whose objects are finite sign sequences ε. Morphisms from ε to ε' are $\mathbf{k}$-linear combinations of diagrams of decorated half-intervals and outer arcs, see Figure 2.1.8, left diagram. An arc is *outer* if it has both endpoints on the boundary of the diagram (*i.e.*, among the signed points of ε and ε'). A half-interval has one floating (inner) endpoint and one endpoint on the boundary of the diagram (outer endpoint). Each floating endpoint is either decorated by $i \in \{1, \dots, k\}$ or undecorated. Some evaluation rules for floating cobordisms are given in Figures 2.1.4, 2.1.6, 2.1.7.

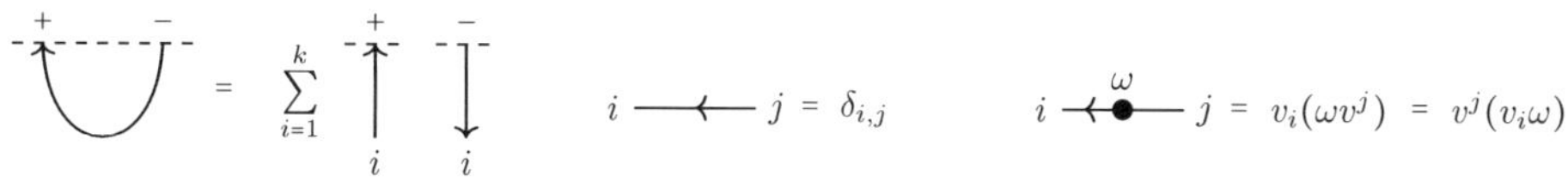

FIGURE 2.1.6. Left: surgery formula $\mathrm{id} = \sum_{i=1}^{k} v^i \otimes v_i$ in the symmetric monoidal category $\mathcal{C}'_{\alpha_I}$. Center: floating i,j-interval evaluates to $(v_i, v^j) = \delta_{i,j}$. Right: a more general floating i,j-interval with the ω dot and its evaluation.

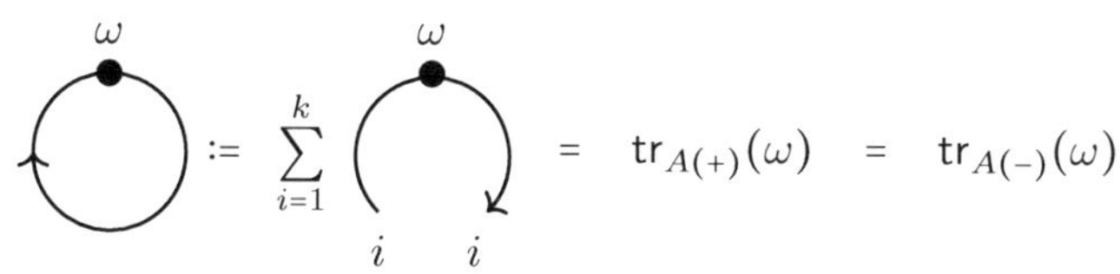

FIGURE 2.1.7. An ω-decorated circle evaluates to the trace of operator ω on $A(+)$ and on $A(-)$.

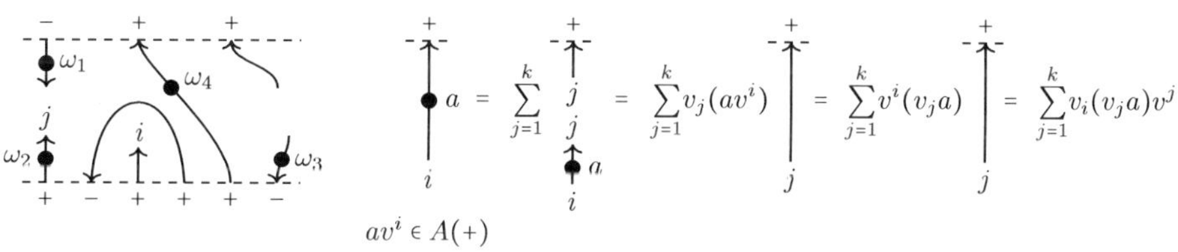

FIGURE 2.1.8. Left: A diagram for a morphism in $\mathcal{C}_{\alpha_I}$. Right: simplifying an a-decorated half-interval, $a \in \Sigma$ (this works, more generally, for any word ω in place of a letter a). Here $v_j(av^i) = (v_j, av^i) = (v_j a, v^i) = v^i(v_j a)$.

To summarize these rules, we observe that in the category $\mathcal{C}'_{\alpha_I}$ arcs and half-intervals can be decorated by words $\omega \in \Sigma^*$. Half-intervals can be decorated by both a label i at the floating (inner) endpoint and words ω. Floating intervals and circles (these appear upon composition of morphisms) are evaluated using the following rules (see rules in Figure 2.1.5 on the right, Figures 2.1.6 and 2.1.7):

- A floating interval with unlabelled endpoints and decorated by word ω evaluates to $\alpha_I(\omega)$. Alternatively, an interval with one or two unlabelled endpoints is evaluated by first converting unlabelled endpoints into a linear combination of labelled endpoints, see equation (9), or, more generally via Figure 2.1.5 by writing $\langle\omega|$ as a linear combination of $v_1, \ldots, v_k$ and $|\omega\rangle$ as a linear combination of $v^1, \ldots, v^k$.
- An interval with labelled endpoints i,j and decorated by a word ω evaluates to $v^i(v_j\omega)$, see Figure 2.1.6.
- A circle decorated by ω evaluates to the trace of ω on $A(+)$ or $A(-)$, see Figure 2.1.7.

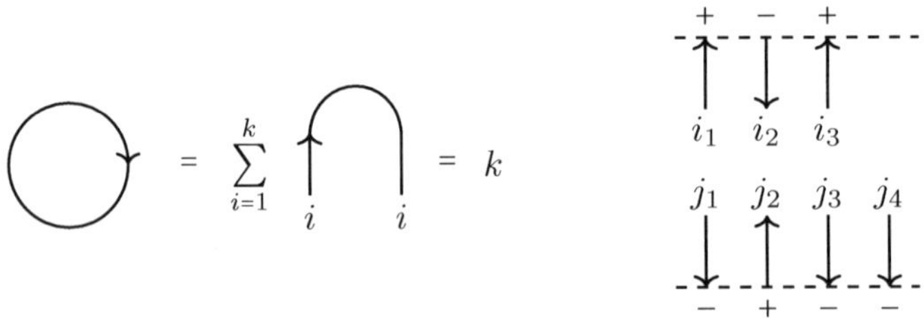

FIGURE 2.1.9. Left: an undecorated circle evaluates to $k = \dim A(+)$ in $\mathcal{C}'_{\alpha_{\mathsf{I}}}$. Right: a basis element of $\mathsf{Hom}_{\mathcal{C}'_{\alpha_{\mathsf{I}}}}(-+--,+-+)$, $i_1,\dots,j_4 \in \{1,\dots,k\}$.

With these evaluation rules at hand, we apply the universal construction to build the category $\mathcal{C}'_{\alpha_{\mathsf{I}}}$. The evaluation rule for decorated circles makes decomposition of the identity in Figure 2.1.6 hold. Consequently, any outer arc reduces to a linear combination of half-intervals, with endpoint decorated by $i \in \{1,\dots,k\}$. Half-intervals with dots are further reduced to linear combinations of endpoint-decorated dotless half-intervals.

Composing an undecorated outer arc with a possibly decorated half-interval results in a half-interval with the same decoration. Composing two half-intervals results in a floating interval, which is then evaluated via α_{I}. An undecorated outer arc can be written as a linear combination of pairs of half-intervals via Figure 2.1.6 relations. In particular, an undecorated circle evaluates to $k = \dim A(+)$, see Figure 2.1.9 left.

The result is a symmetric monoidal $\mathbf{k}$-linear category $\mathcal{C}'_{\alpha_{\mathsf{I}}}$. It has easily describable hom spaces. A basis of $\mathsf{Hom}(\varepsilon,\varepsilon')$ is given by drawing the unique diagram of half-intervals ending at all signs of ε and ε' and adding all possible labels $i \in \{1,\dots,k\}$ to each inner endpoint of the diagram, see an example in Figure 2.1.9 on the right. In particular,

$$\dim_{\mathbf{k}}(\mathsf{Hom}(\varepsilon,\varepsilon')) = k^{|\varepsilon|+|\varepsilon'|}, \tag{12}$$

where $|\varepsilon|$ is the length of the sequence ε. This category is a one-dimensional TQFT with defects, in the sense that the state space of the concatenation of sequences is the tensor product of state spaces for individual sequences:

$$A(\varepsilon\varepsilon') \cong A(\varepsilon)\otimes A(\varepsilon'), \quad A(\varepsilon) \cong A(\epsilon_1)\otimes A(\epsilon_2)\otimes\cdots\otimes A(\epsilon_n), \tag{13}$$

where $\varepsilon = \epsilon_1\epsilon_2\cdots\epsilon_n$, $\epsilon_i \in \{+,-\}$.

We emphasize that the category $\mathcal{C}'_{\alpha_{\mathsf{I}}}$ depends only on the interval evaluation α_{I}. Evaluation of decorated circles is computed as the trace of the action of Σ^* on the state space $A(+)$ associated to α_{I}.

Denote by $\mathsf{Kar}(\mathcal{C}'_{\alpha_{\mathsf{I}}})$ the Karoubi envelope of $\mathcal{C}'_{\alpha_{\mathsf{I}}}$ given by allowing finite direct sums of objects of $\mathcal{C}'_{\alpha_{\mathsf{I}}}$ and then passing to the idempotent closure. Category $\mathsf{Kar}(\mathcal{C}'_{\alpha_{\mathsf{I}}})$ is an additive symmetric monoidal $\mathbf{k}$-linear category.

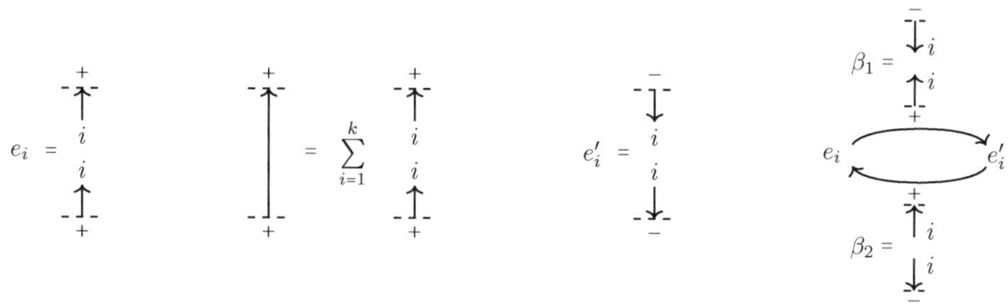

FIGURE 2.1.10. Left to right: idempotent $e_i \in \mathsf{End}_{\mathcal{C}_\alpha}(+)$, idempotent decomposition of id_+, idempotent $e_i' \in \mathsf{End}_{\mathcal{C}_\alpha}(-)$, equivalence of idempotents e_i, e_i' (morphisms β_1, β_2 are written at top, respectively, bottom arrows).

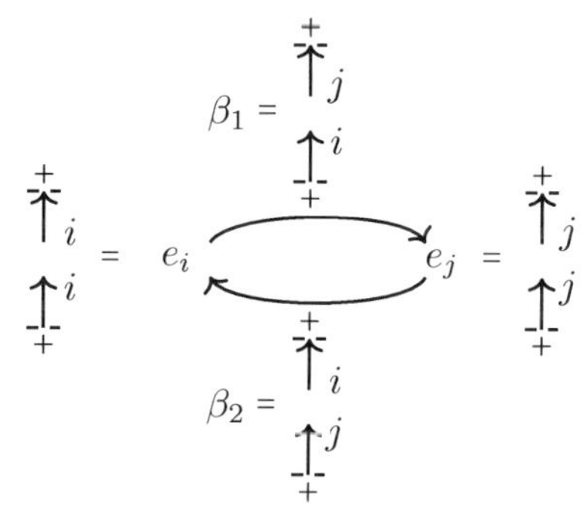

FIGURE 2.1.11. Equivalence of idempotents e_i and e_j and isomorphism of objects $(+, e_i), (+, e_j)$, with morphisms β_1, β_2 written next to top and bottom arrows, respectively.

Consider idempotents $e_i \in \mathsf{End}_{\mathcal{C}'_{\alpha_I}}(+)$ given by a pair of i-labelled half intervals, $i = 1, \dots, k$, see Figure 2.1.10 on the left. These are mutually-orthogonal idempotents giving a decomposition of the identity id_+ endomorphism

$$\mathsf{id}_+ = e_1 + e_2 + \ldots + e_k, \qquad e_i e_j = \delta_{i,j} e_i, \tag{14}$$

see Figure 2.1.10 on the right. There is a similar decomposition of the identity for the dual object – via idempotents $e_1', \dots, e_k'$, see Figure 2.1.10.

Note that the dual simple object $(-, e_i')$, see Figure 2.1.10 in the middle, is isomorphic to $(+, e_i)$, via the pair of morphisms show in that figure on the right. Recall that idempotent endomorphisms e, e' are *equivalent* (and corresponding objects of the Karoubi envelope are isomorphic) if there exist two-way composable morphisms β_1, β_2 such that $e = \beta_2\beta_1$ and $e' = \beta_1\beta_2$. For idempotents e_i, e_i' these two morphisms are written next to the arrows between these idempotents in Figure 2.1.10 on the right.

Furthermore, objects $(+, e_i)$ and $(+, e_j)$ are isomorphic, for $i \neq j$, $1 \leq i, j \leq k$, via the morphisms shown in Figure 2.1.11.

The endomorphism rings of objects $(+, e_i)$ and $(-, e_i')$ are the ground field $\mathbf{k}$, and these $2k$ objects, over $i = 1, \ldots, k$, are pairwise isomorphic. The generating object $+$ is the sum of k of them. This quickly leads to the following result that $\mathsf{Kar}(\mathcal{C}'_{\alpha_\mathsf{I}})$ generated by $(+, e_i)$, $(-, e_i')$ over $i = 1, \ldots, k$ is equivalent to the tensor category $\mathbf{k}\text{-}\mathsf{vect}$ of finite-dimensional $\mathbf{k}$-vector spaces.

PROPOSITION 2.1. The Karoubi envelope of $\mathcal{C}'_{\alpha_\mathsf{I}}$ is equivalent, as an additive symmetric monoidal category, to the category of finite-dimensional $\mathbf{k}$-vector spaces:

$$\mathsf{Kar}(\mathcal{C}'_{\alpha_\mathsf{I}}) \cong \mathbf{k}\text{-}\mathsf{vect}. \tag{15}$$

PROOF. The equivalence is given by the functor that takes each $(+, e_i)$ to a one-dimensional vector space V_i, each $(-, e_i')$ to its dual V_i^* and takes $+$ to the k-dimensional space $V = V_1 \oplus \ldots \oplus V_k$. □

The proposition tells us that $\mathsf{Kar}(\mathcal{C}'_{\alpha_\mathsf{I}})$ has a very simple structure. The complexity of noncommutative rational power series α_I is hidden in the action of Σ^* on $A(+) \cong V$. More generally, given a TQFT taking values in $\mathbf{k}$-vector spaces, the target category of that TQFT is $\mathbf{k}\text{-}\mathsf{vect}$ or some variation of it, so a version of the above proposition holds as well.

The interesting question here is to *explicitly* compute the circular evaluation $\alpha_\mathsf{I}^{\mathsf{tr}}$ given a rational interval evaluation α_I or, equivalently, rational noncommutative power series α_I. We obtain $\alpha_\mathsf{I}^{\mathsf{tr}}$ by considering $A(+)$ and the action of $\mathbf{k}\Sigma^*$ on it, so that the coefficient at ω of the noncommutative circular series of $\alpha_\mathsf{I}^{\mathsf{tr}}$ is the trace of ω on $A(+)$, see (10), and the generating function of the circular evaluation is

$$Z_{\alpha_\mathsf{I}}^{\mathsf{tr}} := \sum_{\omega \in \Sigma^*} \mathsf{tr}_{A(+)}(\omega)\,\omega. \tag{16}$$

For more than one variable, Z_{α_I} is a noncommutative power series in elements of Σ. Circular evaluation $\alpha_\mathsf{I}^{\mathsf{tr}}$ gives a canonical extension of α_I to a TQFT with defects. This extension is unique, in appropriate sense. It is straightforward to write down in the 1-variable case, see Section 2.2.

Given a $\mathbf{k}[\Sigma^*]$-module M, finite-dimensional over $\mathbf{k}$, its characteristic function $\chi(M)$, defined in [**RRV99**], is given by

$$\chi(M) := \sum_{\omega \in \Sigma^*} \mathsf{tr}_M(\omega)\omega. \tag{17}$$

Expression (16) is the characteristic function of $\mathbf{k}\Sigma^*$-module $A(+)$.

2.2. One-variable case and comparison to two-dimensional theory. Consider the above construction in the case of a single variable, $\Sigma = \{a\}$. Then there is only one type of a dot, necessarily labelled a, and n dots on an interval can be denoted by a single dot labelled n. The interval evaluation in encoded by a one-variable power series

$$Z_\mathsf{I}(T) := \sum_{n \geq 0} \alpha_{\mathsf{I},n} T^n. \tag{18}$$

A well-known theorem, see [**Kho20a**, Proposition 2.1] and [**Kho20b**, Theorem 2.3], says that $Z_\mathsf{I}(T)$ is a rational series ($A(+)$ is finite-dimensional) if and only if it is a rational function,

$$Z_\mathsf{I}(T) = \frac{P(T)}{Q(T)}, \tag{19}$$

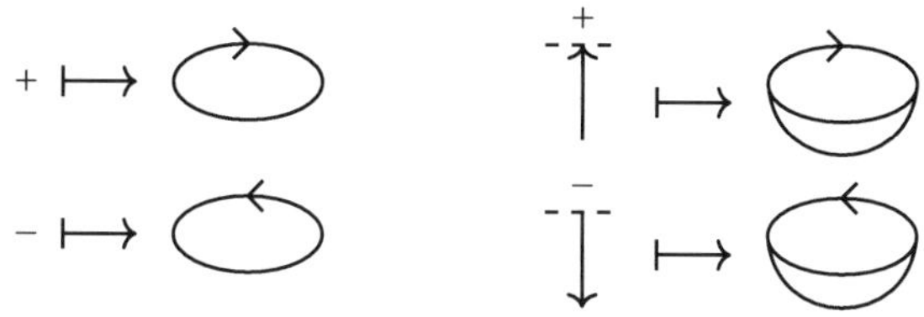

FIGURE 2.2.1. Going from a 1D to a 2D topological theory. Signed points become oriented circles and half-intervals turn into oriented disks.

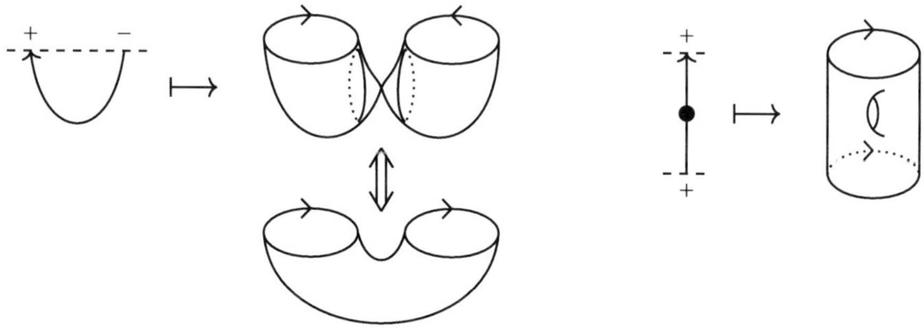

FIGURE 2.2.2. Left: an arc is converted into an annulus. Due to the opposite orientations at the arc's ends, the standard embedding of two oppositely oriented circles into $\mathbb{R}^2$ extends only to an immersion, not an embedding, of an annulus into the lower half-space. This is not an issue for us due to considering 2D cobordisms, not embedded 2D cobordisms. Right: An arc with a dot is converted to a torus with two punctures.

for some polynomials $P(T), Q(T)$, with $Q(0) \neq 0$.

We would like to explicitly compute $\alpha_{\mathsf{I}}^{\mathsf{tr}}$ in this case, given the generating function above (equivalently, given interval evaluation α_{I}).

The state space $A(+)$ can be identified with the state space $A(1)$ in [**Kho20b**, Section 2] of a circle in the 2D topological theory [**Kho20b**] associated to the same generating function $Z_{\mathsf{I}}(T)$. In that 2D topological theory, closed connected oriented surface of genus n evaluates to $\alpha_{\mathsf{I},n} \in \mathbf{k}$, while in our 1D defect topological theory an interval with n dots evaluates to $\alpha_{\mathsf{I},n}$, see the correspondence in Figure 2.2.3 right.

The reason is that there is a functor from the category of (oriented) dotted one-cobordisms to the category of (oriented) two-cobordisms. This functor sends $+$ (and $-$) to an oriented circle. It sends a half-interval to a disk, a dotless arc to an annulus, and an outer arc with a single dot to a two-holed torus, see Figures 2.2.1 and 2.2.2.

There is a bijection between homeomorphism classes of decorated connected dotted 1-manifolds with boundary $+$ and connected oriented surfaces with boundary

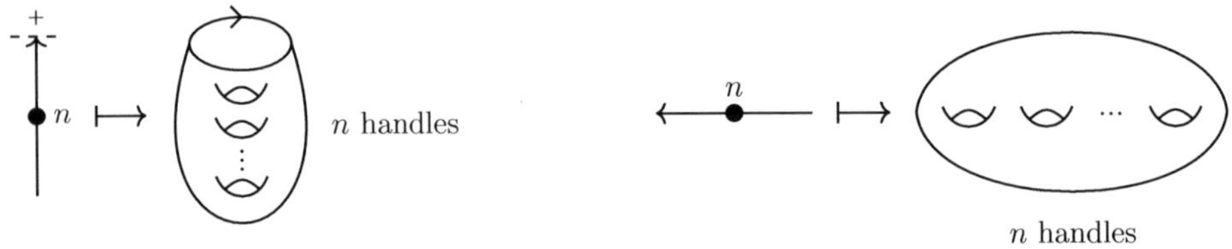

FIGURE 2.2.3. Half-interval with n dots becomes a genus n surface with one boundary circle. An n-dotted interval is transformed into a closed genus n surface.

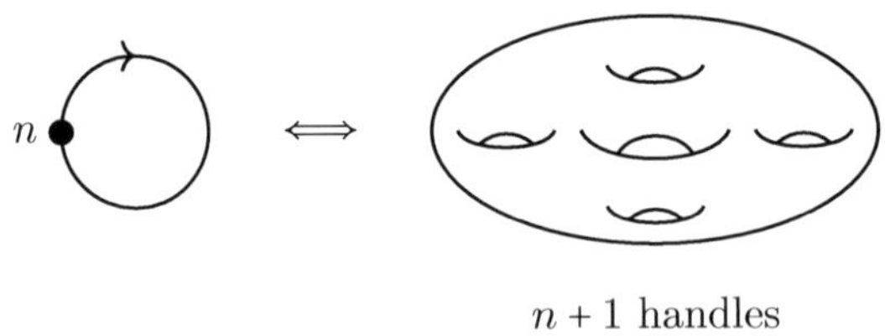

FIGURE 2.2.4. A circle with n dots is mapped to a genus $(n+1)$ surface.

$\mathbb{S}^1$, see Figure 2.2.3 left, where n-dotted half-interval corresponds to a genus n surface with one boundary circle.

This leads to a natural isomorphism of state spaces

$$A(+) \cong A(\mathbb{S}^1) \tag{20}$$

of the 1D topological theory with defects with a rational generating function in (18) for interval evaluation and the 2D topological theory with the same generating function. Notice also a canonical isomorphism $A(+) \cong A(-)$ sending upward-oriented half-interval with n dots to the downward-oriented half-interval with n dots, $n \geq 0$ (this isomorphism exists when $|\Sigma| = 1$ and otherwise requires α_{I} to be invariant under word reversal).

One can look to compare the state spaces for these two theories (in two different dimensions) beyond a single point and a circle. In the above thickening construction, a circle with n dots corresponds to a genus $n+1$ closed surface, see Figure 2.2.4, so for the best match we pick the circular series in the 1D theory to be

$$Z_{\alpha_\circ}(T) = TZ_{\alpha_{\mathsf{I}}}(T), \quad \alpha_{\circ,n+1} = \alpha_{\mathsf{I},n}, \ n \geq 0. \tag{21}$$

Then there is a natural $\mathbf{k}$-linear map

$$A(+-) \xrightarrow{\psi_{+-}} A(2) \tag{22}$$

from the state space of $+-$ in the 1D theory to that of two circles in the 2D theory, with the generating functions (18), (21) for the 1D theory and (18) for the 2D theory, given by the above thickening of dotted 1D cobordisms to 2D cobordisms. This map respects evaluations and is, in fact, an isomorphism, so that $A(+-) \cong A(2)$ in the two theories.

More generally, for any sign sequence ε there is natural map

$$A(\varepsilon) \xrightarrow{\psi_\varepsilon} A(|\varepsilon|) \tag{23}$$

extending to a monoidal functor ψ between corresponding categories for 1D and 2D evaluations. Map ψ_ε is not surjective, for instance, for $|\varepsilon| = 3$ and the constant evaluation function $Z_{\mathsf{I}}(T) = \beta \in \mathbf{k}$, $\beta \neq 0$. Element of $A(3)$ which is the 3-holed sphere is not in the image of ψ_ε, for any length 3 sign sequence ε. Also, ψ_{++} is not, in general, surjective, with the annulus element of $A(2)$ not in its image.

We now come back to the problem of computing $\alpha_{\mathsf{I}}^{\mathsf{tr}}$ given α_{I}, for one-element $\Sigma = \{a\}$. Let us first examine two special cases.

- The generating function $Z_{\mathsf{I}}(T) = \sum_{i=0}^k a_i T^i$ is a polynomial of degree k. Then $\dim A(+) = k+1$ and the operator of multiplication by a is nilpotent. Consequently, circular evaluation $\alpha_\circ$ is the constant function, taking value 0 on any nonzero power of a, with the generating function $Z_\circ(T) = k+1$.
- The generating function is a reduced fraction of the form

$$Z_{\mathsf{I}}(T) = \frac{f(T)}{(\lambda - T)^k}, \quad k \geq 1, \quad \lambda \neq 0, \quad \deg(f(T)) < k. \tag{24}$$

  Then $A(+)$ is a cyclic $\mathbf{k}[a]$-module isomorphic to $\mathbf{k}[a]/((\lambda a - 1)^k)$, where we quotient by the reciprocal polynomial of $(\lambda - a)^k$, see [**Kho20a**]. Note that trace of a^m on this quotient space does not depend on $f(T)$ above, subject to the conditions in (24). Substituting $u = \lambda a - 1$, so that $a = \frac{1}{\lambda}(u+1)$, the trace of a^m on $\mathbf{k}[u]/(u^k)$ is given by

$$\mathsf{tr}(a^m) = \lambda^{-m} k. \tag{25}$$

A rational function, over an algebraically closed field $\mathbf{k}$, has a unique partial fraction decomposition

$$Z_{\mathsf{I}}(T) = \frac{P(T)}{Q(T)} = \sum_{i=1}^r \frac{f_i(T)}{(\lambda_i - T)^{k_i}} + f_0(T), \quad \lambda_i \neq 0, \ \deg(f_i(T)) < k_i, \ i = 1, \dots, r. \tag{26}$$

Then the trace circular series associated with this generating function is

$$\begin{aligned} Z_{\alpha_{\mathsf{I}}}^{\mathsf{tr}}(T) &= \deg(f_0) + 1 + \sum_{m\geq 0} \left(\sum_{i=1}^r k_i \lambda_i^{-m} T^m \right) \\ &= \deg(f_0) + 1 + \sum_{i=1}^r k_i \sum_{m\geq 0} (\lambda_i^{-1} T)^m \\ &= \deg(f_0) + 1 + \sum_{i=1}^r \frac{k_i}{1 - \lambda_i^{-1} T} \end{aligned} \tag{27}$$

If $Z_{\mathsf{I}}(T)$ in (26) is a proper fraction, that is, $f_0(T) = 0$, we set $\deg(f_0) + 1 = 0$ in (27). Note that the characteristic polynomial for the trace series is a divisor of the characteristic polynomial for the original series.

If eigenvalues of a on $A(+)$ are $\mu_1, \dots, \mu_n$, listed with multiplicities, then

$$Z_{\alpha_{\mathsf{I}}}^{\mathsf{tr}}(T) = \sum_{i=1}^n \frac{1}{1 - \mu_i T}, \tag{28}$$

which is [**RRV99**, Example 2.6].

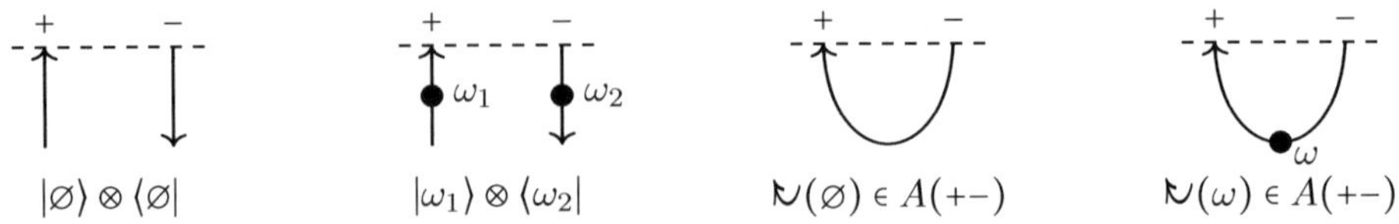

FIGURE 2.3.1. A spanning set for the module $A(+-)$. First and third figure (from the left) are a generator $|\varnothing\rangle \otimes \langle\varnothing|$ of $A(+) \otimes A(-)$ and the unit element 1 of $A(+-)$, respectively.

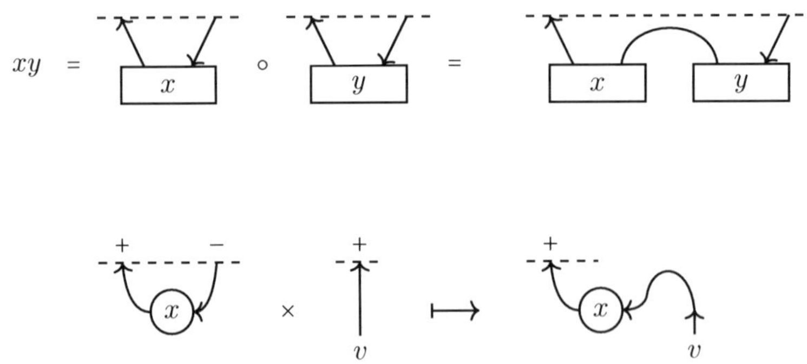

FIGURE 2.3.2. Top: multiplication in $A(+-)$. Bottom: action of $A(+-)$ on $A(+)$. Bottom: action of $A(+-)$ on $A(+)$.

It is an interesting problem to explicitly write down noncommutative trace series $Z^{\mathrm{tr}}_{\alpha_{\mathsf{I}}}(\Sigma)$ associated with an arbitrary rational noncommutative series $Z_{\alpha_{\mathsf{I}}}(\Sigma)$ when the number of variables $|\Sigma|$ is greater than one.

2.3. A topological theory when a circular series is added. To build a more general monoidal category, we additionally pick a circular rational series $\alpha_\circ$, see [**Kho20a, IK22**]. Here $\alpha = (\alpha_{\mathsf{I}}, \alpha_\circ)$ is a pair: a rational noncommutative series α_{I} and a circular rational noncommutative series $\alpha_\circ$. We build category $\mathcal{C}_\alpha$ from it as in Section 2.1 by evaluating floating decorated intervals and circles via α_{I} and $\alpha_\circ$ correspondingly and applying the universal construction to derive further relations on linear combinations of cobordisms with outer boundary, see also [**Kho20a, Kho20b, KS20**]. As before, objects of $\mathcal{C}_\alpha$ are finite sign sequences ε.

In the category $\mathcal{C}_\alpha$, state spaces $A(+), A(-)$ depend only on α_{I} and they are spanned by elements $|\omega\rangle$, $\omega \in \Sigma^*$, respectively $\langle\omega|$, $\omega \in \Sigma^*$. State space $A(+-)$ is spanned by diagrams of two types:

(1) pairs of decorated half-intervals with opposite orientations,
(2) decorated outer arcs,

see Figure 2.3.1.

$A(+-)$ is naturally a unital associative finite-dimensional algebra, with the unit element given by an outer arc with the trivial decoration and with the multiplication shown in Figure 2.3.2. Algebra $A(+-)$ acts on $A(+)$ on the left, see Figure 2.3.2.

Denote by $\mathcal{I}$ the subspace of $A(+-)$ spanned by diagrams of type (1), see Figure 2.3.1 second from left picture. This subspace is a two-sided ideal of $A(+-)$

and a unital $\mathbf{k}$-algebra with the unit element $1' = \sum_{i=1}^{k} v^i \otimes v_i$ shown in Figure 2.1.6 on the right hand side of the equality. Note that the equality $1' = 1$ fails unless $\alpha_\circ = \alpha_\mathsf{I}^{\mathsf{tr}}$. In general, the right hand side diagram is the unit element of $A(+-)$ and the left hand side is the idempotent $1'$.

There is a natural algebra isomorphism

$$\mathcal{I} \cong A(+) \otimes A(-) \cong \mathsf{End}(A(+)) \tag{29}$$

coming from the faithful action of $\mathcal{I}$ on $A(+)$, given by restricting the action from that of $A(+-)$.

The kernel $\mathcal{K}$ of the action of $A(+-)$ on $A(+)$ is a two-sided ideal of $A(+-)$, complementary to $\mathcal{I}$, giving a direct product decomposition

$$A(+-) \cong \mathcal{I} \times \mathcal{K}. \tag{30}$$

In this decomposition both terms on the right are unital $\mathbf{k}$-algebras, with the unit element of $\mathcal{K}$ given by the image of $1 \in A(+-)$ under the projection, that is, by

$$1_\mathcal{K} := 1 - 1' = 1 - \sum_{i=1}^{k} v^i \otimes v_i. \tag{31}$$

In particular, $1_\mathcal{K}$ generates $\mathcal{K}$ as an $A(+-)$-bimodule.

Denote by U the subspace of $A(+-)$ spanned by diagrams of type (2), that is, by decorated arcs connecting $+$ and $-$ boundary points, see the picture on the right in Figure 2.3.1. Recall that we denote by $\cup(\omega)$ the arc decorated by ω, see Figure 2.3.1 on the right. U is a unital subalgebra of $A(+-)$ and there are algebra inclusions

$$U \subset A(+-) \supset A(+) \otimes A(-) \cong \mathcal{I}. \tag{32}$$

Subalgebra U surjects onto $\mathcal{K}$ upon projection to the second term in the direct product (30), and there is a short exact sequence

$$0 \longrightarrow U \cap \mathcal{I} \longrightarrow U \longrightarrow \mathcal{K} \longrightarrow 0. \tag{33}$$

The first term

$$U' := U \cap \mathcal{I} \tag{34}$$

is a two-sided ideal of U. Elements in U' are linear combinations of decorated arcs (elements of U) that decompose in $A(+-)$ into linear combinations of pairs of half-intervals. These decompositions are unique as elements of $A(+) \otimes A(-) \cong \mathcal{I} \subset A(+-)$.

This data carries a triple of discrete invariants:

$$(\dim A(+), \dim(U'), \dim \mathcal{K}), \tag{35}$$

which are three non-negative integers. Note that $\dim A(-) = \dim A(+)$, $\dim \mathcal{I} = (\dim A(-))^2$ and $\dim(U') \leq (\dim A(+))^2$.

REMARK 2.2. The natural inclusion $A(+) \otimes A(-) \subset A(+-)$ is an isomorphism if and only if $\mathcal{K} = 0$. This is exactly the case when there is a decomposition of the identity, that is, when the undecorated arc $1 \in A(+-)$ lies in $\mathcal{I} \cong A(+) \otimes A(-)$, that is, when 1 is *decomposable* (also when $1 = 1'$, see earlier). In this case $A(+-) \cong A(+) \otimes A(-)$ and, more generally, $A(\varepsilon\varepsilon') \cong A(\varepsilon) \otimes A(\varepsilon')$ for any sign sequences $\varepsilon, \varepsilon'$.

Equivalently, the category $\mathcal{C}_\alpha$ gives a TQFT rather than just a topological theory (with defects) if and only if $\mathcal{K} = 0$. This is possible for a unique rational circular series $\alpha_\circ = \alpha_\mathsf{I}^{\mathsf{tr}}$ associated with α_I and with the identity decomposition determined by α_I, see formula (11). Circular evaluation $\alpha_\circ$ is then the trace of

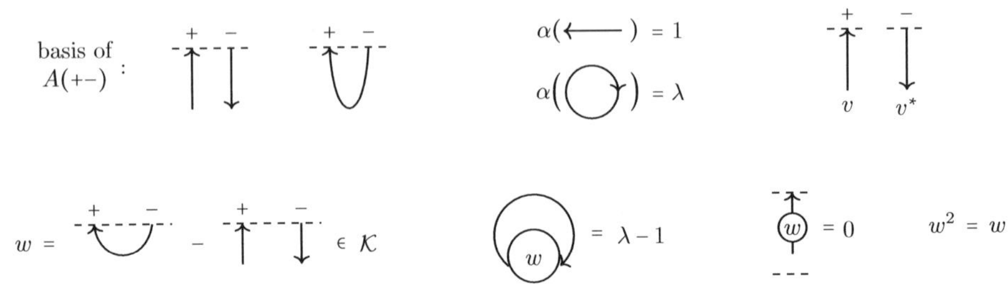

FIGURE 2.3.3. Example 2.3. Top, left to right: a basis of $A(+-)$, evaluation α, vectors v and v^*. Bottom, left to right: spanning vector w of $\mathcal{K}$ and its properties.

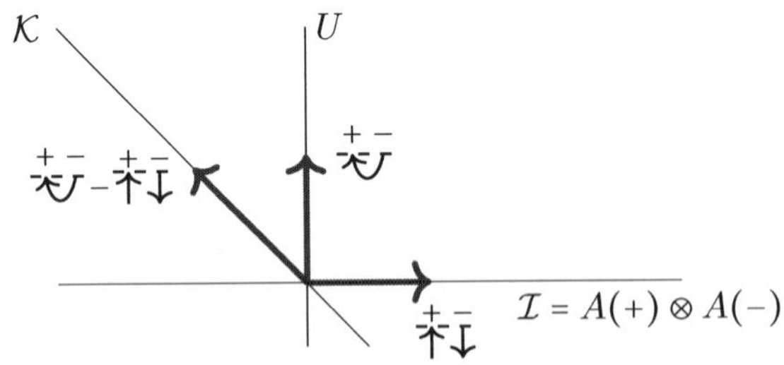

FIGURE 2.3.4. Example 2.3. One-dimensional subspaces $\mathcal{I}$, U and $\mathcal{K}$ in $A(+-)$, with $\dim A(+-) = 2$.

action of words on $A(+)$, see formula (16), with $\alpha_\circ(\omega) = \alpha_{\mathsf{I}}^{\mathsf{tr}}(\omega) = \mathsf{tr}_{A(+)}(\omega)$ for $\omega \in \Sigma^*$. The resulting theory $\alpha = (\alpha_{\mathsf{I}}, \alpha_\circ)$ is a TQFT.

The Boolean version of the identity decomposition (which requires $A(+)$ to be a distributive semilattice) is considered in [**IK22**].

EXAMPLE 2.3. Consider the case when Σ is empty. Then there are only two closed connected cobordisms, undecorated interval and circle. Suppose they evaluate to 1 and $\lambda \neq 1$, respectively, see Figure 2.3.3, top middle. Then $A(+)$ and $A(-)$ are one-dimensional, with basis vectors v and v^*, respectively, see Figure 2.3.3, top right. The space $A(+-)$ is two-dimensional, with the basis shown in Figure 2.3.3, top left. The subalgebra $\mathcal{K}$ is one-dimensional, $\mathcal{K} = \mathbf{k}w$, with a basis element w shown in Figures 2.3.3, bottom left, and 2.3.4. The algebra U is one-dimensional, with the unit element (an arc) as the basis element, see Figure 2.3.4.

In this example $U \cap \mathcal{I} = U \cap \mathcal{K} = 0$, and projection of U onto $\mathcal{K}$ along $\mathcal{I}$ is an algebra isomorphism, See Figure 2.3.4. The triple of parameters is $(1,0,1)$. There is algebra decomposition $A(+-) \cong \mathcal{I} \times \mathcal{K} = \mathbf{k}(1-w) \times \mathbf{k}w$.

When $\Sigma = \{a\}$ is a one-element set, decoration of an interval or a circle is determined by the number $n \geq 0$ of dots on it. If an n-dotted interval evaluates to $\alpha_{\mathsf{I},n}$ and n-dotted circle evaluates to $\alpha_{\circ,n}$, the evaluation is encoded by a pair of one-variable power series in a variable T:

$$Z_{\mathsf{I}}(T) := \sum_{n\geq 0} \alpha_{\mathsf{I},n} T^n, \quad Z_\circ(T) := \sum_{n\geq 0} \alpha_{\circ,n} T^n, \tag{36}$$

FIGURE 2.3.5. Connected closed cobordisms when Σ has cardinality one and their evaluations $\alpha_{\mathsf{I},n}$ and $\alpha_{\circ,n}$.

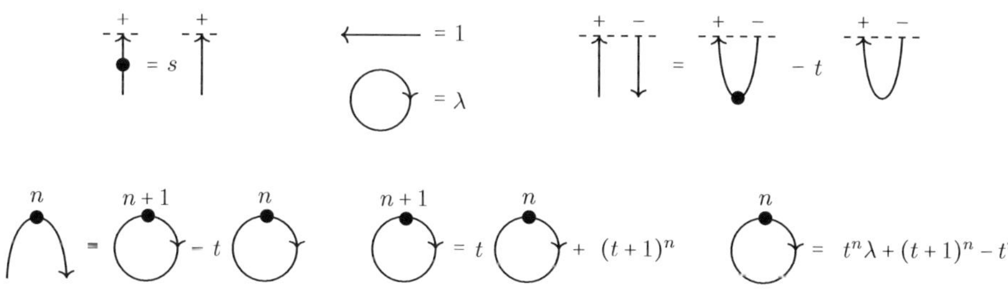

FIGURE 2.3.6. Evaluation (38) and skein relations for that topological theory; $s = t+1$.

see Figure 2.3.5. A simple modification of a result from [**Kho20a**] shows the following.

PROPOSITION 2.4. A one-variable evaluation $\alpha = (\alpha_{\mathsf{I}}, \alpha_\circ)$ is rational (the hom spaces in the category $\mathcal{C}_\alpha$ are finite-dimensional) if and only if both $Z_{\mathsf{I}}(T)$ and $Z_\circ(T)$ are rational functions in T, *i.e.*,

$$Z_{\mathsf{I}}(T) = \frac{P_{\mathsf{I}}(T)}{Q_{\mathsf{I}}(T)}, \qquad Z_\circ(T) = \frac{P_\circ(T)}{Q_\circ(T)}. \tag{37}$$

It is also easy to see that α is a rational evaluation if and only if the state spaces $A(+), A(+-)$ are finite-dimensional.

EXAMPLE 2.5. Let us construct an example with parameters $(1,1,1)$ as in (35). Since $\dim(U) = 2$, take $\Sigma = \{a\}$. Since $\dim A(+) = 1$, the dot acts by some scalar s on the endpoint, see Figure 2.3.6. Pick evaluation of undecorated interval and circle to be 1 and $\lambda \neq 1$, respectively, and introduce parameter t for the skein relation in Figure 2.3.6 top right.

Attaching half-interval at the top right endpoint in each diagram of the relation implies $s = t+1$. Closing up the skein relation by an arc with n dots gives an inductive formula for a circle with n dots. Generating functions (37) are

$$Z_{\mathsf{I}}(T) = \frac{1}{1-(t+1)T}, \qquad Z_\circ(T) = \frac{\lambda - 1}{1-tT} + \frac{1}{1-(t+1)T}. \tag{38}$$

Inductive skein relation to reduce the number of dots is shown in Figure 2.3.7 on the left. Figure 2.3.7 on the right shows the unit element and the basis vector of algebra $\mathcal{K} \cong \mathbf{k}$, equal to $1 - 1'$.

We get a decomposition $A(+-) \cong \mathcal{I} \times \mathcal{K} \cong \mathbf{k} \times \mathbf{k}$ of $A(+-)$ into the product of two copies of the ground field.

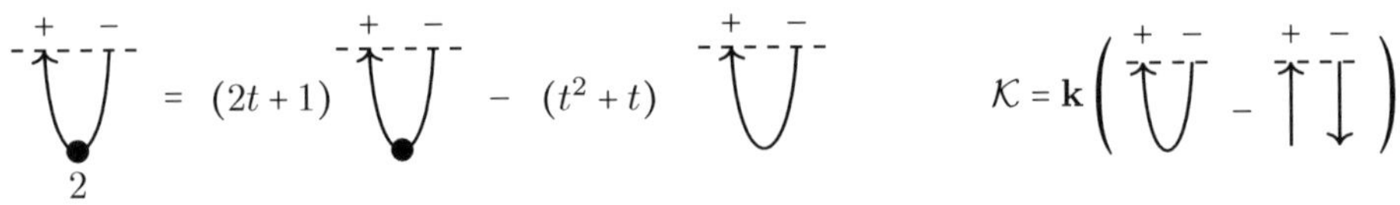

FIGURE 2.3.7. Evaluation (38). Left: we have the relation $a^2 = (2t+1)a-(t^2+t)$, which can also be written as $(a-t)(a-t-1)=0$. Right: $\mathcal{K}$ is a one-dimensional space.

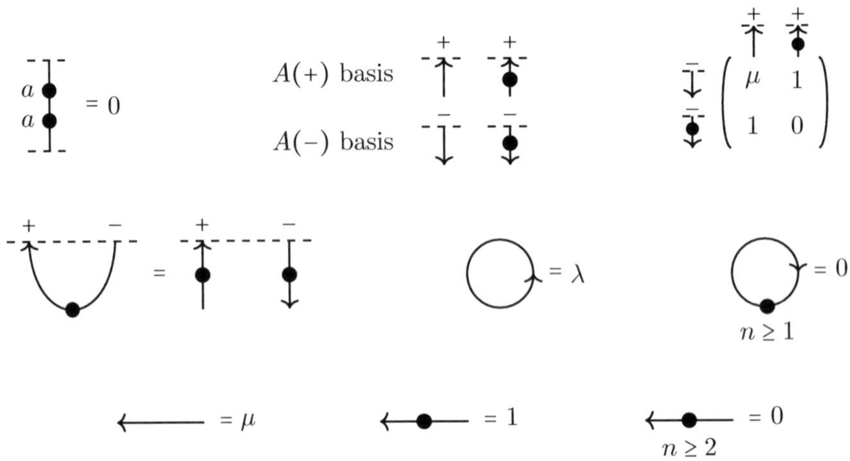

FIGURE 2.3.8. Example 2.6. Evaluation for the generating functions in (39), some relations and bases. Top row: $a^2 = 0$, bases of $A(+)$ and $A(-)$, the matrix of bilinear pairing. Second row: simplification of a dotted arc and circle evaluations. Third row: interval evaluations.

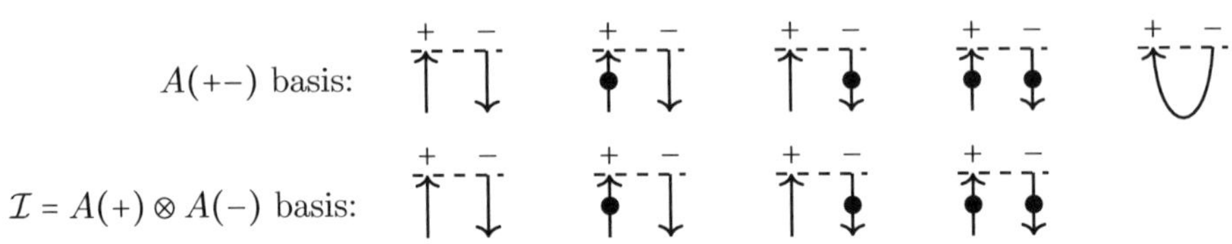

FIGURE 2.3.9. Example 2.6. Bases of $A(+-)$ and $A(+)\otimes A(-)$ for the evaluation (39). In the decomposition $A(+-)\simeq \mathcal{K}\times \mathcal{I}$ factor $\mathcal{K}$ is one-dimensional, spanned by the vector in Figure 2.3.12.

EXAMPLE 2.6. Let us give an example with $\dim A(+) = 2$, $\dim(U') = 1$, $\dim(\mathcal{K}) = 1$ and $U' = I\cap U$, a nilpotent ideal in U. Let $\Sigma = \{a\}$ and make a nilpotent, $a^2 = 0 \in U$, see Figure 2.3.8. Then $\{\langle\varnothing|, \langle a|\}$ is a basis of $A(+)$ and for the bilinear pairing $A(+)\times A(-)\longrightarrow \mathbf{k}$ we can choose the one in Figure 2.3.8, with a parameter $\mu\in\mathbf{k}$. Additionally, choose a relation reducing dot on an arc to a pair of dotted half-intervals. Evaluation α is then shown in Figure 2.3.8 second and bottom rows.

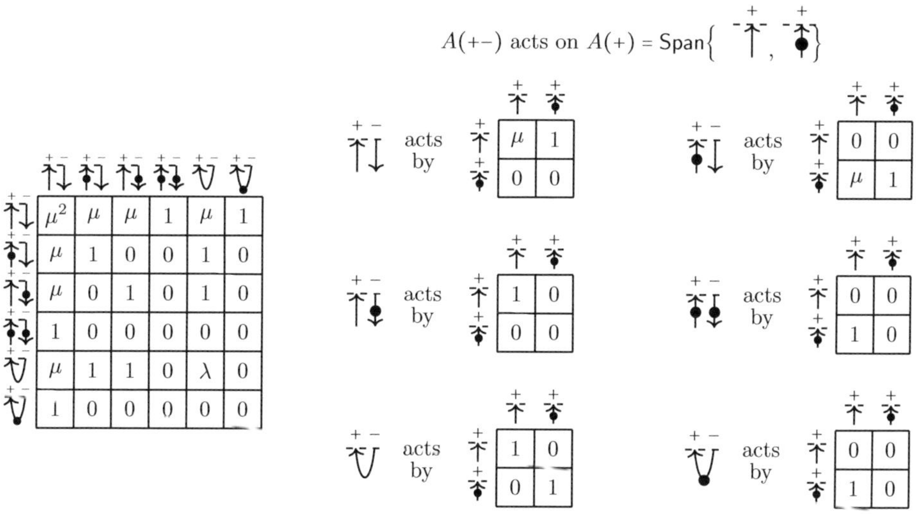

FIGURE 2.3.10. Pairing on $A(+-)$ for Example 2.6. Left: matrix of bilinear pairing. Right: action of $A(+-)$ on $A(+)$.

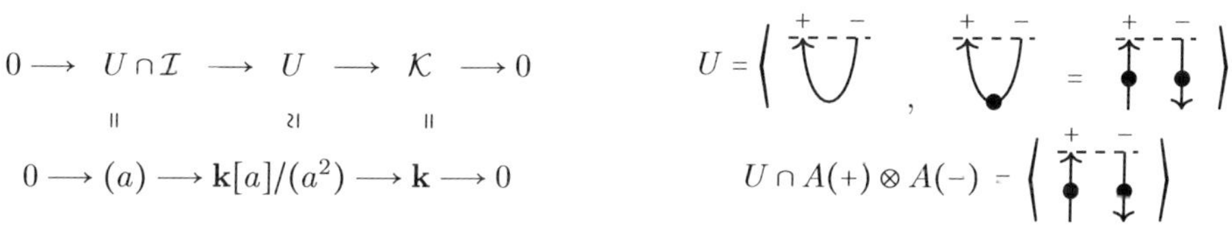

FIGURE 2.3.11. In this short exact sequence, $\dim U = 2$, $\dim \mathcal{K} = 1$, $\dim \mathcal{I} = 4$, and $\dim U \cap \mathcal{I} = 1$. The ideal (a) is nilpotent since $a^2 = 0$. Algebra U is not semisimple and the sequence does not split.

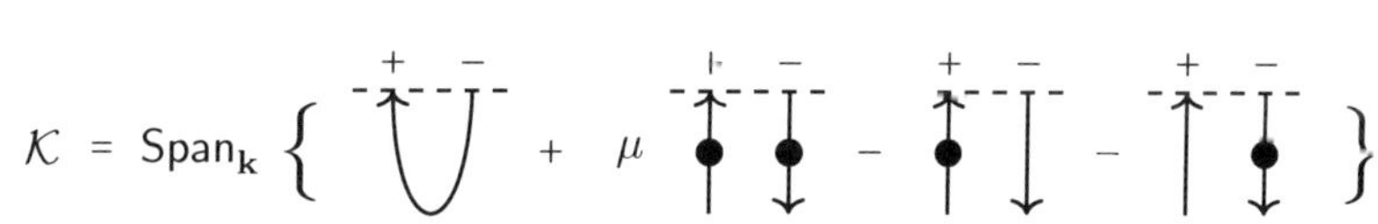

FIGURE 2.3.12. The unit element of $\mathcal{K}$, Example 2.6, with $\dim(\mathcal{K}) = 1$.

Space $A(+-)$ is spanned by the six vectors, with the pairing given in Figure 2.3.10 (rows 4 and 6 are equal and columns 4 and 6 are equal). Dropping the last row and column gives us a basis of $A(+-)$ of cardinality 5. See Figure 2.3.9.

The structure of the short exact sequence (33) for this example is shown in Figure 2.3.11 left. The generating functions are

$$Z_{\mathsf{I}}(T) = \mu + T, \qquad Z_{\circ}(T) = \lambda, \tag{39}$$

see also Figure 2.3.8.

2.4. Karoubi envelope decomposition for arbitrary rational α. We now go back to the case of arbitrary Σ of finite cardinality and a rational pair

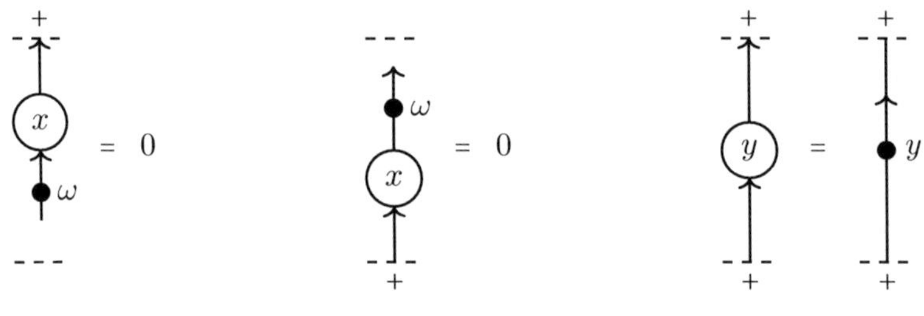

FIGURE 2.4.1. Left: elements $x \in \mathcal{K}$ annihilate $A(+)$ and $A(-)$. Right: two possible notations for $y \in A(+-)$. Also see Figure 2.4.5.

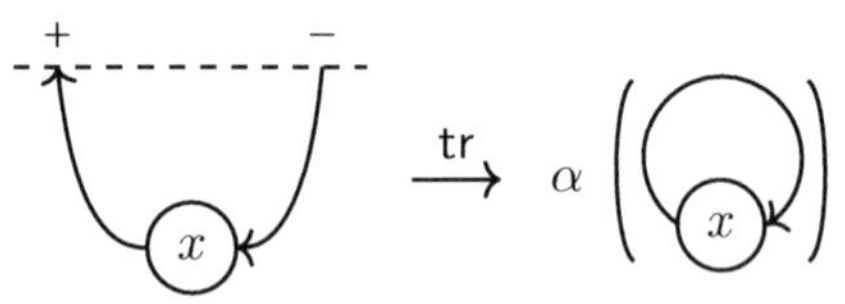

FIGURE 2.4.2. Trace map on $\mathcal{K} \subset A(+-)$.

$\alpha = (\alpha_{|}, \alpha_{\circ})$. Recall the direct product decomposition of $A(+-)$ into $\mathcal{K}$ and the matrix ring $\mathcal{I}$, and the formula for the unit element of $\mathcal{K}$:

$$A(+-) \cong \mathcal{I} \times \mathcal{K}, \quad \mathcal{I} \cong \mathsf{End}_{\mathbf{k}}(A(+)), \quad 1_{\mathcal{K}} = 1 - \sum_{i=1}^{k} v^i \otimes v_i. \tag{40}$$

Elements of $\mathcal{K}$ placed on arcs act trivially on $A(+)$ and $A(-)$, see Figure 2.4.1. In the earlier examples, elements of $\mathcal{K}$ are shown in Figures 2.3.3, 2.3.7, 2.3.12, where in each case $\dim \mathcal{K} = 1$.

Denote by

$$p \,:\, U \subset A(+-) \xrightarrow{p_{\mathcal{K}}} \mathcal{K} \tag{41}$$

the composition of the inclusion of algebra U into $A(+-)$ and projection $p_{\mathcal{K}}$ onto $\mathcal{K}$ along $\mathcal{I}$. The latter map can be written as either left or right multiplication by $1_{\mathcal{K}}$,

$$p_{\mathcal{K}}(y) = 1_{\mathcal{K}}\, y = y 1_{\mathcal{K}}, \tag{42}$$

and, hiding the inclusion, we can write $p(x) = 1_{\mathcal{K}}\, x = x 1_{\mathcal{K}}$ for $x \in U$.

Furthermore, the surjection $\mathbf{k}\Sigma^* \longrightarrow U$ can be composed with p above. Denote the resulting algebra homomorphism by

$$p_* : \mathbf{k}\Sigma^* \longrightarrow \mathcal{K}, \quad p_*(\omega) = 1_{\mathcal{K}}\omega = \omega 1_{\mathcal{K}}, \quad \omega \in \Sigma^*, \tag{43}$$

where, when multiplying by $1_{\mathcal{K}}$, we view ω as an element of U or $A(+-)$.

Introduce a a trace map tr on $\mathcal{K}$ by the circular closure and evaluation, see Figure 2.4.2,

$$\mathsf{tr} \,:\, \mathcal{K} \longrightarrow \mathbf{k}. \tag{44}$$

Note that, in general, both $\alpha_{|}$ and $\alpha_{\circ}$ are used for the evaluation, since elements of $\mathcal{K}$ are linear combinations of elements of U and $A(+) \otimes A(-)$. The circular closure on elements on U, respectively $A(+) \otimes A(-)$, is computed via $\alpha_{\circ}$, respectively $\alpha_{|}$.

The following relation between traces holds:

$$\mathsf{tr}(p_*(\omega)) = \alpha_\circ(\omega) - \alpha_\mathsf{I}^{\mathsf{tr}}(\omega), \quad \omega \in \Sigma^*, \tag{45}$$

where

$$\alpha_\mathsf{I}^{\mathsf{tr}}(\omega) := \mathsf{tr}_{A(+)}(\omega). \tag{46}$$

Recall that evaluation $\alpha_\mathsf{I}^{\mathsf{tr}}$ depends only on the interval evaluation α_I and is given by the trace of words on $A(+)$.

Note that $\mathsf{tr}(\omega) = \alpha_\circ(\omega)$, since we view ω as an element of either $\mathbf{k}\Sigma^*$ or U, via the projection onto the latter. In formula (45) projection p_* appears, which introduces an additional term.

REMARK 2.7. If comparing to formula (11), recall that there circular series $\alpha_\circ$ was picked to depend on α_I and give a 1D TQFT, while here we are considering arbitrary rational $\alpha_\mathsf{I}, \alpha_\circ$.

Extending linearly from ω to elements of $\mathbf{k}\Sigma^*$, we get

$$\mathsf{tr}(p_*(z)) = \alpha_\circ(z) - \alpha_\mathsf{I}^{\mathsf{tr}}(z), \qquad z \in \mathbf{k}\Sigma^*, \tag{47}$$

PROPOSITION 2.8. Trace tr in formula (44) turns $\mathcal{K}$ into a symmetric Frobenius algebra with the unit element $1_\mathcal{K}$.

A Frobenius algebra is called *symmetric* if $\mathsf{tr}(xy) = \mathsf{tr}(yx)$ for any elements x, y.

PROOF. The trace (44) on $\mathcal{K}$ is symmetric, since the circular closure is symmetric. The pairing $A(+-) \otimes A(+-) \longrightarrow \mathbf{k}$ is non-degenerate, and elements of $\mathcal{K}$ are orthogonal to those of $\mathcal{I} = A(+) \otimes A(-)$ in this pairing. Consequently and in view of the direct product (hence direct sum) decomposition (40), any nonzero element of $\mathcal{K}$ can be paired to some element of $\mathcal{K}$ to get a nontrivial evaluation, implying that the trace is non-degenerate. □

REMARK 2.9. Vice versa, any symmetric Frobenius algebra (B, tr_B) can be obtained from some rational evaluation $\alpha = (\alpha_\mathsf{I}, \alpha_\circ)$ in this way. For that, choose a set Σ of generators of algebra B and form circular series $\alpha_\circ(\omega) = \mathsf{tr}_B(\omega)$, viewing $\omega \in \Sigma^*$ as an element of B via the monoid homomorphism $\Sigma^* \longrightarrow B$, where B is naturally a monoid under multiplication. Set the interval evaluation α_I identically to zero, $\alpha_\mathsf{I}(\omega) = 0$ for any word ω. Then $A(+) = 0 = A(-)$, the trace $\alpha_\mathsf{I}^{\mathsf{tr}} = 0$, and $\mathcal{K} = A(+-) \cong B$ with the trace tr. Similar examples can be produced with $A(+) \neq 0$.

Pair (B, tr_B) as in Remark 2.9 gives rise to monoidal category $\mathcal{C}_B$ obtained via the universal construction as follows. First consider the category $\mathcal{C}'_B$ with objects – finite sign sequences and morphisms finite $\mathbf{k}$-linear combinations of one-dimensional cobordisms with defects. Floating endpoints are not allowed this time, and defects are labelled by elements of B, see Figure 2.4.3 on the left. Concatenation and addition of defects corresponds to multiplication and addition in B, and decorated circles are evaluated via the trace on B, see Figure 2.4.3.

Since B is not, in general, commutative, it is essential to require strands to be oriented, to make sense of the concatenation formula in Figure 2.4.3 (second diagram from left). The resulting category $\mathcal{C}'_B$ can be thought as a decorated version of the oriented Brauer category and is known as the *Frobenius–Brauer category* [**SS22, MS22, Sav21**]. Next, we apply the universal construction to $\mathcal{C}'_B$ (since trace tr_B allows to evaluate any closed diagram) to get the quotient category,

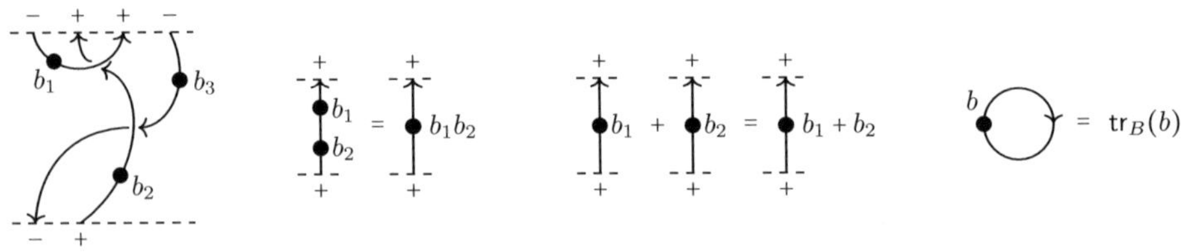

FIGURE 2.4.3. Left: a diagram for a morphism from $-+$ to $-++-$ in categories $\mathcal{C}'_B$ and $\mathcal{C}_B$. Middle: concatenation and addition relations. Right: evaluation of a decorated circle.

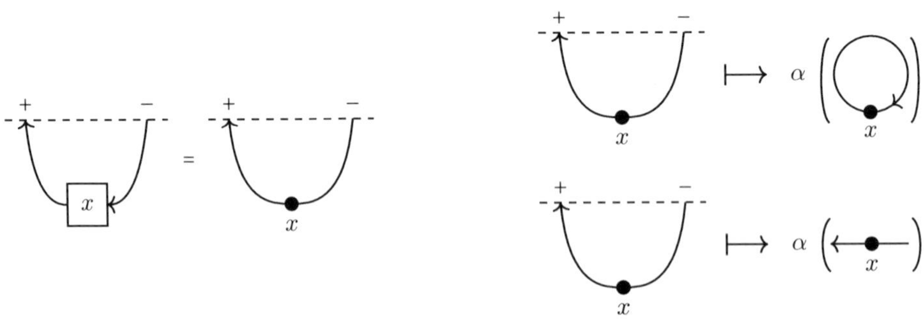

FIGURE 2.4.4. The two trace maps on $A(+-)$.

denoted $\mathcal{C}_B$. Equivalently, one can define $\mathcal{C}_B$ as the quotient of $\mathcal{C}'_B$ by the ideal of negligible morphisms. Trace on B is omitted from our notations for these two categories, but both $\mathcal{C}'_B$ and $\mathcal{C}_B$ depend on it as well.

REMARK 2.10. Algebra $A(+-)$ carries two trace maps, see Figure 2.4.4. This pair of maps is nondegenerate on $A(+-)$, in a suitable sense. In the above construction only the first trace map is considered.

REMARK 2.11. It is interesting that starting with a 1-dimensional theory with defects one obtains a symmetric Frobenius algebra $\mathcal{K}$, since the latter describes a 2D TQFT for thin surfaces, see [**KQ20**, page 19] as well as [**LP08, LP09, MS06, Lau06, KQR21**], hinting at a sort of dimensional lifting. We discuss this later in the paper, in Section 3.

Algebra $A(+-)$ has an idempotent decomposition (40), which we can write as

$$1 = 1_{\mathcal{K}} + \sum_{i=1}^{k} v^i \otimes v_i, \tag{48}$$

with each $v^i \otimes v_i$, $1 \le i \le k$ and $1_{\mathcal{K}}$ together constituting $k+1$ mutually orthogonal idempotents. There are natural algebra isomorphisms

$$\mathsf{End}_{\mathcal{C}_\alpha}(+) \cong A(+-) \cong \mathsf{End}_{\mathcal{C}_\alpha}(-)^{\mathrm{op}}. \tag{49}$$

given by bending strands, see Figures 2.4.5 and 2.4.6.

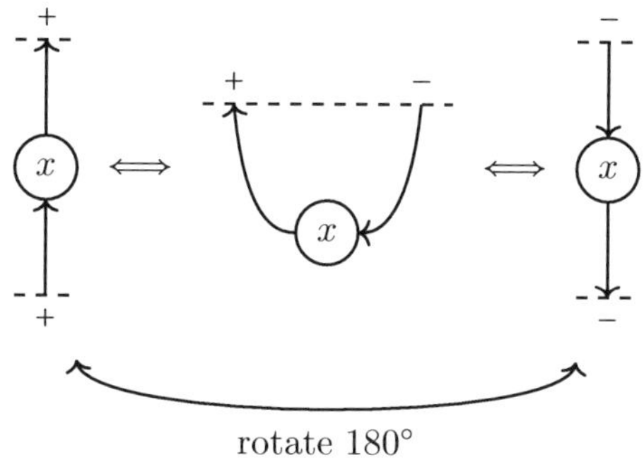

FIGURE 2.4.5. Strand bending leads to isomorphisms (49).

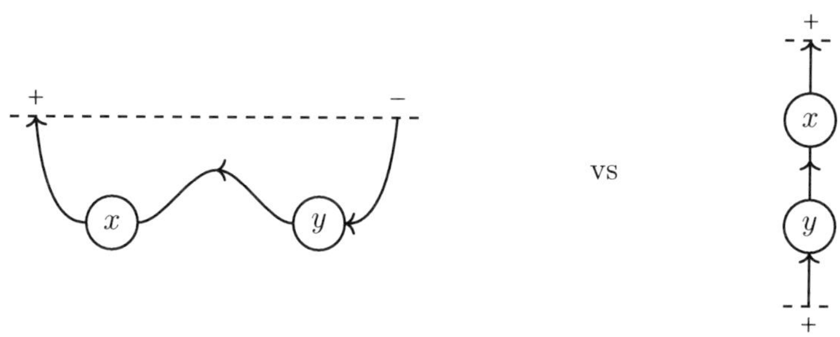

FIGURE 2.4.6. Multiplications in $A(+-)$ and $\mathsf{End}_{\mathcal{C}_\alpha}(+)$ match under the isomorphism (49).

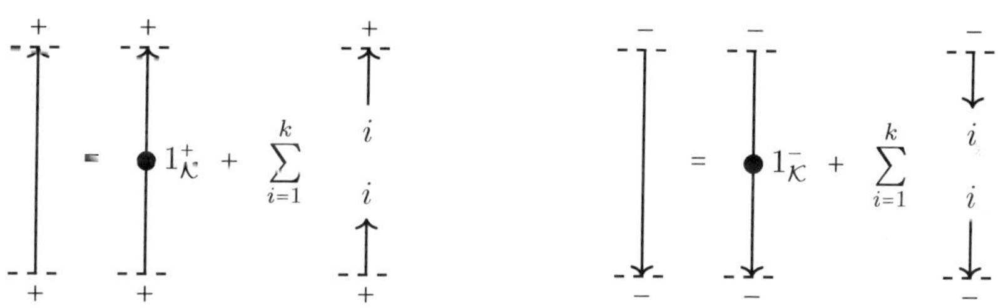

FIGURE 2.4.7. Idempotent decompositions for id_+ and id_-.

The corresponding orthogonal idempotent decompositions in $\mathsf{End}_{\mathcal{C}_\alpha}(+)$ and $\mathsf{End}_{\mathcal{C}_\alpha}(-)^{\mathrm{op}}$ are given by

$$\mathsf{id}_+ = 1^+_{\mathcal{K}} + \sum_{i=1}^{k} v^i \otimes v_i^*, \qquad \mathsf{id}_- = 1^-_{\mathcal{K}} + \sum_{i=1}^{k} v_i \otimes v_*^i, \tag{50}$$

see Figure 2.4.7

Consider the Karoubi envelope $\mathsf{Kar}(\mathcal{C}_\alpha)$ of $\mathcal{C}_\alpha$, also denoted $\underline{\mathcal{DC}}_\alpha$ in [**Kho20a**]. Denote by $e_0^+, e_1^+, \dots, e_k^+$ the idempotents in the endomorphism ring of $+$, see (50), so that $e_0^+ = 1^+_{\mathcal{K}}$ and $e_i = v^i \otimes v_i^*$, and use the same notation but with $-$ instead of

+ for the − object, so that

$$\mathsf{id}_+ = e_0^+ + \ldots + e_k^+, \quad \mathsf{id}_- = e_0^- + \ldots + e_k^- \tag{51}$$

Objects + and − in the category $\mathsf{Kar}(\mathcal{C}_\alpha)$ are isomorphic to the direct sum of objects

$$+ \cong \oplus_{i=0}^k (+, e_i^+), \quad - \cong \oplus_{i=0}^k (-, e_i^-). \tag{52}$$

For $0 \le i \le k$ objects $(+, e_i^+)$ and $(-, e_i^-)$ are dual. Furthermore, we have

PROPOSITION 2.12. Objects $(+, e_i^+)$ and $(-, e_i^-)$ are isomorphic, for $i \ge 1$. Objects $(+, e_i^+)$ and $(+, e_j^+)$ are isomorphic, for $i, j \ge 1$. Each of these objects is isomorphic to the identity object $\mathbf{1}$ of $\mathsf{Kar}(\mathcal{C}_\alpha)$.

PROOF. It is enough to set up pairs of morphisms between the corresponding objects of $\mathcal{C}_\alpha$ such that the compositions are the corresponding idempotents. These morphisms are shown in Figure 2.1.11 for objects $(+, e_i^+)$ and $(+, e_j^+)$. For objects $(+, e_i^+)$ and $(-, e_i^-)$ the morphisms are shown in Figure 2.1.10 on the right. Note also that $\mathsf{End}_{\mathsf{Kar}(\mathcal{C}_\alpha)}((+, e_i^+)) = \mathbf{k}$. Isomorphisms between each of $(+, e_i^+)$ and $(-, e_i^-)$ and $\mathbf{1}$ are given by the half-intervals (up oriented for +, down oriented for −) with the top or bottom boundary + or −. The identity object $\mathbf{1}$ is represented by the empty 0-manifold $\varnothing_0$ (by the empty sign sequence). □

COROLLARY 2.13. $2k$ objects $(+, e_1^+), \ldots, (+, e_k^+)$ and $(-, e_1^-), \ldots, (-, e_k^-)$ are pairwise isomorphic, and each is isomorphic to $\mathbf{1}$. For any of these objects X the endomorphism ring $\mathsf{End}_{\mathsf{Kar}(\mathcal{C}_\alpha)}(X) \cong \mathbf{k}$.

Denote by $\mathcal{C}'$ the full tensor additive subcategory of $\mathsf{Kar}(\mathcal{C}_\alpha)$ generated by these $2k$ objects and $\mathbf{1}$. It is a rigid category.

PROPOSITION 2.14. The category $\mathcal{C}'$ is tensor (symmetric monoidal) equivalent to the category $\mathbf{k}\text{–}\mathsf{vect}$ of finite-dimensional $\mathbf{k}$-vector spaces with the standard tensor structure. It is Karoubi-closed.

PROOF. This is immediate since monoidal generators of $\mathcal{C}'$ are all equivalent to $\mathbf{1}$. The unit object $\mathbf{1} \in \mathsf{Ob}(\mathsf{Kar}(\mathcal{C}_\alpha))$ generates a full monoidal subcategory of $\mathsf{Kar}(\mathcal{C}_\alpha)$ which is monoidal equivalent to $\mathbf{k}\text{–}\mathsf{vect}$. □

Recall the complementary idempotent $e_0^+ = 1_{\mathcal{K}}^+$ to $e_{>0}^+ := e_1^+ + \ldots + e_k^+$ in id_+, see Figure 2.4.7. Likewise, $e_0^- = 1_{\mathcal{K}}^-$ is the complementary idempotent to $e_{>0}^- := e_1^- + \ldots + e_k^-$ in id_-. Consider the corresponding objects of $\mathsf{Kar}(\mathcal{C}_\alpha)$:

$$X_0^+ := (+, e_0^+), \quad X_0^- := (-, e_0^-), \quad X_{>0}^+ := (+, e_*^+), \quad X_{>0}^- := (-, e_*^-). \tag{53}$$

Then the first pair of objects is *monoidal orthogonal* to the second pair in the following strong sense. For any $n > 0$, $m \ge 0$

$$\mathsf{Hom}_{\mathsf{Kar}(\mathcal{C}_\alpha)}((X_0^+ \oplus X_0^-)^{\otimes n}, (X_{>0}^+ \oplus X_{>0}^-)^{\otimes m}) = 0, \tag{54}$$

$$\mathsf{Hom}_{\mathsf{Kar}(\mathcal{C}_\alpha)}((X_{>0}^+ \oplus X_{>0}^-)^{\otimes m}, (X_0^+ \oplus X_0^-)^{\otimes n}) = 0, \tag{55}$$

that is, the space of homs between any nonempty finite tensor product of X_0^+ and X_0^- and a finite tensor product of $X_{>0}^+$ and $X_{>0}^-$ is zero. In particular, there is only the 0 morphism between any nonempty finite tensor product of X_0^+ and X_0^- and $\mathbf{1}$.

Equations (54), (55) follow from the orthogonality between elements of $\mathcal{K}$ and elements of $A(+)$, $A(-)$ shown in Figure 2.4.1 on the left and center.

Objects X_0^+ and X_0^- are dual. Denote by $\widetilde{\mathcal{C}}_\alpha$ the full monoidal additive and Karoubi-closed subcategory of $\mathsf{Kar}(\mathcal{C}_\alpha)$ generated by these two objects (and object

1). We now relate this category to the Frobenius–Brauer category $\mathcal{C}'_{\mathcal{K}}$ associated to the Frobenius algebra $(\mathcal{K}, \mathsf{tr}_{\mathcal{K}})$ and its negligible quotient $\mathcal{C}_{\mathcal{K}}$, see earlier.

Recall that to a symmetric Frobenius **k**-algebra (B, tr_B), we have assigned a category $\mathcal{C}'_B$ (the Frobenius–Brauer category) of 1-cobordisms with dots decorated by elements of B subject to relations in Figure 2.4.3. Then category $\mathcal{C}_B$ is the quotient of $\mathcal{C}'_B$ via the universal construction for the evaluation of closed B-decorated 1-manifold via tr_B. Equivalently, $\mathcal{C}_B$ is the *gligible quotient* of $\mathcal{C}'_B$, the quotient by the 2-sided ideal of negligible morphisms.

Now specialize to the symmetric Frobenius algebra $(\mathcal{K}, \mathsf{tr}_{\mathcal{K}})$ associated to the evaluation α. We have canonical **k**-algebra isomorphisms $\mathsf{End}(X_0^+) \cong \mathcal{K} \cong \mathsf{End}(X_0^-)$.

PROPOSITION 2.15. These isomorphisms extend to a monoidal and full functor

$$\mathcal{F}_0 \; : \; \mathcal{C}'_{\mathcal{K}} \longrightarrow \mathsf{Kar}(\mathcal{C}_\alpha) \tag{56}$$

taking the object $+ \in \mathsf{Ob}(\mathcal{C}'_{\mathcal{K}})$ to $X_0^+ = (+, e_0^+)$ and object $-$ to $X_0^- = (-, e_0^-)$.

PROOF. The functor $\mathcal{F}_0$ takes **1** to **1**. It takes an arc carrying a dot labelled $x \in \mathcal{K}$ to the arc with the same label, which is now viewed as a morphism in $\mathsf{Kar}(\mathcal{C}_\alpha)$ between products of X_0^+ and X_0^-. For instance, The arc in Figure 2.4.4 on the left, for $x \in \mathcal{K}$, can be viewed as a morphism in $\mathcal{C}'_{\mathcal{K}}$ from **1** (the identity object in $\mathcal{C}'_{\mathcal{K}}$) to the object $+-$ and, alternatively, as a morphism in $\mathsf{Kar}(\mathcal{C}_\alpha)$ from **1** (the identity object in $\mathsf{Kar}(\mathcal{C}_\alpha)$) to the object $X_0^+ \otimes X_0^-$.

Earlier computations, including orthogonality relations (54), (55) imply that $\mathcal{F}_0$ is well-defined and surjective on morphisms (a full functor). □

Category $\mathsf{Kar}(\mathcal{C}_\alpha)$ is the Karoubi envelope of $\mathcal{C}_\alpha$, the latter defined via the universal construction for the evaluation α. Restricting evaluation α to closures of elements of $\mathcal{K} \subset A(+-)$ results in the trace $\mathsf{tr}_{\mathcal{K}}$ on $\mathcal{K}$ used in the construction of the category $\mathcal{C}'_{\mathcal{K}}$ and its gligible quotient $\mathcal{C}_{\mathcal{K}}$ (quotient via the universal construction). The two evaluations – in categories $\mathcal{C}_\alpha$ and $\mathcal{C}'_{\mathcal{K}}$ – match, for the above inclusion $\mathcal{K} \subset A(+-)$. Consequently, we obtain the following statement.

PROPOSITION 2.16. Functor $\mathcal{F}_0$ in (56) factors through the gligible quotient $\mathcal{C}_{\mathcal{K}}$ of $\mathcal{C}'_{\mathcal{K}}$ and induces a fully-faithful and monoidal functor

$$\mathcal{F} \; : \; \mathcal{C}_{\mathcal{K}} \longrightarrow \mathsf{Kar}(\mathcal{C}_\alpha) \tag{57}$$

In particular, functor $\mathcal{F}_0$ factorizes as the composition

$$\mathcal{C}'_{\mathcal{K}} \longrightarrow \mathcal{C}_{\mathcal{K}} \stackrel{\mathcal{F}}{\longrightarrow} \mathsf{Kar}(\mathcal{C}_\alpha). \tag{58}$$

where the first arrow is the gligible quotient functor.

Functor $\mathcal{F}$ induces a monoidal functor

$$\mathsf{Kar}(\mathcal{F}) \; : \; \mathsf{Kar}(\mathcal{C}_{\mathcal{K}}) \; \longrightarrow \mathsf{Kar}(\mathcal{C}_\alpha). \tag{59}$$

PROPOSITION 2.17. Functor $\mathsf{Kar}(\mathcal{F})$ is an equivalence of categories.

PROOF. Recall the complementary object $X_{>0}^+$ to X_0^+ in $+ \in \mathsf{Ob}(\mathsf{Kar}(\mathcal{C}_\alpha)$, so that $+ \cong X_0^+ \oplus X_{>0}^+$. Likewise, $X_{<0}^-$ is complementary to X_0^- with respect to the object $-$ of $\mathsf{Kar}(\mathcal{C}_\alpha)$. Objects $X_{>0}^+$ and $X_{<0}^-$ are both isomorphic to the direct sum of k copies of the object **1**. Consequently, the category $\mathsf{Kar}(\mathcal{C}_\alpha)$ is equivalent to the Karoubi closure of the monoidal subcategory generated by X_0^+ and X_0^-. Taking earlier propositions into consideration completes the proof. □

We now explain the meaning of these results. Starting with a rational evaluation $\alpha = (\alpha_{\mathsf{I}}, \alpha_{\circ})$ we formed the decomposition $A(+-) \cong \mathcal{I} \times \mathcal{K}$, with $\mathcal{I} \cong A(+) \otimes A(-)$ and the complementary factor $\mathcal{K}$ orthogonal, in a suitable sense, to $A(+)$ and $A(-)$.

Algebra $\mathcal{K}$ is symmetric Frobenius, with the nondegenerate trace given by the closure operation in Figure 2.4.2. Note also the trace formula (45) for the trace computed on the projection from $\mathbf{k}\Sigma^* \subset A(+-)$ onto $\mathcal{K}$ as the difference $\alpha_{\circ} - \alpha_{\mathsf{I}}^{\mathsf{tr}}$.

Symmetric Frobenius algebras (Frobenius algebras B with a symmetric trace, $\mathsf{tr}(xy) = \mathsf{tr}(yx), x, y \in B$) are also called just *symmetric* algebras. To the symmetric algebra $(\mathcal{K}, \mathsf{tr}_{\mathcal{K}})$ we assign the Frobenius–Brauer category $\mathcal{C}'_{\mathcal{K}}$ and its gligible quotient $\mathcal{C}_{\mathcal{K}}$. Then there is a natural fully-faithful monoidal functor $\mathcal{F} : \mathcal{C}_{\mathcal{K}} \longrightarrow \mathsf{Kar}(\mathcal{C}_{\alpha})$, see Proposition 2.16, inducing an equivalence (59) of Karoubi envelopes of $\mathcal{C}_{\mathcal{K}}$ and $\mathcal{C}_{\alpha}$.

Conceptually, passing to the Karoubi envelope $\mathsf{Kar}(\mathcal{C}_{\alpha})$ allows one to reduce the consideration to the Frobenius algebra $\mathcal{K}$ together with the trace $\mathsf{tr}_{\mathcal{K}}$ on it. Decorated half-arc morphisms go between $\mathbf{1}$ and summands of + and − objects which are equivalent to $\mathbf{1}$ and can be ignored in the Karoubi envelope. In a sense, no genuinely new idempotents appear when considering half-intervals. Beyond the identity object $\mathbf{1}$ we only need the objects X_0^+, X_0^- coming from $\mathcal{K}$, tensor products of these objects and summands of these tensor products to describe a category equivalent to $\mathsf{Kar}(\mathcal{C}_{\alpha})$.

General one-variable case. Let us specialize again to a single variable to generalize examples 2.5–2.6. Consider a general one-variable case $\Sigma = \{a\}$, with rational evaluation functions as in (37), (36). State space $A(+)$ depends on $Z_{\mathsf{I}}(T)$ only, and let

$$Z_{\mathsf{I}}(T) = \frac{P_{\mathsf{I}}(T)}{Q_{\mathsf{I}}(T)}, \quad n_{\mathsf{I}} = \deg(P_{\mathsf{I}}(T)), \quad m_{\mathsf{I}} = \deg(Q_{\mathsf{I}}(T)), \tag{60}$$

for polynomials $P_{\mathsf{I}}(T), Q_{\mathsf{I}}(T)$ with $Q_{\mathsf{I}}(0) \neq 0$. Consider the polynomial

$$g_{\mathsf{I}}(T) \ := \ T^{m_{\mathsf{I}}} Q_{\mathsf{I}}(1/T)\, T^{\max(0,n_{\mathsf{I}}-m_{\mathsf{I}}+1)} \ = \ T^{m_{\mathsf{I}}+\max(0,n_{\mathsf{I}}-m_{\mathsf{I}}+1)} Q_{\mathsf{I}}(1/T), \tag{61}$$

the product of the reciprocal polynomial of $Q_{\mathsf{I}}(T)$ and a power of T. Then $g_{\mathsf{I}}(T)$ is the characteristic polynomial for the action of a on $A(+)$, see [**Kho20b, KOK22**] for the equivalent case of the state space of a circle in a two-dimensional topological theory. In particular,

$$\dim A(+) \ = \ \dim A(-) \ = \ \deg(g_{\mathsf{I}}(T)) \ = \ m_{\mathsf{I}} + \max(0, n_{\mathsf{I}} - m_{\mathsf{I}} + 1). \tag{62}$$

and $A(+)$ is a cyclic $\mathbf{k}[a]$-module given by

$$A(+) \ \cong \ \mathbf{k}[a]/(g_{\mathsf{I}}(a)). \tag{63}$$

Note that for any rational series

$$Z(T) = \frac{P(T)}{Q(T)}, \quad n = \deg(P(T)), \quad m = \deg(Q(T)), \quad Q(0) \neq 0, \tag{64}$$

we can form the corresponding polynomial

$$g_Z(T) \ := \ T^m Q(1/T)\, T^{\max(0,n-m+1)} \ = \ T^{m+\max(0,n-m+1)} Q(1/T), \tag{65}$$

which is the characteristic polynomial for the action of a on the state space $A(+)$ associated to the series $Z(T)$.

In the direct product decomposition $A(+-) \cong \mathcal{I} \times \mathcal{K}$ the factor $\mathcal{I} \cong A(+) \otimes A(-)$ can be understood from this data. To handle $\mathcal{K}$, let us start with the potentially larger algebra U. These algebras fit into a diagram of surjective homomorphisms

$$\mathbf{k}[a] \xrightarrow{p_0} U \xrightarrow{p} \mathcal{K}, \quad p_* : \mathbf{k}[a] \longrightarrow \mathcal{K}, \quad p_* := p\, p_0, \tag{66}$$

where map p_0 sends a^n to an arc with n dots (see Figure 2.3.1 on the right and specialize $\omega = a^n$), and p is the composition of the inclusion of U into $A(+-)$ and projection onto $\mathcal{K}$, see (41). Finally, let $p_* := p \circ p_0$, see (43).

Algebra U is naturally a quotient $U \cong \mathbf{k}[a]/(g(a))$, for some polynomial $g(a)$. A polynomial $f(a)$ is 0 in U if and only if it (1) acts trivially on $A(+)$ (equivalently, $f(a) \in (g_{\mathsf{I}}(a))$) and (2) belongs to the kernel of the bilinear form on U given by $(f_1, f_2) = \mathsf{tr}(f_1 f_2)$, with the trace given by closing an arc into into a circle and evaluating it via $\alpha_\circ$, see the top right diagram in Figure 2.4.4.

Evaluation $\alpha_\circ$ is encoded by a rational series

$$Z_\circ(T) = \frac{P_\circ(T)}{Q_\circ(T)}, \quad n_\circ = \deg(P_{\mathsf{I}}(T)), \quad m_\circ = \deg(Q_{\mathsf{I}}(T)), \tag{67}$$

with $Q_\circ(0) \neq 0$. Form the polynomial (as in (61))

$$g_\circ(T) := T^{m_\circ} Q_\circ(1/T)\, T^{\max(0, n_\circ - m_\circ + 1)} = T^{m_\circ + \max(0, n_\circ - m_\circ + 1)} Q_\circ(1/T), \tag{68}$$

Then the principal ideal $(g_\circ(a)) \subset \mathbf{k}[a]$ is the kernel of the above bilinear form, and U is naturally the quotient of $\mathbf{k}[a]$ by the intersection of the two principal ideals $(g_{\mathsf{I}}(a))$ and $(g_\circ(a))$. Let $g_\alpha(a) := \mathsf{lcm}(g_{\mathsf{I}}(a), g_\circ(a))$ be the lcm of these two polynomials. We get a canonical isomorphism

$$U \cong \mathbf{k}[a]/(g_\alpha(a)) \cong \mathbf{k}[a]/(\mathsf{lcm}(g_{\mathsf{I}}(a), g_\circ(a))). \tag{69}$$

Coming back to $\mathcal{K}$, recall that it is a quotient of U and $\mathbf{k}[a]$, see (66) and (43), via quotient maps p and $p\,p_0$, respectively.

Form the trace series $Z_{\mathsf{I}}^{\mathsf{tr}} := Z_{\alpha_{\mathsf{I}}}^{\mathsf{tr}}(T)$ of $Z_{\alpha_{\mathsf{I}}}(T)$, see formulas (26)-(28), and the difference of the two series

$$Z_{\circ\mathsf{I}}(T) := Z_\circ(T) - Z_{\mathsf{I}}^{\mathsf{tr}}(T). \tag{70}$$

Let $g_{\circ\mathsf{I}}(T)$ be the characteristic polynomial for the action of a associated to these series, see (65).

The relation (45) on the trace, reproduced below,

$$\mathsf{tr}(p_*(\omega)) = \alpha_\circ(\omega) - \alpha_{\mathsf{I}}^{\mathsf{tr}}(\omega), \quad \omega \in \Sigma^*, \tag{71}$$

implies that under the homomorphism p_* in (66) the trace on $\omega = a^n$, viewed as an element of $\mathcal{K}$, is given by the coefficient at T^n of the series $Z_{\circ\mathsf{I}}(T)$ above. Consequently, there is a natural isomorphism

$$\mathcal{K} \cong \mathbf{k}[a]/(g_{\circ\mathsf{I}}(a)), \tag{72}$$

with $g_{\circ\mathsf{I}}(T)$ associated to the series $Z_{\circ\mathsf{I}}(T)$. The quotient map $U \longrightarrow \mathcal{K}$ takes $a \in U$ to $p_*(a) \in \mathcal{K}$. From the isomorphisms

$$U \cong \mathbf{k}[a]/(\mathsf{lcm}(g_{\mathsf{I}}(a), g_\circ(a))), \quad \mathcal{K} \cong \mathbf{k}[a]/(g_{\circ\mathsf{I}}(a)) \tag{73}$$

that respect the surjection, we obtain an inclusion of ideals $(\mathsf{lcm}(g_{\mathsf{I}}(a), g_\circ(a))) \subset (g_{\circ\mathsf{I}}(a))$, so that the one-variable polynomial $g_{\circ\mathsf{I}}(T)$ is a divisor of $\mathsf{lcm}(g_{\mathsf{I}}(T), g_\circ(T))$.

3. From 1D to 2D

3.1. Category of thin flat surfaces and symmetric Frobenius algebras. Factorization $A(+-) \cong \mathcal{I} \times \mathcal{K}$ allows us to understand, as described above, the category $\mathsf{Kar}(\mathcal{C}_\alpha)$. The term $\mathcal{I}$ contributes objects isomorphic to $\mathbf{1}$ to the Karoubi envelope, while the symmetric Frobenius algebra $\mathcal{K}$ leads to a $\mathcal{K}$-decorated Frobenius–Brauer category and its negligible quotient, which is equivalent to $\mathsf{Kar}(\mathcal{C}_\alpha)$.

Thus, from a one-dimensional theory with defects we can extract a key piece of structure, the symmetric Frobenius algebra $(\mathcal{K}, \mathsf{tr}_{\mathcal{K}})$ which, on one hand, describes $\mathsf{Kar}(\mathcal{C}_\alpha)$, and on the other has a strongly two-dimensional flavor, since it gives a tensor functor on a suitable category of 2-dimensional cobordisms with corners.

In this section we review the two-dimensional nature of non-necessarily commutative symmetric algebras (B, tr_B) and their interpretation via tensor functors from the category of *thin flat surfaces with boundary and corners* to $\mathbf{k}$–vect. A version of this correspondence goes back at least to Moore-Segal [**MS06**], also see [**Lau05, Lau06, LP07, LP08, Cap13, KQR21**] and unpublished notes [**KQ20**]. Our modest contribution is to point out that elements of B can be placed as labelled dots (0-dimensional defects) along the inner edges of cobordisms and to provide a neck-cutting formula in this language. We discuss obstacles in the non-semisimple case to extending a thin surface TQFT to an open-closed TQFT as in [**MS06, LP07, LP08, Cap13**]. We also show how a combination of a symmetric Frobenius algebra and an evaluation series for closed cobordisms leads to a universal construction which occupies an intermediate role between open-closed two-dimensional TQFT [**MS06, LP07, LP08**] and general two-dimensional topological theories for surfaces with corners [**KQR21**].

The category TCob_2 of *thin flat surfaces* is defined in [**KQR21**], where it is denoted TFS. It has objects $n \in \mathbb{Z}_+ = \{0, 1, 2, \dots\}$ represented by n disjoint intervals in $\mathbb{R}$ ordered from left to right. Morphisms from n to m in TCob_2 are surfaces S with boundary and corners together with an immersion into the strip $\mathbb{R} \times [0,1]$, see Figure 3.1.1 for an example. A surface inherits an orientation from that of $\mathbb{R} \times [0,1]$. An immersion can have overlaps, that can be perturbed in $\mathbb{R}^2 \times [0,1]$ into an embedding of S into the latter. Two morphisms are equal if the corresponding oriented surfaces are diffeomorphic rel boundary (forgetting the immersion or the embedding). We refer to [**KQR21**] for details.

Category TCob_2 is symmetric monoidal, with the monoidal structure given by placing objects and morphisms next to each other, in parallel. Figure 3.1.2 shows a possible set of generating morphisms of the monoidal category TCob_2. Figure 3.1.3 shows some relations in TCob_2, and we refer to [**Lau05, KQR21**] for a complete set of relations.

The following result is well-known, *c.f.*, [**MS06, LP07, LP08, LP09, KQ20**].

Proposition 3.1. Symmetric monoidal functors $F : \mathsf{TCob}_2 \longrightarrow \mathbf{k}$–$\mathsf{vect}$ are classified by symmetric Frobenius $\mathbf{k}$-algebras (B, tr_B), assigning multiplication, unit, trace, comultiplication and permutation maps to cobordisms in Figure 3.1.1.

Denote the functor associated to (B, tr_B) by F_B. In this notation we suppress the dependence of F_B on the trace map tr_B. Note that the symmetric trace condition comes from the Figure 3.1.4 equality of morphisms in TCob_2.

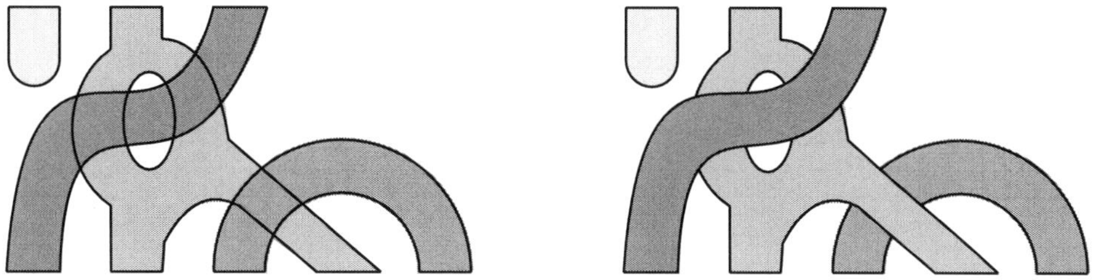

FIGURE 3.1.1. Left: a thin flat surface for a morphism $5 \to 3$ in TCob_2. Right: Same morphism represented by a thin flat surface embedded in $\mathbb{R}^2 \times [0,1]$. Different connected components of a surface are shaded in different ways to make it easier to discern overlapping regions.

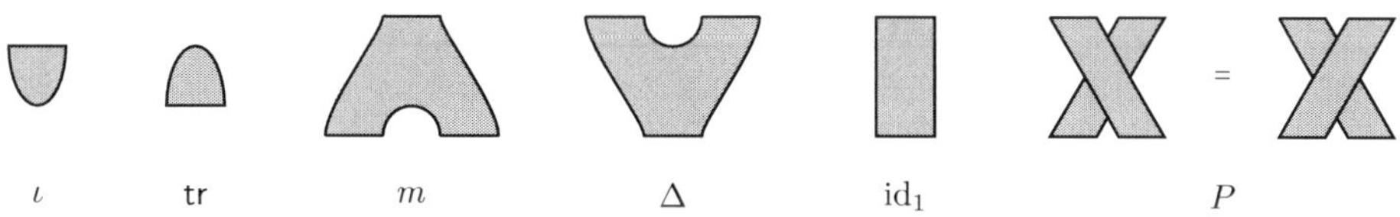

FIGURE 3.1.2. A set of generating morphisms in the symmetric monoidal category TCob_2. The labels next to morphisms denote the structure maps of a symmetric Frobenius algebra associated to a monoidal functor from TCob_2 to $\mathbf{k}$–vect, see Proposition 3.1 below. From left to right, these are the unit map ι, the trace map tr, multiplication m and its dual Δ, identity, and permutation map P.

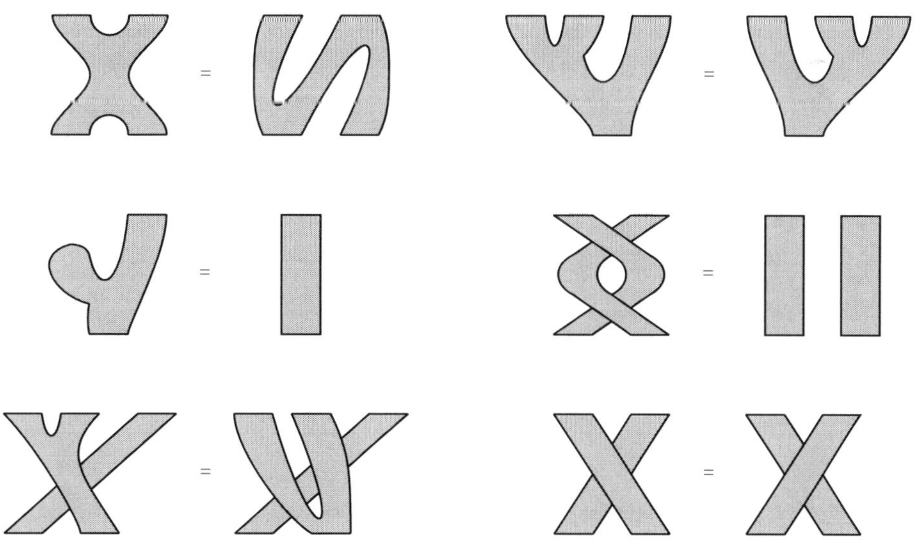

FIGURE 3.1.3. Some relations in TCob_2 on these generating morphisms.

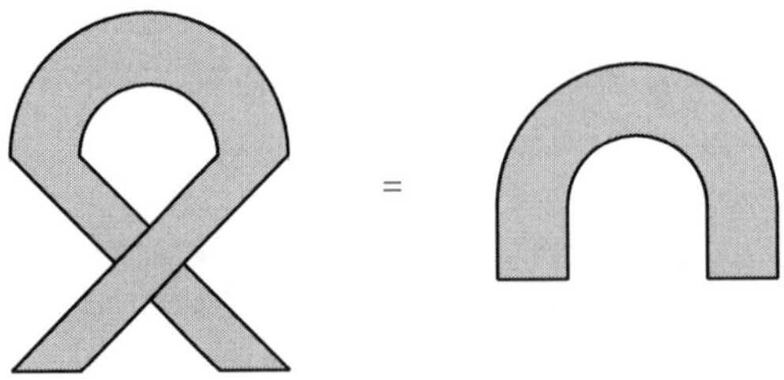

FIGURE 3.1.4. Equality of these two morphisms from 2 to 0 corresponds to the symmetricity of the trace: $\mathsf{tr}(xy) = \mathsf{tr}(yx)$, where $x, y \in B$.

FIGURE 3.1.5. (a) Endomorphism ℓ_a of B, $\ell_a(x) = ax$, (b) Endomorphism r_a of B, $r_a(x) = xa$, (c) Element $a \in B$, (d) Trace map $B \longrightarrow \mathbf{k}$, where $x \mapsto \mathsf{tr}(ax)$.

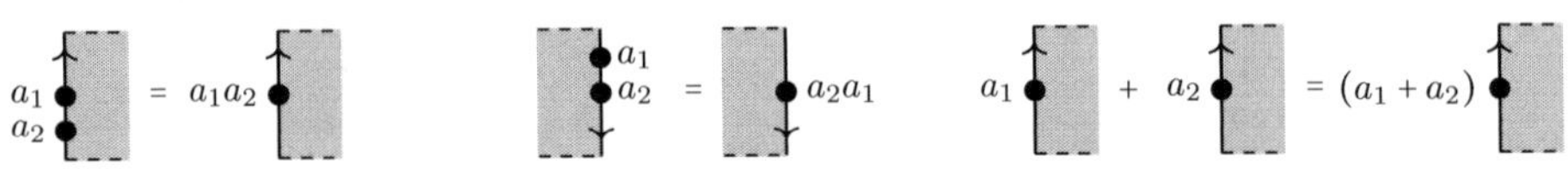

FIGURE 3.1.6. Merging and adding boundary dots.

The more familiar correspondence is that between *commutative* Frobenius algebras and symmetric monoidal functors from the category Cob_2 of oriented surfaces with boundary to $\mathbf{k}-vect$. In that case object n of Cob_2 is represented by n circles, not intervals, and the Frobenius algebra is commutative.

In this case of cobordisms between closed 1-manifolds and commutative Frobenius algebras (A, tr_A) the functor can be enhanced by introducing dots floating on the components of a cobordism and labelled by elements of A. Such dotted cobordism can be evaluated to a linear map between tensor powers of A, with dot a denoting multiplication by a map $m_a : A \longrightarrow A, m_a(x) = ax$.

A similar enhancement exists for tensor functors $F_B : \mathsf{TCob}_2 \longrightarrow \mathbf{k}\text{−}\mathsf{vect}$ as above and we could not find it in the literature. We explain it now.

A dot labelled $a \in B$ on a side boundary denotes the endomorphism of multiplication by a in B. The endomorphism is the left multiplication ℓ_a by a if the local orientation at the dot is up and right multiplication by a if the local orientation at the dot is down, see Figure 3.1.5.

Likewise, an a-labelled dot at the side boundary of a half-disk denotes either the element $a \in B$ or the map $x \mapsto \mathsf{tr}(ax)$ from B to $\mathbf{k}$, see Figure 3.1.5.

Dots can move freely along side boundaries and relations in Figure 3.1.6 hold. Figure 3.1.7 depicts that dots may slide past local maxima and minima.

FIGURE 3.1.7. Dot sliding past local maxima and minima.

FIGURE 3.1.8. Left: the surgery relation. Right: evaluation of a dot on the boundary of disk.

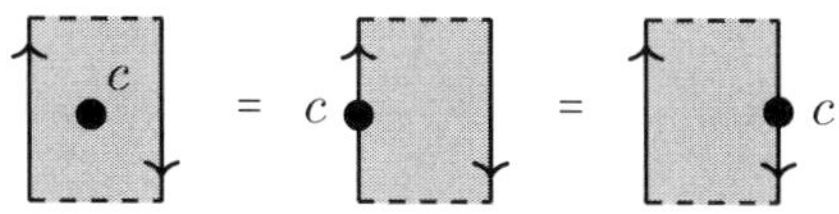

FIGURE 3.1.9. Central element c floating inside a thin surface can move on and off the boundary.

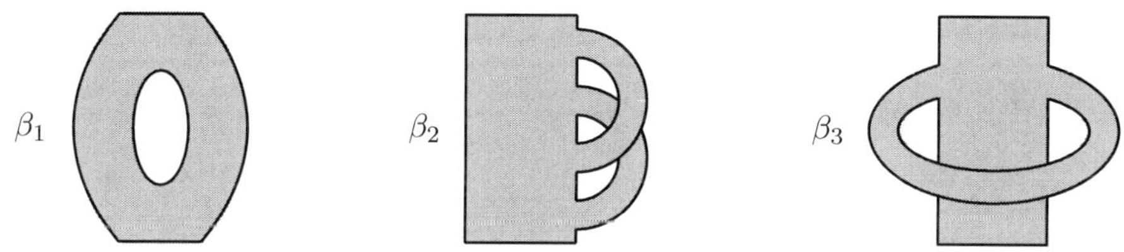

FIGURE 3.1.10. Endomorphisms $\beta_1, \beta_2, \beta_3$ of object 1 in category TCob_2.

Pick a basis $\{x_i\}_{i=1}^n$ of the vector space B and the dual basis $\{y_i\}_{i=1}^n$ of B with respect to the trace form, so that $\mathsf{tr}(x_i y_j) = \delta_{i,j}$. Then the surgery relation holds, see Figure 3.1.8 left, that allows to cut a surface along an interval connecting two side boundary points. In particular, any closed surface in TCob_2 (a surface with empty horizontal boundary, that is, an endomorphism of object $\varnothing_1$), with side boundary decorated by elements of B, can be evaluated to an element of $\mathbf{k}$. Indeed, each connected component of such surface has nonempty side boundary and the surgery relation can be iteratively applied for an evaluation.

One can further augment possible decorations by allowing elements c of the center $Z(B)$ to float inside a surface. Such a floating dot denotes the multiplication by c endomorphism of B. A floating central element c can land onto a side boundary, see Figure 3.1.9.

Recall from [**KQR21**] that category TCob_2 has three commuting endomorphisms $\beta_1, \beta_2, \beta_3$ of object 1, see Figure 3.1.10. These endomorphisms were denoted b_1, b_2, b_3 in [**KQR21**] and satisfy the relation $\beta_1\beta_3 = \beta_3^2$. Consider the functor F_B from TCob_2 to $\mathbf{k}\text{-}\mathsf{vect}$. It takes these endomorphisms to $\mathbf{k}$-linear maps $F_B(\beta_1), F_B(\beta_2), F_B(\beta_3)$.

PROPOSITION 3.2. *$F_B(\beta_1)$ and $F_B(\beta_2)$ are B-bimodule endomorphisms $B \longrightarrow B$, determined by central elements $b_1 := F_B(\beta_1)(1)$ and $b_2 := F_B(\beta_2)(2)$, $b_1, b_2 \in$*

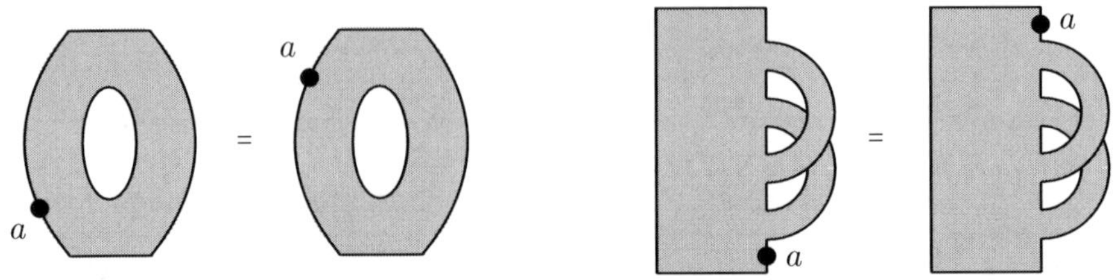

FIGURE 3.1.11. Moving a dot along a side boundary shows that maps $F_B(\beta_1)$, $F_B(\beta_2)$ are B-bimodule maps. Two out of four of these dot moving relations are shown.

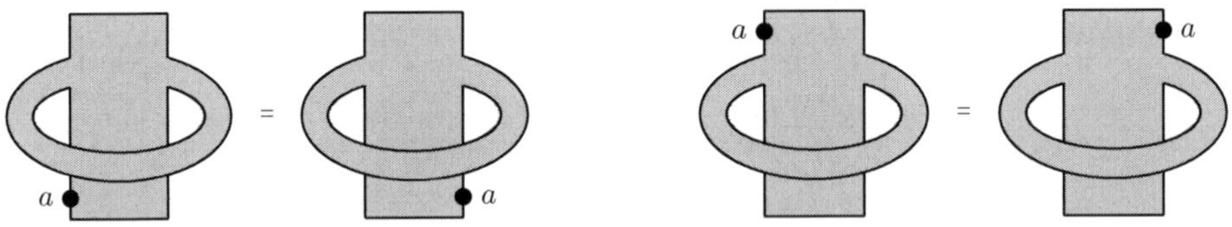

FIGURE 3.1.12. Dot sliding along side boundaries of the β_1 cobordism implies factorization (74) of $F_B(\beta_3)$.

$Z(B)$. Map $F_B(\beta_3)$ can be factored as

$$B \xrightarrow{q_B} [B,B] \xrightarrow{\beta_B} Z(B) \xrightarrow{\iota_B} B \tag{74}$$

for a unique linear map β_B, where q_B is the quotient map by the commutator subspace and ι_B the inclusion of the center into B.

PROOF. Maps $F_B(\beta_1)$ and $F_B(\beta_2)$ are B-bimodule maps, since a-dots can be slid along the side edges of these cobordisms from near a corner at the bottom to the matching corner point at the top, see Figure 3.1.11.

For the map $F_B(\beta_3)$, a dot a near the bottom left corner can be moved to the dot a near the bottom right corner, by dragging it along a side boundary, see Figure 3.1.12 on the left. Consequently, $F_B(\beta_3)(ab) = F_B(\beta_3)(ba)$ for any $a, b \in B$, so that $F_B(\beta_3)$ factors through map q_B.

Likewise, a-dot at the top left corner of β_3 can be moved to the top right corner of β_3, by dragging it along a side boundary, see Figure 3.1.12 on the right. This means $aF_B(\beta_3)(b) = F_B(\beta_3)(b)a$ for any $a, b \in B$, that is, $F_B(\beta_3)(b)$ belongs to the center $Z(B)$ of B. Hence, $F_B(\beta_3)$ admits a factorization (74). □

The map

$$\beta_B : B/[B,B] \longrightarrow Z(B) \tag{75}$$

is associated to a Frobenius algebra B with a symmetric trace.

EXAMPLE 3.3. Consider the group algebra $B = \mathbf{k}[C_p]$ of a prime order cyclic group $C_p = \{1, g | g^p = 1\}$ over a field $\mathbf{k}$ of characteristic p with the trace

$$\mathsf{tr}_B : B \longrightarrow \mathbf{k}, \quad \mathsf{tr}_B(1) = 1, \quad \mathsf{tr}_B(g^i) = 0, \ 0 < i < p. \tag{76}$$

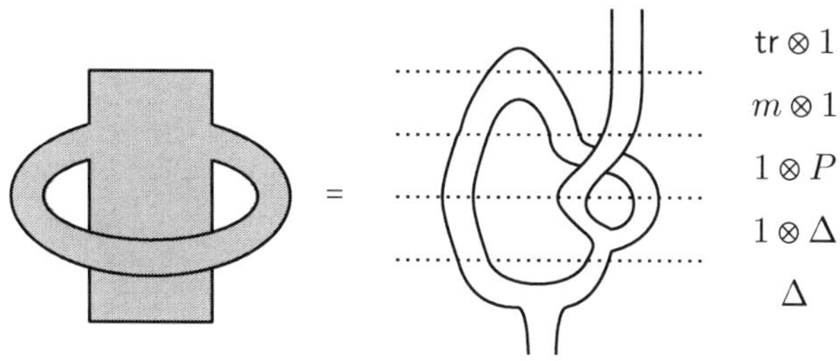

FIGURE 3.1.13. Factorization of the maps β_3 and $F_B(\beta_3)$. Here P is the permutation cobordism.

The comultiplication structure map of this thin surface TQFT is

$$\Delta(g^i) = \sum_{i=0}^{p-1} g^j \otimes g^{i-j}. \tag{77}$$

The map $F_B(\beta_3)$ is the composition

$$(\mathsf{tr}_B \otimes 1)(m \otimes 1)(1 \otimes P)(1 \otimes \Delta)\Delta, \tag{78}$$

see Figure 3.1.13. Computing this map for the group algebra B yields $F_B(\beta_3) = 0$ and $\beta_B = 0$.

Choosing a finite set Σ of labels for the side boundary dots, one obtains a decorated version $\mathsf{TCob}_{2,\Sigma}$ of the thin flat surface category, where side boundaries can carry dots labelled by elements of Σ as in Figures 3.1.5, 3.1.7, for instance. A symmetric Frobenius algebra (B, tr_B) as above together with a map of sets $\Sigma \longrightarrow B$ determines a symmetric monoidal functor $\mathsf{TCob}_{2,\Sigma} \longrightarrow \mathbf{k}\text{-}\mathsf{vect}$.

EXAMPLE 3.4. For Frobenius B in Example 3.3 and $\Sigma = \{g\}$, the relations for the resulting functor are shown in Figure 3.1.14, where a dot labelled $i \in \mathbb{Z}/p$ denotes multiplication by g^i at that position (a dot labelled 0 can be erased).

3.2. Open-closed TQFTs and topological theories.

Open-closed 2D TQFTs and knowledgeable Frobenius algebras. The usual category of oriented cobordisms between closed one-manifolds and the category of thin flat surfaces TCob_2 can be unified into the category of open-closed 2D cobordisms [**MS06, LP08, Laz01, Lau06**]. In that category OCCob_2 objects are compact oriented 1-manifolds with boundary, thus finite unions of intervals and circles. Morphisms are diffeomorphism classes (rel boundary) of oriented surfaces with boundary and corners. These surfaces have *side boundary* and corner points where side boundaries meet top and bottom boundary intervals. We refer to [**LP08**] for details. As a set of generating morphisms one can take the union of

- standard generating morphisms for thin surfaces in Figure 3.1.2 that correspond to the structure maps of a symmetric Frobenius algebra, and the permutation map P of two intervals,
- standard generating morphisms for the usual category of two-dimensional oriented cobordisms. These morphisms correspond to the structure maps in a commutative Frobenius algebra and the transposition of two circles,

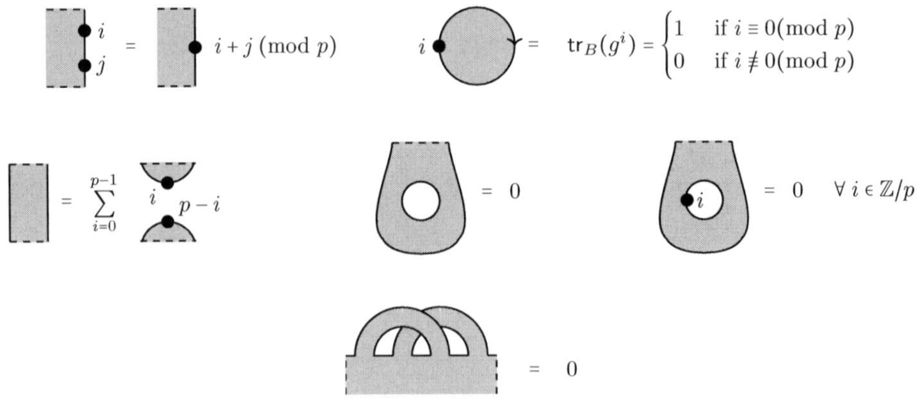

FIGURE 3.1.14. Boundary dots and relations for the thin surface TQFT in Examples 3.3, 3.4. The relation in the top right corner is the trace formula for B. The second and third relations in the second row and the bottom row relation follow from the surgery formula (first relation in the second row) and $\mathsf{char}(\mathbf{k}) = p$.

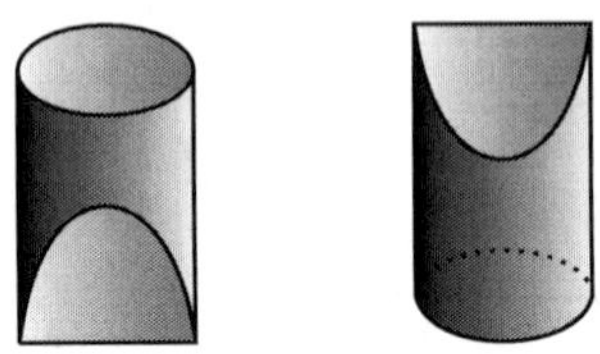

$\jmath$ $\jmath^*$

FIGURE 3.2.1. Zipper and cozipper cobordisms $\jmath$ and $\jmath^*$ in OCCob_2.

- Zipper and cozipper cobordisms, see Figure 3.2.1, and the transposition of an interval and an circle, similar to the transposition P of two intervals in Figure 3.1.2.

A 2D TQFT for the open-closed category consists of a symmetric Frobenius algebra (B, tr_B), which is the state space of an interval, and a commutative Frobenius algebra (C, tr_C), describing the state space of a circle. These two Frobenius algebras are subject to the following interactions:

(1) There exists a trace-respecting coalgebra homomorphism $\jmath : B \longrightarrow C$ (the *zipper homomorphism*), see Figure 3.2.1 left. Its dual is the *cozipper* map $\jmath^* : C \longrightarrow B$, see Figure 3.2.1 right. That $\jmath$ is a coalgebra homomorphism comes from diffeomorphisms (rel boundary) in the top row of Figure 3.2.2; $\jmath$ intertwines comultiplication and the counit maps of the two coalgebras. The dual (cozipper) map intertwines the two multiplications and takes the unit element of C to the unit element of B, thus it is an algebra homomorphism. These properties of $\jmath^*$ correspond to the relations given by rotating those in the top row of Figure 3.2.2 by 180°.

(2) Maps $\jmath, \jmath^*$ are subject to
   - The *knowledge* relation, shown on the left in the second row in 3.2.2,

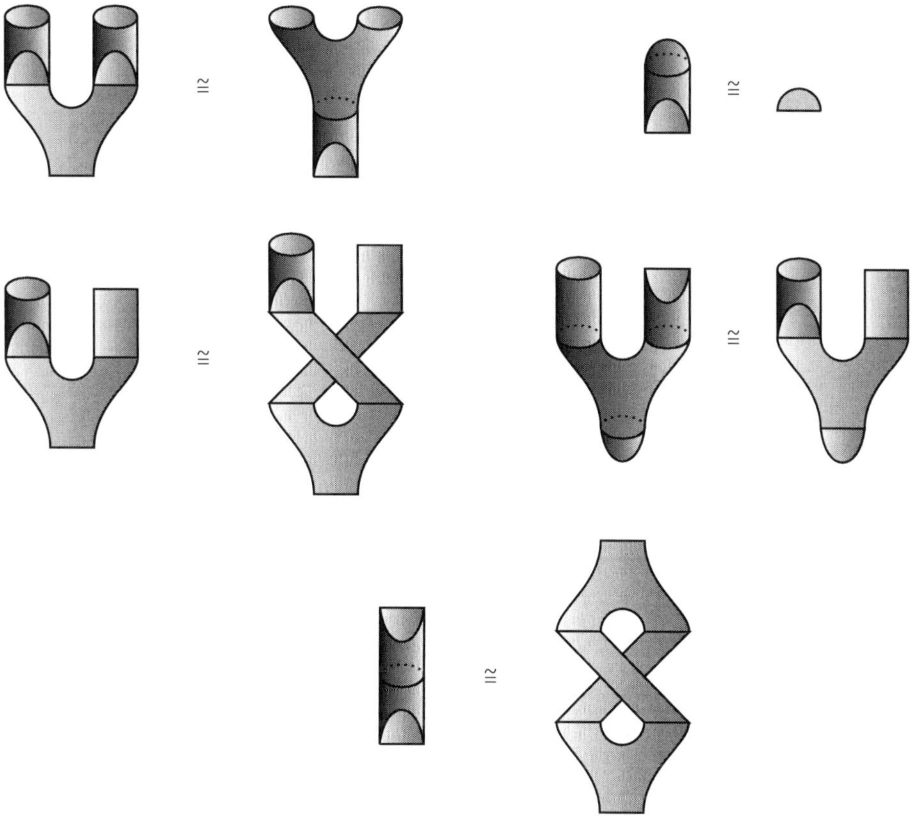

FIGURE 3.2.2. Defining relations in the category OCCob_2 that involve zipper and cozipper homomorphisms.

- The duality between the zipper and the cozipper, shown on the right in the second row of 3.2.2,
- The Cardy condition, see the third row in 3.2.2.

REMARK 3.5. Lauda and Pfeiffer [**LP08, LP09**] view a cobordism as a morphism from the top boundary to the bottom boundary, while our convention is the opposite. They also denote a symmetric Frobenius algebra by A instead of our B, and zipper morphism by ι instead of $\jmath$, so that, for instance, the zipper morphism is $C \overset{\iota}{\longrightarrow} A$ in [**LP08, LP09**] and $B \overset{\jmath}{\longrightarrow} C$ in the present paper.

Such pairs (B, C) are called *knowledgeable Frobenius algebras*, see [**LP09**, Definition 2.2]. Examples of many such pairs in the literature are build from a *strongly separable* symmetric Frobenius algebra B. An algebra is called strongly separable if the trace form $(a,b)_\ell = \mathsf{tr}_A(L_a \circ L_b)$ is non-degenerate, where L_a is the operator $A \longrightarrow A$ of left multiplication by a.

A strongly separable symmetric Frobenius algebra B extends to a knowledgeable Frobenius algebra (B, C) by taking $C = Z(A)$ to be the center of A, see [**LP07, LP09**]. A strongly separable algebra is necessarily semisimple.

EXAMPLE 3.6. Consider the thin surface TQFT and the Frobenius algebra B in Examples 3.3, 3.4. One can look to extend this open TQFT F to an open-closed

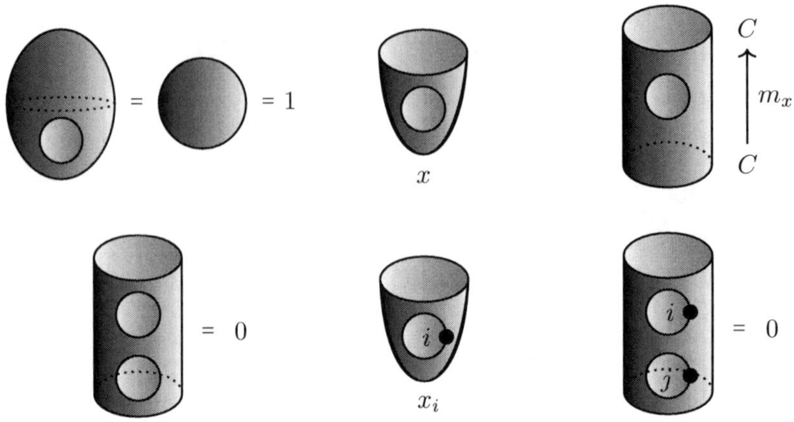

FIGURE 3.2.3. Top row, left to right: evaluation of a one-holed sphere; element $x \in C$ is given by one-holed disk; endomorphism $m_x(c) = xc$ of multiplication by x. Bottom row, left to right: $x^2 = 0$ (cobordism for x^2 contains the cobordism in the center of Figure 3.1.14 as a subcobordism), element $x_i \in C$, products $x_i x_j = 0$.

TQFT (to a knowledgeable Frobenius algebra (B, C)). Then the commutative Frobenius algebra C associated to a circle should have a distinguished element $x \in C$, $x = \jmath(1_B)$ shown in Figure 3.2.3 in the top row and given by a cup with a hole in it. It is the image of the identity element 1_B of B under the zipper map $\jmath$. The trace $\mathsf{tr}_C(x) = 1$ since a dotless disk evaluates to 1, so that $x \neq 0 \in C$. On the other hand, $x^2 = 0$ since the corresponding cobordism contains a subsurface shown in the middle of the second row in Figure 3.1.14, which evaluates to 0. Thus, C contains a subalgebra $\mathbf{k}[x]/(x^2)$.

More generally, C contains elements $x_i = \jmath(g^i)$, $i \in \mathbb{Z}/p$, with $x = x_0$, see Figure 3.2.3 bottom row. We have
(79)
$$x_i x_j = 0,\ i, j \in \mathbb{Z}/p, \quad \Delta_C(x_i) = \sum_{k=0}^{p-1} x_{i+k} \otimes x_{p-k}, \quad \mathsf{tr}_C(x_i) = \delta_{i,0}, \quad \jmath^*(x_i) = 0,\ i \in \mathbb{Z}/p.$$

These elements (or their linear combinations) may potentially be 0 for $i \neq 0$. The trace on C must be nondegenerate, pairwise products of x_i's are 0 and $\mathsf{tr}_C(x_i) = 0$ for $i \neq 0$. To avoid introducing more generators for C, we further assume that $x_i = 0$ for $i \neq 0$ and look to complete B to a knowledgeable Frobenius algebra (B, C) with $C \cong \mathbf{k}[x]/(x^2)$. Then the zipper and the cozipper maps are

$$\jmath : B \longrightarrow C, \jmath(g^i) = 0,\ i \neq 0 \in \mathbb{Z}/p, \ \ \jmath(1_B) = x, \ \ \mathsf{im}(\jmath) = \mathbf{k}\, x,$$
$$\jmath^* : C \longrightarrow B, \jmath^*(1_C) = 1_B, \ \ \jmath^*(x) = 0, \ \ \mathsf{im}(\jmath^*) = \mathbf{k}\, 1.$$

with $\jmath^* \jmath = 0$, map $\jmath$ a coalgebra homomorphism and $\jmath^*$ an algebra homomorphism. Each individual map $\jmath, \jmath^*$ is nonzero but has a one-dimensional image. To define the trace on C, pick a parameter $\lambda \in \mathbf{k}$:

$$\mathsf{tr}_C(1) = \lambda, \quad \mathsf{tr}_C(x) = 1. \tag{80}$$

The multiplication ($x^2 = 0$) and trace on C determine the comultiplication

$$\Delta_C(1) = 1 \otimes x + x \otimes 1 - \lambda x \otimes x, \quad \Delta_C(x) = x \otimes x. \tag{81}$$

The remaining open-closed TQFT relations, as shown in Figure 3.2.2, are straightforward to check.

In this knowledgeable Frobenius algebra, both B and C are nonsemisimple algebras. The handle (or punctured torus) element of C equals $m_C \circ \Delta_C(1) = 2x$, and a closed surface of genus g evaluates to $\alpha_{0,g}$, where

$$\alpha_{0,0} = \lambda, \quad \alpha_{0,1} = 2, \quad \alpha_{0,g} = 0 \text{ for } g > 1 \tag{82}$$

(recall that the coefficients belong to a field $\mathbf{k}$ of characteristic p). Evaluation $\alpha_{m,g}$ of a connected surface of genus g with m side boundary circles is

$$\alpha_{1,0} = 1, \quad \alpha_{m,g} = 0 \text{ if } m > 0 \text{ and } g \neq 0. \tag{83}$$

Thus, in this evaluation, at most three coefficients: $\alpha_{0,0}, \alpha_{0,1}, \alpha_{1,0}$ are nonzero.

More generally, $\jmath^*\jmath = 0$ if and only if $\alpha_{m,g} = 0$ for all $m \geq 1, g \geq 1$ and $m \geq 2, g \geq 0$.

Example 3.7 in [**LP09**] is somewhat similar to the present example, with the algebra B isomorphic to the one above (and $\mathsf{char}(\mathbf{k}) = p$), algebra $C = \mathbf{k}[x]/(x^2 - ht - t)$ two-dimensional and nonsemisimple when parameters h, t satisfy $h^2 = 4t$, but with different zipper and trace maps. In particular, $\jmath^*\jmath \neq 0$ in that example ($\iota\iota^* \neq 0$ in the notations of [**LP09**]).

In the above example of a knowledgeable Frobenius pair (B, C) both B and C are nonsemisimple and the maps β_B and $F_B(\beta_3)$, see (74) and (75), are zero, so that $\jmath^*\jmath = 0$.

Frobenius algebras that appear in link homology and categorification are typically nonsemisimple, which creates an obstacle to merging link homology with open-closed 2D TQFTs, where a vast majority of examples is built from semisimple Frobenius algebras. This obstruction is discussed in [**LP09**] and [**Cap13**]. One well-known way out of this is a functorial extension of link homology to tangles [**Kho02**] and then to tangle cobordisms [**Kho06, BN05**].

This discrepancy between semisimple Frobenius algebras common in open-closed TQFTs in dimension two and rather special nonsemisimple Frobenius algebras that give rise to link cobordism TQFTs in dimension four and categorification of quantum invariants is an interesting phenomenon that is not fully understood.

Combining a symmetric Frobenius algebra with the universal construction for closed surfaces. Even when symmetric Frobenius B does not extend to a knowledgeable Frobenius (B, C) it is possible to extend B to a functor from OCCob_2 to the category of vector spaces but with a weaker axioms than that of a TQFT. This can be achieved by combining the Frobenius structure (B, tr_B) with the universal construction.

Symmetric Frobenius algebra (B, tr_B) gives a thin flat surface TQFT. In particular, it evaluates any connected oriented surface $S_{n,g}$ with $n \geq 1$ boundary components and g handles to a number $\alpha_{B,n,g} \in \mathbf{k}$. This number can be computed by viewing the surface as an endomorphism of the identity object 0 of TCob_2 and computing the element of $\mathbf{k}$ it goes to under the functor F_B. Alternatively, one can use the surgery relation in Figure 3.1.8 and other relations in Figures 3.1.8, 3.1.6.

Doing the universal construction in the modification of TCob_2 where side boundaries are decorated by generators of B results in the TQFT F_B.

To extend to all oriented surfaces, we choose an evaluation $\alpha_{0,g} \in \mathbf{k}$ of a closed oriented surface of genus g for all $g \geq 0$. It is convenient to require that the generating function

$$Z_0(T) := \sum_{g\geq 0} \alpha_{0,g} T^g \tag{84}$$

is rational. With this additional choice evaluations of all surfaces $S_{n,g}$, $n, g \geq 0$ of genus g with n boundary circles are defined.

The universal construction can be applied to the category OCCob_2 of open-closed cobordisms. To get a better match with state spaces build from (B, tr_B) for cobordisms that have corners, it is convenient to pick a set W of generators of B and allow these generators to float on side boundaries of cobordisms. One obtains a minor modification, denoted $\mathsf{OCCob}_2(W)$ of the category OCCob_2. Endomorphisms of the 0 object of OCCob_2 that come from surfaces with decorated side boundaries are then evaluated via (B, tr_B) while evaluations of closed surfaces are encoded in the generating function (84).

Let us do the universal construction for $\mathsf{OCCob}_2(W)$ evaluating surfaces with side boundary their possible W-decorations via (B, tr_B) and closed surfaces via coefficients of (84). The resulting category and a functor is an extension of the thin surface TQFT associated with (B, tr_B). Objects of $\mathsf{OCCob}_2(W)$ are finite unions $I^{\sqcup k} \sqcup (\mathbb{S}^1)^{\sqcup m}$ of intervals and circles. Denote the state space of that one-manifold by $A(k, m)$. Then the surgery formula still applies near each interval component, and the state space simplifies via the isomorphism

$$A(k, m) \cong B^{\otimes k} \otimes A(0, m). \tag{85}$$

That is, the state space is isomorphic to the tensor product of B's, one for each interval, and the state space of m circles. The latter state space contains a subquotient isomorphic to the state space of m circles in the closed 2D topological theory with the generating function (84), as studied in [**Kho20b, KS20, KOK22**]. The state space $A(0, m)$ may be strictly bigger than the latter state space, due to the presence of surfaces that bound m circles at the top but have side circles (such surfaces can be viewed as morphisms in $\mathsf{OCCob}_2(W)$ from the identity object 0 to m circles).

This universal construction based on (B, tr_B) and rational power series (84) occupies an intermediate position between open-closed TQFTs and the more general universal construction for surfaces with boundary and corners studied in [**KQR21**]. In the present case (B, tr_B) allows to evaluate surfaces of all genera with at least one side boundary circle and produce a TQFT (as long as we add additional observables on the boundary lines for generators of B) for these surfaces, then extend to cobordisms that may have top and bottom boundary circles and closed surfaces via (84). Without enlarging OCCob_2 to $\mathsf{OCCob}_2(W)$ the resulting state space of the union of k intervals could be only a subspace of $B^{\otimes k}$.

REMARK 3.7. Going back to the category TCob_2 of thin flat surfaces, one can further introduce one-dimensional interval defects that connect two points on the boundary of a surface. These defects are labelled by endomorphisms of B (by $\mathbf{k}$-linear maps $B \to B$), see Figure 3.2.4.

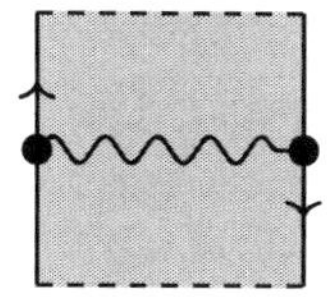

FIGURE 3.2.4. A line defect connecting two boundary points can be labelled by an element of $\mathsf{End}_{\mathbf{k}}(B)$.

REMARK 3.8. Categorifications of the Heisenberg algebra come from the study of natural transformations on compositions of the induction and restriction functors between symmetric groups or Hecke algebras [**Kho14, LS13**]. More general categorifications of the Heisenberg algebra [**BSW22, Sav19, RS17, BSW21**] add elements of a Frobenius algebra as decorations on strands of diagrams in those graphical calculi. The neck-cutting formula in Figure 3.1.8 on the left is called *the Frobenius skein relation* in that case and is referred to as *teleportation* in [**BSW22**].

A symmetric Frobenius algebra gives a TQFT for thin flat surfaces (two-dimensional objects), which is one of the indications that various Heisenberg algebra categorifications should admit reformulations via a suitable graphical calculus of foam-like objects in $\mathbb{R}^3$ rather than graphs (or intersecting decorated lines) in $\mathbb{R}^2$.

4. Embeddings into a 1D TQFT and dimensional lifting

Embeddings into a 1D TQFT: semisimplicity restriction. It is natural to ask under what conditions on $\mathcal{K}$ and the trace $\mathsf{tr}_{\mathcal{K}}$ is the corresponding one-dimensional theory of arcs with $\mathcal{K}$-defects and the circle evaluation given by $\mathsf{tr}_{\mathcal{K}}$ embeddable into a one-dimensional TQFT. Oriented 1D TQFTs are described by finite-dimensional vector spaces V, with the state space of $+-$ oriented 0-manifold isomorphic to $V \otimes V^*$. When viewed as an algebra under the composition in Figure 2.4.6 it is naturally isomorphic to the endomorphism or the matrix algebra $\mathsf{End}_{\mathbf{k}}(V) \cong M_n(\mathbf{k})$.

Suppose given a symmetric Frobenius algebra (B, tr_B). It gives rise to the category of arcs with B-defects and circles evaluated via tr_B and the negligible quotient of that category. A monoidal functor from either of these two categories into a 1D TQFT given by V is described by a homomorphism of algebras $\phi : B \longrightarrow M_n(\mathbf{k})$ that converts trace tr_B to the usual trace on the matrix algebra $M_n(\mathbf{k})$, that is $\mathsf{tr}(\phi(a)) = \mathsf{tr}_B(a), \forall a \in B$. That is, ϕ must intertwine the two traces. Then ϕ is necessarily an inclusion and $\mathsf{tr}_B(1) = n = \dim(V)$.

Consider the Jacobson radical $J \subset B$. Then J is a two-sided nilpotent ideal, $J^n = 0$, and B/J is semisimple. Any element $x \in J$ is nilpotent and $\phi(x)$ is a nilpotent matrix, so that $\mathsf{tr}(\phi(x)) = 0$. Consequently, $\mathsf{tr}_B(x) = 0$, for all elements x in the Jacobson radical. Nondegeneracy of tr_B implies that $J = 0$, so that B is semisimple, and we obtain the following result.

PROPOSITION 4.1. A one-dimensional topological theory with B-labelled defects associated to (B, tr_B) can be embedded into a one-dimensional TQFT over $\mathbf{k}$–vect only if B is semisimple.

We see that tr_B can come from a trace on a matrix algebra in the above way only if B is a semisimple $\mathbf{k}$-algebra, while in applications (for instance, to link homology) we most often encounter cases when B is not semisimple.

Furthermore, assuming that B is semisimple, only few traces on B correspond to embeddings into one-dimensional TQFTs. Namely, $B \cong \prod_{i=1}^{k} \mathsf{Mat}_{n_i}(D_i)$ is then isomorphic to the product of matrix algebras over finite-dimensional division rings D_i over F and representation V has the form $V \cong \oplus_{i=1}^{k} V_i$, where

$$V_i \cong (D_i^{n_i})^{r_i} \tag{86}$$

is the sum of r_i copies of the column representation $D_i^{n_i}$ of the matrix algebra $\mathsf{Mat}_{n_i}(D_i)$. Under this isomorphism,

$$\mathsf{tr}_B = \sum_{i=1}^{k} r_i \, \mathsf{tr}_i, \tag{87}$$

where tr_i is the trace on $\mathsf{Mat}_{n_i}(D_i)$ with values in $\mathbf{k}$ which is the composition of the matrix algebra trace and the map $\mathsf{tr}_i' : D_i \longrightarrow \mathbf{k}$ which is the trace of left multiplication in D_i viewed as a $\mathbf{k}$-vector space.

For example, suppose that the division ring D_i is commutative, thus it is a field F such that $\mathbf{k} \subset F$ is a finite extension and that $n_i = 1$. Any non-zero $\mathbf{k}$-linear map $\varepsilon : F \longrightarrow \mathbf{k}$ turns F into a commutative Frobenius $\mathbf{k}$-algebra, but only the trace map $\mathsf{tr}_{\mathbf{k}} : F \longrightarrow \mathbf{k}$ and its multiples $r \, \mathsf{tr}_{\mathbf{k}}$, $r \in \mathbb{N}$ (further assuming that $F/\mathbf{k}$ is separable) come from embeddings into a 1D TQFT.

Dimensional liftings. We see that trace-preserving embeddings of symmetric Frobenius algebras into matrix algebras are scarce. At the same time, a symmetric Frobenius algebra (B, tr_B) gives rise to a 2D TQFT for thin surfaces as explained earlier. A one-dimensional TQFT with defects α produces a two-dimensional TQFT, restricted to thin surfaces, via the symmetric Frobenius algebra $(\mathcal{K}, \mathsf{tr}_{\mathcal{K}})$. This dimensional lifting from one to two dimension can be very loosely compared to the Drinfeld center of a monoidal category (and the Drinfeld double of a Hopf algebra). Monoidal categories are naturally two-dimensional structures, with morphisms often represented by planar diagram. The Drinfeld center of a monoidal category is a braided monoidal category, providing invariants of braids and lifting the structure one dimension up, from two to three dimensions. Likewise, the Drinfeld double of a Hopf algebra converts a two-dimensional structure (the category of representations of a Hopf algebra is monoidal) to a three-dimensional structure (a quasitriangular Hopf algebra, with the category of representations being a braided monoidal category).

The graphical nature of a monoidal category $\mathcal{C}$ is that of planar networks of morphisms between tensor products of objects of $\mathcal{C}$. Such planar networks can be thought of as defects in the two-dimensional theory of the underlying plane $\mathbb{R}^2$. Drinfeld's center and doubling constructions lift these "two-dimensional theories with defects" to three-dimensional theories.

Of course, the above discussion and comparison of dimensional liftings is highly informal.

Acknowledgments. We would like to thank Aaron Lauda and Vladimir Retakh for illuminating discussions. M.S.I. was partially supported by Naval Academy Research Council (Jr. NARC) Fellowship over the summer. M.K. gratefully acknowledges partial support via NSF grants DMS-1807425, DMS-2204033 and Simons Collaboration Award 994328.

References

[BHMV95] C. Blanchet, N. Habegger, G. Masbaum, and P. Vogel, *Topological quantum field theories derived from the Kauffman bracket*, Topology **34** (1995), no. 4, 883–927, DOI 10.1016/0040-9383(94)00051-4. MR1362791

[BN05] Dror Bar-Natan, *Khovanov's homology for tangles and cobordisms*, Geom. Topol. **9** (2005), 1443–1499, DOI 10.2140/gt.2005.9.1443. MR2174270

[BR90] Jean Berstel and Christophe Reutenauer, *Zeta functions of formal languages*, Trans. Amer. Math. Soc. **321** (1990), no. 2, 533–546, DOI 10.2307/2001573. MR998123

[BSW21] Jonathan Brundan, Alistair Savage, and Ben Webster, *Foundations of Frobenius Heisenberg categories*, J. Algebra **578** (2021), 115–185, DOI 10.1016/j.jalgebra.2021.02.025. MR4234799

[BSW22] Jonathan Brundan, Alistair Savage, and Ben Webster, *Quantum Frobenius Heisenberg categorification*, J. Pure Appl. Algebra **226** (2022), no. 1, Paper No. 106792, 50 pp., DOI 10.1016/j.jpaa.2021.106792. MR4262074

[Cap13] Carmen Caprau, *Twin TQFTs with Frobenius algebras*, J. Math., posted on 2013, Art. ID 407068, 25 pp., DOI 10.1155/2013/407068. MR3101321

[GIK+23] Paul Gustafson, Mee Seong Im, Remy Kaldawy, Mikhail Khovanov, and Zachary Lihn, *Automata and one-dimensional TQFTs with defects*, arXiv preprint arXiv:2301.00700 (2023), 1–36.

[IK22] Mee Seong Im and Mikhail Khovanov, *Topological theories and automata*, arXiv preprint arXiv:2202.13398 (2022), 1–70.

[IKO23] Mee Seong Im, Mikhail Khovanov, and Victor Ostrik, *Universal construction in monoidal and non-monoidal settings, the Brauer envelope, and pseudocharacters*, arXiv preprint arXiv:2303.02696 (2023), 1–59.

[IZ22] Mee Seong Im and Paul Zimmer, *One-dimensional topological theories with defects and linear generating functions*, Involve **15** (2022), no. 2, 319–331, DOI 10.2140/involve.2022.15.319. MR4462160

[Kho02] Mikhail Khovanov, *A functor-valued invariant of tangles*, Algebr. Geom. Topol. **2** (2002), 665–741, DOI 10.2140/agt.2002.2.665. MR1928174

[Kho04] Mikhail Khovanov, *sl(3) link homology*, Algebr. Geom. Topol. **4** (2004), 1045–1081, DOI 10.2140/agt.2004.4.1045. MR2100691

[Kho06] Mikhail Khovanov, *An invariant of tangle cobordisms*, Trans. Amer. Math. Soc. **358** (2006), no. 1, 315–327, DOI 10.1090/S0002-9947-05-03665-2. MR2171235

[Kho14] Mikhail Khovanov, *Heisenberg algebra and a graphical calculus*, Fund. Math. **225** (2014), no. 1, 169–210, DOI 10.4064/fm225-1-8. MR3205569

[Kho20a] ———, *Decorated one-dimensional cobordisms and tensor envelopes of noncommutative recognizable power series*, arXiv preprint arXiv:2010.05730 (2020), 1–33.

[Kho20b] ———, *Universal construction of topological theories in two dimensions*, arXiv preprint arXiv:2007.03361 (2020), 1–56.

[KK20] Mikhail Khovanov and Nitu Kitchloo, *A deformation of Robert-Wagner foam evaluation and link homology*, arXiv preprint arXiv:2004.14197 (2020), 1–58.

[KOK22] Mikhail Khovanov, Victor Ostrik, and Yakov Kononov, *Two-dimensional topological theories, rational functions and their tensor envelopes*, Selecta Math. (N.S.) **28** (2022), no. 4, Paper No. 71, 68 pp., DOI 10.1007/s00029-022-00785-z. MR4450143

[KQ20] Mikhail Khovanov and You Qi, *Introduction to categorification: Section 2, Topological Quantum Field Theories*, Lecture notes: Cob_n and n-d TQFT (2020), 1–22.

[KQR21] Mikhail Khovanov, You Qi, and Lev Rozansky, *Evaluating thin flat surfaces*, Comm. Math. Phys. **385** (2021), no. 3, 1835–1870, DOI 10.1007/s00220-021-04011-3. MR4284003

[KS20] Mikhail Khovanov and Radmila Sazdanovic, *Bilinear pairings on two-dimensional cobordisms and generalizations of the Deligne category*, arXiv preprint arXiv:2007.11640 (2020), 1–15.

[KS21] ———, *Diagrammatic categorification of the Chebyshev polynomials of the second kind*, J. Pure Appl. Algebra **225** (2021), no. 6, Paper No. 106592, 23 pp., https://doi.org/10.1016/j.jpaa.2020.106592.

[Lau05] Aaron Lauda, *Frobenius algebras and planar open string topological field theories*, arXiv preprint arXiv:math/0508349 (2005), 1–66.

[Lau06] ______, *Open-Closed Topological Quantum Field Theory and Tangle Homology*, Ph.D. thesis, University of Cambridge, 2006, pp. 1–169.

[Laz01] C. I. Lazaroiu, *On the structure of open-closed topological field theory in two dimensions*, Nuclear Phys. B **603** (2001), no. 3, 497–530, DOI 10.1016/S0550-3213(01)00135-3. MR1839382

[LP07] Aaron D. Lauda and Hendryk Pfeiffer, *State sum construction of two-dimensional open-closed topological quantum field theories*, J. Knot Theory Ramifications **16** (2007), no. 9, 1121–1163, DOI 10.1142/S0218216507005725. MR2375819

[LP08] Aaron D. Lauda and Hendryk Pfeiffer, *Open-closed strings: two-dimensional extended TQFTs and Frobenius algebras*, Topology Appl. **155** (2008), no. 7, 623–666, DOI 10.1016/j.topol.2007.11.005. MR2395583

[LP09] Aaron D. Lauda and Hendryk Pfeiffer, *Open-closed TQFTS extend Khovanov homology from links to tangles*, J. Knot Theory Ramifications **18** (2009), no. 1, 87–150, DOI 10.1142/S0218216509006793. MR2490019

[LS13] Anthony Licata and Alistair Savage, *Hecke algebras, finite general linear groups, and Heisenberg categorification*, Quantum Topol. **4** (2013), no. 2, 125–185, DOI 10.4171/QT/37. MR3032820

[Mei21] Ehud Meir, *Interpolations of monoidal categories and algebraic structures by invariant theory*, arXiv preprint arXiv:2105.04622 (2021), 1–31.

[MS06] Gregory W. Moore and Graeme Segal, *D-branes and K-theory in 2D topological field theory*, arXiv preprint arXiv:hep-th/0609042 (2006), 1–88.

[MS22] Alexandra McSween and Alistair Savage, *Affine oriented Frobenius Brauer categories*, Comm. Algebra **51** (2023), no. 2, 742–756, DOI 10.1080/00927872.2022.2113401. MR4532823

[MV07] Marco Mackaay and Pedro Vaz, *The universal* $\mathfrak{sl}_3$*-link homology*, Algebr. Geom. Topol. **7** (2007), 1135–1169, DOI 10.2140/agt.2007.7.1135. MR2336253

[RRV99] Vladimir Retakh, Christophe Reutenauer, and Arkady Vaintrob, *Noncommutative rational functions and Farber's invariants of boundary links*, Differential topology, infinite-dimensional Lie algebras, and applications, Amer. Math. Soc. Transl. Ser. 2, vol. 194, Amer. Math. Soc., Providence, RI, 1999, pp. 237–246, DOI 10.1090/trans2/194/11. MR1729366

[RS17] Daniele Rosso and Alistair Savage, *A general approach to Heisenberg categorification via wreath product algebras*, Math. Z. **286** (2017), no. 1-2, 603–655, DOI 10.1007/s00209-016-1776-9. MR3648512

[RW20] Louis-Hadrien Robert and Emmanuel Wagner, *A closed formula for the evaluation of foams*, Quantum Topol. **11** (2020), no. 3, 411–487, DOI 10.4171/qt/139. MR4164001

[Sav19] Alistair Savage, *Frobenius Heisenberg categorification*, Algebr. Comb. **2** (2019), no. 5, 937–967, DOI 10.5802/alco.73. MR4023572

[Sav21] ______, *Affine oriented Frobenius Brauer categories*, Presentation Slides (2021), 1–33.

[SS22] Saima Samchuck-Schnarch, *Frobenius Brauer Categories*, uOttawa Research (2022), 1–107, PhD Thesis, University of Ottawa.

Department of Mathematics, United States Naval Academy, Annapolis, Maryland 21402

Email address: meeseongim@gmail.com

Department of Mathematics, Columbia University, New York, New York 10027

Email address: khovanov@math.columbia.edu

Contemporary Mathematics
Volume **791**, 2024
https://doi.org/10.1090/conm/791/15875

A deformation of Robert-Wagner foam evaluation and link homology

Mikhail Khovanov and Nitu Kitchloo

ABSTRACT. We consider a deformation of the Robert-Wagner foam evaluation formula, with an eye toward a relation to formal groups. Integrality of the deformed evaluation is established, giving rise to state spaces for planar $GL(N)$ MOY graphs (Murakami-Ohtsuki-Yamada graphs). Skein relations for the deformation are worked out in details in the $GL(2)$ case. These skein relations deform $GL(2)$ foam relations of Beliakova, Hogancamp, Putyra and Wehrli. We establish the Reidemeister move invariance of the resulting chain complexes assigned to link diagrams, giving us a link homology theory.

CONTENTS

1. Introduction

1.1. MOY graphs and quantum invariants for level one representation. Foams are 2-dimensional combinatorial CW-complexes, often with extra decorations, embedded in $\mathbb{R}^3$. They naturally appear [**Kh2, KRo2, MV1, MSV, QR, RWd**] in the study of link homology theories that categorify quantum $\mathfrak{sl}_N$ or $\mathfrak{gl}_N$ link invariants for level one representations when $N \geq 3$.

Reshetikhin-Turaev-Witten invariants [**RT, W**] of oriented links L in the 3-sphere $\mathbb{S}^3$ depend on the choice of a simple Lie algebra $\mathfrak{g}$ and an irreducible representation of $\mathfrak{g}$ associated to each component of L. When $\mathfrak{g} = \mathfrak{sl}_N$ and the components are labelled by level one representations of $\mathfrak{sl}_N$, the Reshetikhin-Turaev-Witten invariant $P(L) \in \mathbb{Z}[q, q^{-1}]$ can be written [**MOY**] as a linear combinations of terms $P(\Gamma) \in \mathbb{Z}_+[q, q^{-1}]$ over trivalent oriented planar graphs Γ with edges labelled by integers between 1 to N. $P(\Gamma)$ is known as the Murakami-Ohtsuki-Yamada or *MOY invariant* of Γ.

2020 *Mathematics Subject Classification.* Primary 57K18, 14L05, 18N25.

An edge labelled a corresponds to the identity intertwiner of $\Lambda_q^a V$, the latter a quantum group representation which q-deforms the a-th exterior power of the fundamental representation of the Lie algebra $\mathfrak{sl}_N$. At this point it is convenient to shift from $\mathfrak{sl}_N$ to $\mathfrak{gl}_N$, and view $\Lambda_q^a V$ as a representation of $U_q(\mathfrak{gl}_N)$ rather than that of $\mathfrak{sl}_N$. This change will be more essential at the categorified level of homological invariants rather than for uncategorified quantum invariants, taking values in $\mathbb{Z}[q, q^{-1}]$.

Oriented labelled graphs Γ are built out of trivalent vertices that correspond to suitably scaled inclusion and projection of $\Lambda_q^{a+b} V$ into and out of the tensor product $\Lambda_q^a V \otimes \Lambda_q^b V$, see Figure 1.11.

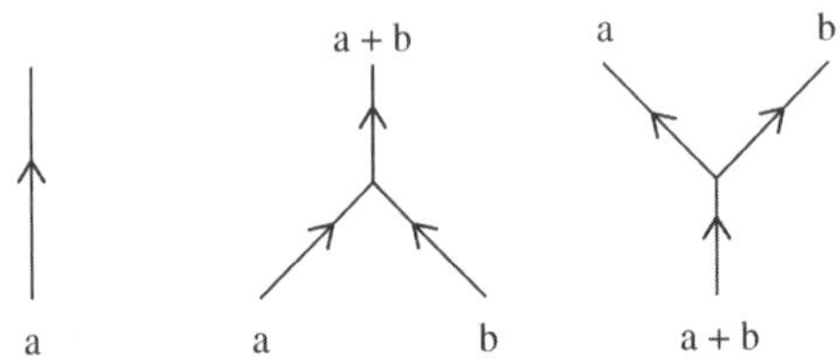

FIGURE 1.11. Generating diagrams for $GL(N)$ MOY graphs. They correspond to the identity intertwiner on $\Lambda_q^a V$ and projection and inclusion (up to scaling) between $\Lambda_q^a V \otimes \Lambda_q^b V$ and $\Lambda_q^{a+b} V$.

Quantum $\mathfrak{gl}_N$ (or MOY) invariant of Γ is given by a suitable convolution of these maps, which for closed graphs Γ results in a Laurent polynomial $P(\Gamma) \in \mathbb{Z}[q, q^{-1}]$ with nonnegative coefficients, see [**MOY**] for integrality and [**RW2**, Appendix 2A] for nonnegativity via a suitable state sum formula. Planar graph invariant $P(\Gamma)$ can be computed either via a state sum formula or inductively via skein relations.

As we mention earlier, $\mathbb{Z}[q, q^{-1}]$-linear combinations of invariants $P(\Gamma)$ give quantum link invariants $P(L)$, when $\mathfrak{g} = \mathfrak{gl}_N$ and components of L are labelled by level one representations, that is, by $\Lambda_q^a V$, over different a's.

The reason for the popularity of this specialization (from $\mathfrak{g}$ to $\mathfrak{gl}_N$ and to level one representations), especially with an eye towards categorification, is the relative simplicity of these formulas compared to the case of general $\mathfrak{g}$ and its representations, where canonical choices of intertwiners associated to graph's vertices are harder to guess, spaces of these intertwiners may be more than one-dimensional, decomposition of a crossing into a linear combinations of planar graphs has more complicated coefficients or may be difficult to select, and evaluations of $P(\Gamma)$ lose positivity, acquire denominators and live in $\mathbb{Q}(q)$ rather than $\mathbb{Z}_+[q, q^{-1}]$. Any such complication makes categorical lifting noticeably harder. An approach to categorification of the Reshetikhin-Turaev-Witten link invariants for an arbitrary $\mathfrak{g}$ and arbitrary representations has been developed by Webster [**We**]. It is an open problem to find a foam-like interpretation of Webster link homology theories and refine them to achieve functoriality under link cobordisms.

1.2. Foams and Robert-Wagner evaluation. The key property of $P(\Gamma)$ is it having non-negative coefficients, that is, taking values in $\mathbb{Z}_+[q, q^{-1}]$, rather than just in $\mathbb{Z}[q, q^{-1}]$, where link invariants $P(L)$ live. In the lifting of $P(L)$ to homology groups, state spaces $\langle \Gamma \rangle$ will be graded, with graded rank (as a free module over the graded ring R_N of symmetric functions, see below) having non-negative coefficients, thus lying in $\mathbb{Z}_+[q,q^{-1}]$, Homology groups $H(L)$ come from complexes of state spaces $\langle\Gamma\rangle$, built from various resolutions Γ of L.

Louis-Hadrien Robert and Emmanuel Wagner discovered a remarkable evaluation formula for $GL(N)$ foams [**RW1**]. Their formula leads to a natural construction of homology groups (or state spaces) for each planar trivalent MOY graph Γ as above.

At the categorified level of this story, Robert-Wagner foam evaluation leads to a state space $\langle\Gamma\rangle$, a graded module over the ring $R_N = \mathbb{Z}[x_1,\dots,x_N]^{S_N}$ of symmetric polynomials in $x_1,\dots,x_N$ with coefficients in $\mathbb{Z}$. Robert and Wagner prove [**RW1**] that the graded R_N-module is free and finitely-generated, of graded rank $P(\Gamma)$.

Thus, graded rank of R_N-module $\langle\Gamma\rangle$ categorifies the quantum $\mathfrak{gl}_N$ invariant (the Murakami-Ohtsuki-Yamada invariant) of these planar graphs. Forming suitable complexes out of these state spaces and taking homology groups leads to bigraded homology theories of links that categorify the HOMFLYPT polynomial and its generalizations to other quantum exterior powers of the fundamental representation [**ETW**], see also earlier approaches [**Y, Wu1, Wu2**] to categorification of $\mathfrak{gl}_N$ link homology with components colored by arbitrary level one representations.

We now recall the details of Robert-Wagner's foam invariant. A $GL(N)$-foam F is a two-dimensional piecewise-linear compact CW-complex F embedded in $\mathbb{R}^3$. Its facets are oriented in a compatible way and labelled by numbers from 0 to N called the *thickness* of a facet (facets of thickness 0 may be removed) with points of three types:

- A regular point on a facet of thickness a.
- A point on a singular edge, which has a neighbourhood homeomorphic to the product of a tripod T and an interval I. The three facets must have thickness $a, b, a+b$ respectively. One can think of thickness a, b facets as merging into the thick facet or vice versa, of the facet of thickness $a+b$ splitting into two thinner facets of thickness a and b.
- A singular vertex where four singular edges meet. The six corners of the foam at the vertex have thickness $a, b, c, a+b, b+c, a+b+c$ respectively.

Neighbourhoods of these three types of points are depicted below.

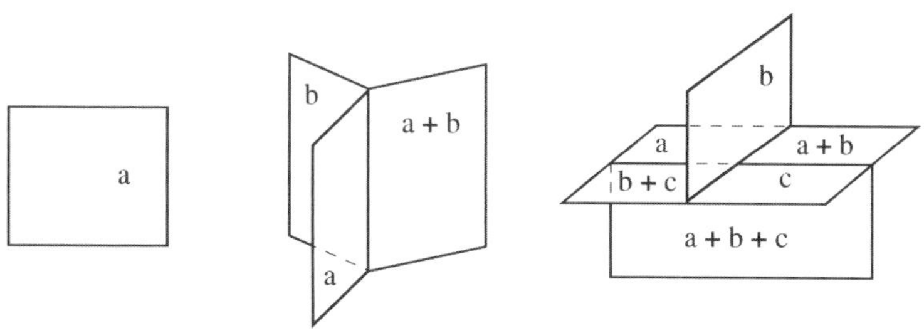

FIGURE 1.21. Three types of points on a foam.

Orientations of facets are compatible at singular edges, see Figure 1.23 below.

A singular vertex can be viewed, see Figure 1.22, as the singular point of the cobordism between two labelled trees that are the two splittings of an edge of thickness $a + b + c$ into edges of thickness a, b, c, respectively. This is a kind of "associativity" cobordism, which is invertible when viewed as an appropriate module map between state spaces associated to MOY planar graphs in the foam theory. We follow the orientation conventions from [**ETW**]. They show compatible

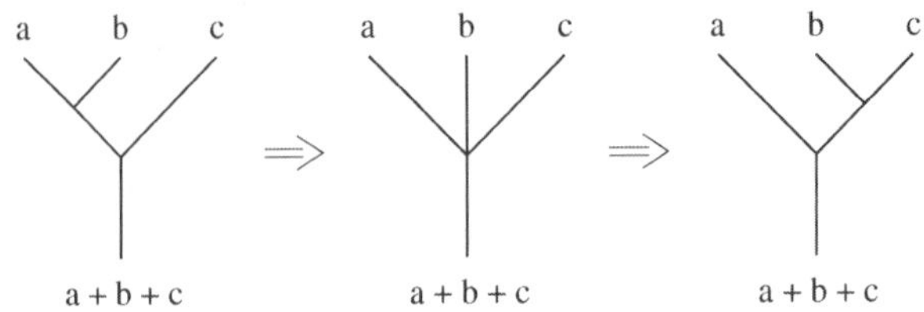

FIGURE 1.22. Cross-sections near a singular vertex.

orientations on facets of thickness a and b attached along a singular edge to a facet of thickness $a+b$. The same diagram shows induced orientations on top and bottom boundaries of foam F. This convention will be used once we pass from closed foams to foams with boundary, viewed as cobordisms between $GL(N)$ MOY graphs.

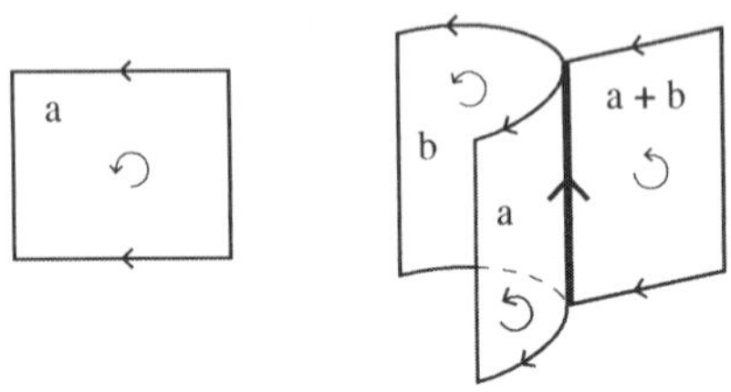

FIGURE 1.23. Orientation conventions from [**ETW**, Figure 1]. An orientation of a facet induces an orientation of its top boundary (if non-empty, for non-closed foams only) by sticking the first vector of the orientation basis up out of the foam. The remaining vector then induces an orientation of the boundary. For the bottom boundary the resulting orientation is reversed. To induce an orientation on a singular circle, approach it with an orientation basis from a thin facet and point the first vector into the thick facet. The second vector then defines an orientation of the singular circle (or a singular arc, if foam is not closed). This is the one convention we choose out of the four possible conventions for inducing orientations on the boundary and on singular lines, given an orientation of a facet.

Facets f of a foam F are the connected components of the set $F \setminus s(F)$, where $s(F)$ is the set of the singular points of F. Thickness of f is denoted $\ell(f)$. The set of facets of F is denoted $f(F)$. A coloring c of F is a map $c : f(F) \longrightarrow 2^{I_N}$ from the set of facets to the set of subsets of $I_N = \{1, \dots, N\}$ such that subset $c(f)$ has cardinality $\ell(f)$ and for any three facets f_1, f_2, f_3 attached to a singular edge with $\ell(f_3) = \ell(f_1) + \ell(f_2)$ equality $c(f_3) = c(f_1) \sqcup c(f_2)$ holds. In other words, the subset for f_3 is the union of subsets for f_1 and f_2. A foam may come with decorations (dots). A dot on a facet f of thickness a represents a homogeneous symmetric polynomial P_f in a variables.

Any coloring c gives rise to closed surfaces $F_i(c)$, $1 \le i \le N$, which are unions of facets f such that $c(f)$ contains i. One also forms symmetric differences $F_{ij}(c) = F_i(c)\Delta F_j(c)$, which are the unions of facets f such that $c(f)$ contains exactly one element of the set $\{i, j\}$. Surfaces $F_{ij}(c)$, $i \neq j$ are closed orientable as well.

Rogert-Wagner evaluation $\langle F, c\rangle_{RW}$ of a foam on a coloring c is

$$\langle F, c\rangle_{RW} = (-1)^{s(F,c)} \frac{P(F,c)}{Q(F,c)}, \tag{1.1}$$

where

$$\begin{aligned} s(F,c) &= \theta^+(c) + \sum_{i=1}^{N} i\chi(F_i(c))/2\ , \\ \theta^+(c) &= \sum_{i<j} \theta^+_{ij}(c), \\ P(F,c) &= \prod_{f\in f(F)} P_f(c), \\ Q(F,c) &= \prod_{1\le i<j\le N} (x_i - x_j)^{\chi(F_{ij}(c))/2}. \end{aligned}$$

Here $\theta^+_{ij}(c)$ counts the number of circular seams on the surface $F_{ij}(c)$ along which the cyclic order of the three attached facets (with i but not j in the coloring c, with j but not i in the coloring, and with i, j in the coloring) is one of the two types, called *positive* type. Positivity is determined by the left hand rule with the direction of the thumb along the positive orientation, on turning the fingers of the left hand from the facet with color i to the one with color j for $i < j$.

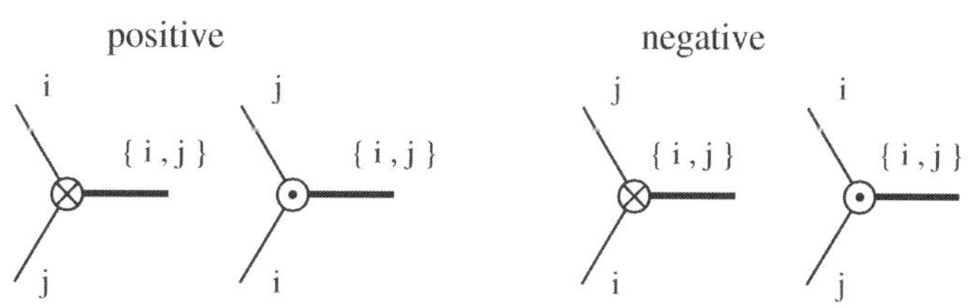

FIGURE 1.24. θ^+_{ij} counts the number of positive (i, j)-circles, $i < j$.

$\chi(S)$ denotes the Euler characteristic of the surface S. Term $P_f(c)$ is the product of symmetric functions associated to the dots on the facet f, in variables x_k, where k runs over the elements of $c(f)$.

Now define

$$\langle F\rangle_{RW} \ = \ \sum_{c} \langle F, c\rangle_{RW}, \tag{1.2}$$

the sum over all colorings c of F. We refer the readers to [**RW1**] for more details on $GL(N)$-foams and their evaluations.

Later in the paper we deform Robert-Wagner evaluation and for the most part work with the deformation. To keep notations light, we use Robert-Wagner's notation $\langle F\rangle$ to denote deformed evaluation and denote their original one by $\langle F\rangle_{RW}$.

One of the first key results of Robert and Wagner [**RW1**] is that $\langle F\rangle_{RW}$ is a (symmetric) polynomial rather than just a rational function of $x_1,\dots,x_N$, thus an element of R_N.

Generic intersection of a foam F with a plane $\mathbb{R}^2$ in $\mathbb{R}^3$ will result in an oriented planar graph Γ, which is exactly a $GL(N)$ MOY graph. It is straightforward to introduce foams with boundary. With the evaluation $\langle F\rangle_{RW}$ for closed foams at hand, one can now define the state space $\langle\Gamma\rangle_{RW}$ of Γ as a graded R_N-module freely generated by symbols $\langle F\rangle_{RW}$ of foams F from the empty graph to Γ, modulo the relations that $\sum_k a_k\langle F_k\rangle_{RW} = 0$ for foams F_k from $\emptyset$ to Γ and $a_k \in R_N$ iff for any foam G from Γ to the empty graph $\sum_k a_k\langle GF_k\rangle_{RW} = 0$. Here GF_k is a closed foam, the gluing or composition of G and F_k along Γ. Robert-Wagner state spaces (or homology) of graphs Γ are then used as building blocks for link homology groups [**ETW**].

As an informal remark, we want to point out that the foams considered in the Robert-Wagner construction [**RW1**] should really be called $GL(N)$*-foams*. For $SL(N)$*-foams* one would want to allow seamed edges along which three facets of thickness a,b,c with $a+b+c=N$ or $a+b+c=2N$ meet and allow singular vertices along which such seam edges interact. Robert and Wagner [**RW1**, Section 2.3] briefly discuss how to extend their evaluation to such foams.

Fundamental applications of Robert-Wagner foam evaluation are developed in [**ETW, RW2, RW3**], with more clearly on the way, see also [**KR, Bo**]. Foam evaluation in the limit $N\to\infty$ and restricted to foams in special position provides a connection between foams and Soergel and singular Soergel bimodules [**RW2, KRW**]. Other approaches to Soergel and singular Soergel bimodules via foams [**Vz, MV1, QR, RWd, Wd**] do not use foam evaluation, utilizing instead matrix factorizations, more direct foam computations in $N=2,3$ cases, and other methods. Earlier, an extension of the Kapustin-Li formula was proposed for foam evaluation [**KRo2**], but due to its more implicit nature was not easy to apply [**MSV**].

Looking beyond foam evaluation, both foams as they are used in link homology and spin foams [**Ba**] have "foam" in their names, but we don't know if there is a relation between the two theories beyond this observation. Also see Natanzon [**Nt**] and the follow-up papers for yet another direction in the foam theory.

1.3. Formal groups as a motivation. In this paper we propose a deformation of the Robert-Wagner evaluation formula, motivated by algebraic topology and generalized cohomology theories related to formal groups. Link homology theories in the $SL(2)$ case have been lifted to spectra by Lipshitz and Sarkar [**LS1, LS2**] and Hu, Kriz and Kriz [**HKK**]. More recently, a lifting of bigraded $GL(N)$ link homologies as well as the triply-graded homology to equivariant spectra has been constructed by the second author [**K1, K2**].

Application of generalized cohomology theories to these spectra results in new homological link invariants as well as cohomological operations on them. A purely combinatorial or algebraic description of these homological invariants is clearly desirable, and modifying foam theory and foam evaluation may be a natural first step in this direction.

$GL(N)$ foams are closely related to Grassmannians and partial and full complex flag varieties. A family of cohomology theories known as complex oriented cohomology theories is related to these varieties as well, and to deformations of the formula for the first Chern class in singular cohomology of the tensor product of line bundles $c_1(L_1 \otimes L_2) = c_1(L_1) + c_1(L_2)$ to formulas

$$c_1(L_1 \otimes L_2) = \mathscr{F}(c_1(L_1), c_1(L_2))$$

that hold for the first Chern class invariant in these geralized cohomology theories, where

$$\mathscr{F}(x, y) = x + y + \sum_{i+j>1} a_{i,j} x^i y^j$$

is, in general, a power series in x, y with coefficients in the ground ring. Such a power series admits rich internal structure, making it a *Formal group law.* In Section 3 we shall study formal group laws in detail, but let us briefly point out some relevant structure in this introduction. Among the relations satisfied by $\mathscr{F}(x,y)$ is the associativity relation, which leads to polynomial relations on $a_{i,j}$ which admit a universal solution with one generator for each $k = i + j - 1$, $k \geq 1$. This solution is hard to write down explicitly, and most manipulations with general formal group laws are implicit [**Ha, St**] (see section 3 for examples).

With formal group law $\mathscr{F}(x,y)$ at hand, one defines $-_{\mathscr{F}} x$ or $[-1]x$ as the power series $-x + \dots$ which solves the equation $\mathscr{F}(x, [-1]x) = 0$, and forms the power series $x -_{\mathscr{F}} y = \mathscr{F}(x, -_{\mathscr{F}} y)$, also denoted $x[-1]y$:

$$x[-1]y \;=\; x -_{\mathscr{F}} y \;:=\; \mathscr{F}(x, -_{\mathscr{F}} y).$$

This expression deforms $x - y$, so that $x -_{\mathscr{F}} y = x - y +$ higher order terms. One can show $x -_{\mathscr{F}} y = (x - y)q(x, y)$ for an invertible element $q(x, y)$ of a suitable power series ring. We write $x - y = p(x, y)(x -_{\mathscr{F}} y)$ where $p(x, y)q(x, y) = 1$, and use $p(x, y) = q(x, y)^{-1}$ in our computations.

From the standpoint of algebraic topology, $x -_{\mathscr{F}} y$ represents the Euler class (in the cohomology theory corresponding to the formal group law $\mathscr{F}(x, y)$) of the line bundle $\mathcal{L}_1 \otimes \mathcal{L}_2^*$, where $\mathcal{L}_1$ and $\mathcal{L}_2$ represent the tautological line bundles over the product space $\mathbb{CP}^\infty \times \mathbb{CP}^\infty$. In other words, the expression $q(x, y)$ should be interpreted as the *relative Euler class* for the bundle $\mathcal{L}_1 \otimes \mathcal{L}_2^*$, in the sense that one compares the Euler classes in the cohomology theory corresponding to $\mathscr{F}(x, y)$, to the standard Euler class in singular cohomology. In this context, products of the form $q(x_{i_1}, x_{j_1})\, q(x_{i_2}, x_{j_2}) \dots q(x_{i_k}, x_{j_k})$ (which we will come across often in this paper) may be interpreted as the relative Euler class of the *direct sum* of the line bundles corresponding to each factor.

Robert-Wagner foam evaluation formulas contain powers of $x_i - x_j$ in the denominator, and a natural idea would be to carefully replace them with $x_i -_{\mathscr{F}} x_j$. We pursue a variant of this idea in this paper. Similar replacements have already been considered for various formulas in the theory of symmetric functions, including the Weyl formula for the Schur function, see [**NN1, NN2, Na**] and references therein. Foam evaluation specializes to the Weyl formula for the Schur function in the case of the so-called theta-foam and its natural generalizations.

On the algebraic topology side, the expressions $x_i - x_j$ for $1 \leq i \neq j \leq N$ have a natural meaning as the Euler classes in singular cohomology for the roots α of $GL(N)$ (we have included both positive and negative roots). In particular, the deformation $x_i -_{\mathscr{F}} x_j$ can be interpreted as the Euler class of α in an exotic cohomology theory corresponding to the formal group law $\mathscr{F}(x,y)$. We may therefore speculate that the corresponding deformed foam evaluation formula is obtained by applying an exotic cohomology theory to a (hitherto undefined) homotopy type. The existence of such a homotopy type for foam evaluations is very compelling given the results by the second author [**K1, K2**]. Since $GL(N)$ has $N(N-1)$ roots α representing the weights $x_i - x_j$ in the standard basis for $1 \leq i \neq j \leq N$, we see that our deformed evaluation formulas will be expressible in terms of $N(N-1)$ parameters given by the relative Euler classes $q(x_i, x_j)$. However, these extra parameters will satisfy certain constraints with coefficients in the algebra of symmetric power series in N-variables (which is the $GL(N)$-equivariant cohomology of a point). This suggests that the possible underlying homotopy type for foam evaluations is built from universal bundles using suitable subsets of roots of $GL(N)$.

The discussion above motivates our deformation using the language of formal group laws and related cohomology theories. We go into this further in section 3. Interestingly however, although motivated by it, our deformation setup will end up not requiring all the constraints on the power series $q(x,y)$ imposed by a formal group law. For instance, we will not require associativity from our analogue of the power series $\mathscr{F}(x,y)$. We therefore take as Ansatz, the series $p(x,y)$, the inverse of $q(x,y)$, with arbitrary coefficients. Most of the information in the coefficients of $q(x,y)$ will turn out to be redundant in our framework, at least in the $GL(2)$ case. However, it is conceivable that one may endow our constructions with the action of cohomology operations which are sensitive to more coefficients in the power series $q(x,y)$.

1.4. Plan of the paper. In section 2, motivated by analogies with formal group laws, we write down a multi-parameter deformation of the Robert-Wagner evaluation of closed $GL(N)$-foams and prove its integrality for any such foam. In section 3 we review formal group laws and corresponding generalizations of the divided difference operators. In section 4 we specialize to $N=2$ and study this deformation, which ultimately adds two more variables, of the $GL(2)$ foam evaluation. Skein relations for the deformed $GL(2)$ foam evaluation are derived in section 4.4. In section 4.5 we work out the ground ring R for the deformed theory, which has four generators E_1, E_2, ρ_0, ρ_1 of degrees $2, 4, -2, 0$, respectively. For comparison, the ground ring for the usual $GL(2)$-equivariant link homology has generators E_1, E_2 (also denoted h, t, up to a minus sign). In section 4.6 we show, unsurprisingly, that the state spaces (or homology) of planar $GL(2)$ webs are free modules over the graded ring R of rank $(q+q^{-1})^k$ over the ground ring R, where k is the number of thin circles in a web. In section 5 we extend the state spaces to homology groups of planar link diagrams and show the invariance under the Reidemeister moves.

Specializing power series $p(x,y)$ to $p(x,y)=1$ recovers the $GL(2)$ foam theory of Beliakova, Hogancamp, Putyra, and Wehrli [**BHPW**]. Simplifying computations in our section 4 to this case gives a foam evaluation approach to their theory.

$GL(2)$ foam theory that comes from this deformation seems very similar to the $SL(2)$ theory as set up by Vogel [**V**] and extended by him to get a strong invariant of tangle cobordisms, without the sign indeterminacy. The relation is given by dropping double facets but remembering singular circles along which the facets attach to the thin surface of a $GL(2)$ foam. $GL(2)$ foam theory also has an overlap with Ehrig-Stroppel-Tubbenhauer's generic $GL(2)$ foams [**EST1**].

We have already mentioned connections to Clark-Morrison-Walker [**CMW**] and Caprau [**Ca1, Ca2**] who have achieved full functoriality of the $SL(2)$ link homology via diagrammatical calculi that employ singular circles on thin surfaces. These circles should be remnants of attached double facets. Caprau's theory is $GL(2)$ equivariant, with variables h and a in place of our E_1 and E_2.

Twisting of $SL(2)$ theories in Vogel [**V**] is related to the deformation via power series $p(x,y)$ in this paper. We plan to elucidate connections to Vogel [**V**], Ehrig-Stroppel-Tubbenhauer [**EST1**], and to Turaev-Turner's rank two Frobenius algebra structures [**TT**] in a follow-up paper and also see whether the $p(x,y)$ deformation corresponds to the twisting [**Kh4, V**] in the $N=2$ and the general case.

2. Deformed evaluation for $GL(N)$ foams

The $GL(N)$ Robert-Wagner formula has denominators of the form $(x_i-x_j)^{\chi_{ij}(c)/2}$, where $\chi_{ij}(c)=\chi(F_{ij}(c))$ is the Euler characteristic of the bicolored surface $F_{ij}(c)$. The expression x_i-x_j can be generalized to $x_i -_{\mathcal{F}} x_j = x_i[-1]x_j$, where F is a formal group law. Unlike the additive case, when $x-y=-(y-x)$, most formal group laws do not satisfy $x[-1]y=-(y[-1]x)$, while those that do are called *symmetric*. Converting $(x_i-x_j)^{\chi_{ij}(c)/2}$ to $(x_i[-1]x_j)^{\chi_{ij}(c)/2}$ to modify the Robert-Wagner formula may be possible, but it would not contain the opposite terms $x_j[-1]x_i$, that perhaps should be present to maintain some symmetry, despite us having fixed a set of positive roots $\{x_i-x_j\}_{i<j}$.

To distribute the exponent $\chi_{ij}(c)/2$ across both terms $x_i -_{\mathcal{F}} x_j$ and $x_j -_{\mathcal{F}} x_i$, we recall the relation [**RW1**, Lemma 2.7] on Euler characteristics

$$\chi(F_{ij}(c)) \;=\; \chi(F_i(c))+\chi(F_j(c))-2\chi(F_{i\cap j}(c)), \tag{2.1}$$

where $F_{i\cap j}(c)$ is the surface, possibly with boundary, consisting of the union of facets that contain colors i and j,

$$F_{i\cap j}(c)=F_i(c)\cap F_j(c).$$

The Euler characteristic $\chi(F_{i\cap j}(c))$ may be odd, due to the presence of boundary, but $2\chi(F_{i\cap j}(c))$ is even, as are the other three terms in the formula (since the other three are Euler characteristics of closed surfaces).

Formula (2.1) simply describes the Euler characteristic of the symmetric difference of two spaces, specialized to the case of surfaces $F_i(c)$ and $F_j(c)$ inside a foam. Using shorthand notations, we can rewrite it as

$$\chi_{ij}(c) \;=\; \chi_i(c)+\chi_j(c)-2\chi_{i\cap j}(c), \tag{2.2}$$

where $\chi_i(c)=\chi(F_i(c))$ and $\chi_{i\cap j}(c)=\chi(F_{i\cap j}(c))$. We can now modify the evaluation formula by changing the (i,j) color pair contribution to the denominator to

$$(x_i[-1]x_j)^{\chi_i(c)/2}(x_j[-1]x_i)^{\chi_j(c)/2}$$

and multiplying the numerator by $(x_i-x_j)^{\chi_{i\cap j}(c)}$, to define

$$\langle F,c\rangle=(-1)^{\theta^+(c)}\prod_{f\in f(F)}P_f(c)\prod_{i<j}\frac{(x_i-x_j)^{\chi_{i\cap j}(c)}}{(x_i[-1]x_j)^{\chi_i(c)/2}(x_j[-1]x_i)^{\chi_j(c)/2}} \tag{2.3}$$

We denote the modified evaluation by $\langle F,c\rangle$ and the original one in [**RW1**] by $\langle F,c\rangle_{RW}$.

Setting aside formal group laws at this point, let us now formally *define* $x[-1]y$ as follows. Choose a commutative graded ring $\mathbf{k}$ and homogeneous elements $\beta_{k,\ell}\in\mathbf{k}$ in degree $-2(k+\ell)$ for all $k,\ell\in\mathbb{Z}_+=\{0,1,2,\dots\}$ such that $(k,\ell)\neq(0,0)$. The element

$$p(x,y) \;=\; 1+\sum_{(k,\ell)\neq(0,0)}\beta_{k,\ell}x^ky^\ell \tag{2.4}$$

belongs to the power sum ring $\mathbf{k}[\![x,y]\!]$. In general, $\beta_{k,\ell}\neq\beta_{\ell,k}$. The element $p(x,y)$ has the inverse $q(x,y)=p^{-1}(x,y)\in\mathbf{k}[\![x,y]\!]$. Define

$$x[-1]y \;=\; q(x,y)(x-y) \;=\; p(x,y)^{-1}(x-y)\in\mathbf{k}[\![x,y]\!]. \tag{2.5}$$

Equivalently, $x-y=p(x,y)(x[-1]y)$. Denote $p_{i,j}=p(x_i,x_j)$ or, interchangeably, p_{ij}, and $q_{ij}=q(x_i,x_j)=p_{ij}^{-1}$. Then

$$x_i[-1]x_j \;=\; (x_i-x_j)p_{ij}^{-1}. \tag{2.6}$$

Note that $x[-1]y=-(y[-1]x)$ iff $p(x,y)=p(y,x)$ iff $\beta_{\ell,k}=\beta_{k,\ell}$ for all k,ℓ. We refer to this as the *symmetric* case.

The universal case is that of the ring

$$\mathbf{k} \;=\; \mathbb{Z}[\beta_{k,\ell}] \tag{2.7}$$

over all k,ℓ as above ($k,\ell\in\mathbb{Z}_+,(k,\ell)\neq(0,0)$). This ring is non-positively graded, with nontrivial homogeneous components in even non-positive degrees $0,-2,-4,\dots$. It is a graded polynomial ring with $k+1$ generators in degree $-2k$ over all $k\geq 1$. The universal symmetric case is when $\beta_{\ell,k}=\beta_{k,\ell}$ are formal variables over all $0\leq k\leq\ell,(k,\ell)\neq(0,0)$.

Convert denominators in (2.3) via (2.6) and combine with a power of x_i-x_j in the numerator to get

$$\frac{(x_i-x_j)^{\chi_{i\cap j}(c)}}{(x_i[-1]x_j)^{\chi_i(c)/2}(x_j[-1]x_i)^{\chi_j(c)/2}}=\frac{(x_i-x_j)^{\chi_{i\cap j}(c)}}{(x_i-x_j)^{\chi_i(c)/2}q_{ij}^{\chi_i(c)/2}(x_j-x_i)^{\chi_j(c)/2}q_{ji}^{\chi_j(c)/2}}=$$
$$\frac{(-1)^{\chi_j(c)/2}p_{ij}^{\chi_i(c)/2}p_{ji}^{\chi_j(c)/2}}{(x_i-x_j)^{\chi_{ij}(c)/2}}$$

Taking the product over all $1\leq i<j\leq N$, the minus signs will combine to $(-1)^{\sum_{j=1}^N(j-1)\chi_j(c)/2}$ and

$$\langle F,c\rangle=(-1)^{s'(F,c)}\prod_{f\in f(F)}P_f(c)\prod_{i<j}\frac{p_{ij}^{\chi_i(c)/2}p_{ji}^{\chi_j(c)/2}}{(x_i-x_j)^{\chi_{ij}(c)/2}}, \tag{2.8}$$

where

$$s'(F,c)=\theta^+(c)+\sum_{j=1}^N(j-1)\chi_j(c)/2. \tag{2.9}$$

We can then define, as before,

$$\langle F\rangle=\sum_c\langle F,c\rangle. \tag{2.10}$$

In the Robert-Wagner formula, due to $\chi_i(c) = \chi(F_i(c))$ being even, one can rewrite the sign as

$$(-1)^{s(F,c)} = (-1)^{\theta^+(c)+\left(\sum_{i=1, i \text{ odd}}^N \chi_i(c)\right)/2} \tag{2.11}$$

versus

$$(-1)^{s'(F,c)} = (-1)^{\theta^+(c)+\left(\sum_{i=1, i \text{ even}}^N \chi_i(c)\right)/2} \tag{2.12}$$

in (2.9). These two signs differ by $(-1)^{\left(\sum_{i=1}^N \chi_i(c)\right)/2}$.

Notice that the sum $\sum_{i=1}^N \chi_i(c)$ does not depend on the coloring c of F and can be computed as a sort of Euler characteristic of F, denoted $\chi(F)$. An open facet f of thickness a contributes $a\chi(f)$, an edge where facets of thickness $a, b, a+b$ meet contributes $-(a+b)$, a vertex along which facets of thickness a, b, c merge in two ways into a facet of thickness $a+b+c$ contributes $a+b+c$. In each case, a generalized 0-, 1- or 2-cell of a foam contributes its Euler characteristic times its thickness, defined as the number of surfaces $F_i(c)$, over all i, that contain that generalized cell. Consequently, the sign difference in the two evaluations is by $(-1)^{\chi(F)/2}$, with

$$\chi(F) = \sum_{i=1}^N \chi(F_i(c)), \quad \text{for any coloring } c. \tag{2.13}$$

Thus, to recover the Robert-Wagner evaluation $\langle F \rangle_{RW}$ from (2.8) and (2.10) one should specialize $p(x, y) = 1$, so that $p_{ij} = 1$ for all $i \neq j$ and scale by the sign,

$$\langle F \rangle_{RW} = (-1)^{\chi(F)/2} \langle F \rangle|_{p(x,y)=1} \tag{2.14}$$

We keep the sign term $(-1)^{\chi(F)/2}$ so that, in the $N = 2$ case, the 2-sphere of thickness one carrying a single dot would evaluate to 1 (upon specializing to $p_{ij} = 1$), as in [**BHPW**], rather than -1, as in [**RW1**]. Adding this sign term is a matter of preference, while we hope that the deformation via $p(x, y)$ will eventually prove significant.

Let us now show that the formula (2.10) given by summing the expressions (2.8) over all colorings gives rise to a symmetric power series that does not involve denominators. We begin with a simple lemma:

LEMMA 2.1. *Given a coloring c, let $p(c)$ denote the expression*

$$p(c) = \prod_{1 \le i < j \le N} p_{ij}^{\chi_i(c)/2} p_{ji}^{\chi_j(c)/2}.$$

Let c' be a coloring obtained from c by a Kempe move relative to 1 and 2 along a connected surface Σ_s, then one has a relation

$$p(c') = p(c)\, p_{1,2,s},$$

where $p_{1,2,s}$ is the expression

$$p_{1,2,s} := \frac{p_{12}^{\chi(F_2\cap\Sigma_s)/2 - \chi(F_1\cap\Sigma_s)/2}}{p_{21}^{\chi(F_2\cap\Sigma_s)/2 - \chi(F_1\cap\Sigma_s)/2}} \prod_{2 < j \le N} \frac{p_{1j}^{\chi(F_2\cap\Sigma_s)/2 - \chi(F_1\cap\Sigma_s)/2}}{p_{2j}^{\chi(F_2\cap\Sigma_s)/2 - \chi(F_1\cap\Sigma_s)/2}}.$$

Notice, in particular, that $p_{1,2,s}$ is an invertible power series starting with 1, *and the transposition σ that switches the variables x_1 and x_2 has the property*

$$\sigma(p_{1,2,s}) = \frac{1}{p_{1,2,s}}.$$

Furthermore, $p_{1,2,s}$ is of the form $p_{1,2,s} = 1 \mod (x_1 - x_2)$.

PROOF. The relation between $p(c')$ and $p(c)$ is straightforward to verify, and is left to the reader. Now, since $p_{1,2,s}$ is an invertible power series, we may consider the expression

$$p_{1,2,s} - \frac{1}{p_{1,2,s}} = \frac{1}{p_{1,2,s}}((p_{1,2,s})^2 - 1).$$

The above expression switches sign under the action of the transposition σ, and is therefore divisible by $(x_1 - x_2)$. We conclude that $(x_1 - x_2)$ divides the expression $(p_{1,2,s})^2 - 1$. Factoring this expression, se see that $(x_1 - x_2)$ must divide $p_{1,2,s} - 1$. □

REMARK 2.2. *Notice that the definition of $p_{1,2,s}$ as a ratio of $p(c')$ and $p(c)$ can be extended to the case when c' is a coloring obtained from c by a $(1,2)$-Kempe move along several connected components*

$$\Sigma_{\underline{s}} := \Sigma_{s_1} \sqcup \ldots \sqcup \Sigma_{s_k}.$$

This expression, $p_{1,2,\underline{s}}$ satisfies a locality property that is crucial for the following theorem

$$p_{1,2,\underline{s}} = p_{1,2,s_1} \cdots p_{1,2,s_k},$$

where $p_{1,2,s_i}$ denotes the ratio of $p(c_i')$ and $p(c)$, with c_i' obtained from c by a $(1,2)$-Kempe move on Σ_{s_i}.

The above lemma allows us to prove:

THEOREM 2.3. *The $GL(N)$-foam evaluation $\langle F \rangle$ is a symmetric power series in the variables $x_1, x_2, \ldots, x_N$. In particular, $\langle F \rangle$ is free of denominators.*

PROOF. The proof of the above theorem is essentially a simple variation on the argument given in [**RW1**, Proposition 2.18]. Consider the expression $\langle F \rangle$. It is clear that it is symmetric in the variables $x_1, x_2, \ldots, x_N$ with possible denominators of the form $(x_i - x_j)^k$. By symmetry, the proof of the theorem will follow if we can show that the denominator $(x_1 - x_2)$ does not appear in $\langle F \rangle$.

Let us decompose the set of colorings of $\langle F \rangle$ into a collection of equivalence classes relative to the colors 1 and 2. Given a coloring c of F, decompose $F_{12}(c)$ into connected components,

$$F_{12}(c) = \mathbf{\Sigma} = \Sigma_1 \cup \Sigma_2 \cup \ldots \cup \Sigma_r.$$

The equivalence class of colorings C_c that contains c consists of colorings of F that can be obtained from c by performing Kempe moves about various connected components $\Sigma_s \subseteq F_{12}(c)$, where we recall that a Kempe move about Σ_s switches the colors 1 and 2 of the facets in Σ_s.

As in [**RW1**, Proposition 2.18], consider the expressions

$$P_{F/\mathbf{\Sigma}}(F,c) = p(c) \prod_{f \text{ not a facet in } \mathbf{\Sigma}} P(c(f)), \qquad \widetilde{Q}(F,c) = \frac{Q(F,c)\prod_{s,k>2}(x_1 - x_k)^{l_{\Sigma_s}(c,k)/2}}{(x_1 - x_2)^{\chi_{12}(c)/2}}$$

where $p(c)$ is as defined in Lemma 2.1, and the integers $l_{\Sigma_s}(c,k)$ are as defined in [**RW1**, Lemma 2.10]. Also, for $1 \leq s \leq r$, define

$$T_s(F,c) = \widehat{P}_{\Sigma_s}(F,c) + (-1)^{\chi(\Sigma_s)/2} p_{1,2,s}\, \sigma(\widehat{P}_{\Sigma_s}(F,c)),$$

where σ is the transposition that swaps x_1 and x_2, the term $p_{1,2,s}$ is as defined in Lemma 2.1, and the expression $\widehat{P}_{\Sigma_s}(F,c)$ is defined as

$$\widehat{P}_{\Sigma_s}(F,c) = \prod_{f \text{ a facet in } \Sigma_s} P(c(f)) \prod_{2<k\leq N} (x_1 - x_k)^{l_{\Sigma_s}(c,k)/2}.$$

Using Lemma 2.1 and Remark 2.2, we may express the foam evaluation $\langle F, c\rangle$ on summing over the equivalence class C_c as

$$\sum_{c' \in C_c} \langle F, c'\rangle = (-1)^{s'(F,c)} \frac{P_{F/\mathbf{\Sigma}}(F,c)}{\widetilde{Q}(F,c)} \prod_{s=1}^{r} (x_1 - x_2)^{-\chi(\Sigma_s)/2} T_s(F,c).$$

Since $\widetilde{Q}(F,c)$ is not divisible by $x_1 - x_2$, it is enough for our purposes to show that the expression $(x_1 - x_2)^{-\chi(\Sigma_s)/2} T_s(F,c)$ does not have a denominator given by a power of $x_1 - x_2$. The only case that is relevant is when Σ_s is a surface of genus zero. It is therefore sufficient to show that $T_s(F,c)$ is divisible by $(x_1 - x_2)$ when Σ_s is a surface of genus zero. In this case, we have

$$T_s(F,c) = \widehat{P}_{\Sigma_s}(F,c) - p_{1,2,s}\, \sigma(\widehat{P}_{\Sigma_s}(F,c)).$$

By Lemma 2.1, recall that $p_{1,2,s}$ is of the form $1 \mod (x_1 - x_2)$. We therefore have

$$T_s(F,c) = \widehat{P}_{\Sigma_s}(F,c) - \sigma(\widehat{P}_{\Sigma_s}(F,c)) \mod (x_1 - x_2).$$

However, the expression $\widehat{P}_{\Sigma_s}(F,c) - \sigma(\widehat{P}_{\Sigma_s}(F,c))$ is also divisible by $x_1 - x_2$ since it switches sign under σ. It follows that $T_s(F,c)$ is divisible by $x_1 - x_2$ whenever Σ_s is a surface of genus zero. The proof of the theorem easily follows on summing $\langle F, c'\rangle$ over all the equivalence classes C_c. □

3. Formal groups and generalized divided difference operators

In this section we study formal group laws and their relationship to topology in some detail. Good references are [**Ha, St**] and the references therein. Due to standard conventions the choice of notation R in this section conflicts with its use in the next section.

Let us begin by recalling the definition of a formal group law. A formal group law defined over a ring R is a power series $\mathscr{F}(x,y)$ with coefficients in R so that $\mathscr{F}(x,y)$ represents a commutative group structure on the formal affine line over R. In other words, one requires $\mathscr{F}(x,y)$ to satisfy the following three properties

$$\mathscr{F}(x,y) = \mathscr{F}(y,x) \qquad \text{commutativity}$$

$$\mathscr{F}(0,x) = \mathscr{F}(x,0) = x \qquad \text{unitarity}$$

$$\mathscr{F}(x,\mathscr{F}(y,z)) = \mathscr{F}(\mathscr{F}(x,y),z) \qquad \text{associativity.}$$

REMARK 3.1. *There is a universal ring known as the Lazard ring which is initial among all rings that support a formal group law. This ring L can be defined to be generated by symbols $a_{i,j}$ where the universal formal group law has the form*

$$\mathscr{F}(x,y) = x + y + \sum_{i,j>0} a_{i,j} x^i y^j.$$

We then impose relations on the generators $a_{i,j}$ that are forced by the relations of commutativity and associativity (the relation for unitarity is built into the form of

$\mathscr{F}(x,y)$). For instance, commutativity implies that $a_{i,j} = a_{j,i}$. The relation for associativity is clearly more involved.

In topology, formal group laws appear when one describes the E-cohomolgy of a space BU(1), where E is any complex oriented cohomology theory and BU(1) denotes the classifying space of the group U(1) (the space BU(1) is equivalent to the infinite projective plane $\mathbb{CP}^\infty$). More precisely, one starts with the observation that $E^*(\mathrm{BU}(1))$ can be expressed as $R[\![x]\!]$, with $R = E^*(pt)$ and x being the first Chern class in cohomological degree 2. The abelian group structure on U(1) induces a map

$$\mathrm{BU}(1) \times \mathrm{BU}(1) \longrightarrow \mathrm{BU}(1).$$

Evaluating this map in E-cohomology then gives rise to the underlying a formal group law $\mathscr{F}_E(x,y)$ for the complex oriented cohomology theory E:

$$R[\![x]\!] = E^*(\mathrm{BU}(1)) \longrightarrow E^*(\mathrm{BU}(1) \times \mathrm{BU}(1)) = R[\![x,y]\!], \qquad x \longmapsto \mathscr{F}_E(x,y).$$

In what follows therefore, we work in the graded setting. So $\mathscr{F}(x,y)$ will denote a formal group law over a graded power series ring $R[\![x,y]\!]$, where R is a graded $\mathbb{Z}$-algebra, and the variables x and y are defined to have degree 2. We assume that $\mathscr{F}(x,y)$ is in homogeneous degree 2, namely

$$\mathscr{F}(x,y) = x + y + \sum_{i,j>0} a_{ij} x^i y^j, \qquad a_{ij} \in R^{2-2(i+j)}.$$

DEFINITION 3.2. *The formal negative of the variable x is defined to be the (unique) power series $[-1]x$ with the property*

$$\mathscr{F}(x,[-1]x) = \mathscr{F}([-1]x,x) = 0.$$

The formal difference $x[-1]y$ is defined as $\mathscr{F}(x,[-1]y)$. It is a power series in two variables x,y that has homogeneous degree 2.

EXAMPLE 3.3. *Let R be the $\mathbb{Z}$-algebra $\mathbb{Z}[\beta]$, where β is in degree -2. The multiplicative formal group law $\mathscr{F}(x,y)$ and its formal difference are given by*

$$\mathscr{F}(x,y) = x + y - \beta xy, \qquad x[-1]y = \frac{x-y}{1-\beta y}, \qquad [-1](x) = -\sum_{i\geq 0} \beta^i x^{i+1}.$$

EXAMPLE 3.4. *Let R be the $\mathbb{Z}$-algebra $\mathbb{Z}[\beta^2]$ as before with β^2 in degree -4. The Lorentz or L-formal group law $\mathscr{F}(x,y)$ and its formal difference are given by*

$$\mathscr{F}(x,y) = \frac{x+y}{1+\beta^2 xy}, \qquad x[-1]y = \frac{x-y}{1-\beta^2 xy}, \qquad [-1](x) = -x.$$

EXAMPLE 3.5. *Let R be the $\mathbb{Z}$-algebra $\mathbb{Z}[\frac{1}{2},\beta^2]$ with β^2 in degree -4. The $\hat{A}$-formal group law $\mathscr{F}(x,y)$ and its formal difference are given by*

$$\mathscr{F}(x,y) = x\sqrt{1+\beta^2(y/2)^2} + y\sqrt{1+\beta^2(x/2)^2},$$

$$x[-1]y = x\sqrt{1+\beta^2(y/2)^2} - y\sqrt{1+\beta^2(x/2)^2}, \qquad [-1](x) = -x.$$

where the radicals are expressed as a power series (with coefficients in $\mathbb{Z}[\frac{1}{2},\beta^2]$) by the formal application of the binomial expansion.

EXAMPLE 3.6. *Let R be the $\mathbb{Z}$-algebra $\mathbb{Z}[\frac{1}{2}, \epsilon, \delta]$ with the degree of δ being -4 and that of ϵ being -8. The Jacobi formal group law $\mathscr{F}(x, y)$ and its formal inverse are given by*

$$\mathscr{F}(x,y) = \frac{x\sqrt{J(y)} + y\sqrt{J(x)}}{1 - \epsilon x^2 y^2}, \quad \textit{where} \quad J(z) = 1 - 2\delta z^2 + \epsilon z^4.$$

$$x[-1]y = \frac{x\sqrt{J(y)} - y\sqrt{J(x)}}{1 - \epsilon x^2 y^2}, \qquad [-1](x) = -x.$$

Examples 3.4 and 3.5 are specializations of 3.6 at the "cusps" described by $\epsilon = \beta^4, \delta = \beta^2$ and $\epsilon = 0, \delta = -\beta^2/8$ respectively.

Let us return to the universal example. In other words, we consider the example of R being the Lazard ring introduced earlier. On introducing a grading on the variables x and y so that the universal formal group law belongs in homogeneous degree 2, the Lazard ring naturally acquires a grading as described earlier. With this grading, the Lazard ring can be shown to be isomorphic to the graded coefficient ring of a complex oriented cohomology theory known as complex cobordism, MU. In other words $R \cong \mathrm{MU}^*(pt)$ as a graded ring.

By the definition of complex cobordism, the elements of $\mathrm{MU}^{-k}(pt)$ are cobordism classes of k-dimensional manifolds endowed with an almost complex structure on their stable normal bundle, and with the ring structure being induced by the cartesian product of manifolds. The ring $\mathrm{MU}^*(pt)$ can be shown to be a polynomial algebra over $\mathbb{Z}$, with one generator in each negative even degree. Working rationally, the generator in degree $-2n$ may be chosen to be the cobordism class of the complex projective space of dimension $2n$, denoted by $[\mathbb{CP}^n]$.

Any formal group law over a $\mathbb{Q}$-algebra is isomorphic to the additive formal group law. This isomorphism is called the logarithm, written as $\log_{\mathscr{F}}(x)$, and is the unique power series with leading term being x, that interpolates the given formal group law $\mathscr{F}(x, y)$ with the additive one $\mathbb{G}_a(x, y) = x + y$.

On extending scalars from $\mathbb{Z}$ to $\mathbb{Q}$, the logarithm in the universal case has an explicit description

$$\log_{\mathrm{MU}}(x) = \sum_{k \geq 0} \frac{[\mathbb{CP}^k]}{k+1} x^{k+1}, \quad \text{so that} \quad \log_{\mathrm{MU}}(\mathscr{F}_{\mathrm{MU}}(x,y)) = \log_{\mathrm{MU}}(x) + \log_{\mathrm{MU}}(y)$$

An immediate corollary of the above description is the following example.

EXAMPLE 3.7. *Let R be the $\mathbb{Z}$-algebra $\mathrm{MU}^*(pt)$. The universal formal group law $\mathscr{F}_{\mathrm{MU}}(x, y)$ and its formal difference are given by*

$$\mathscr{F}_{\mathrm{MU}}(x,y) = \exp_{\mathrm{MU}}(\log_{\mathscr{F}}(x) + \log_{\mathscr{F}}(y)) = \exp_{\mathrm{MU}}(\sum_{k \geq 0} \frac{[\mathbb{CP}^k]}{k+1}(x^{k+1} + y^{k+1})),$$

$$x[-1]y = \exp_{\mathrm{MU}}(\sum_{k \geq 0} \frac{[\mathbb{CP}^k]}{k+1}(x^{k+1} - y^{k+1})) = \exp_{\mathrm{MU}}((x-y)\sum_{k \geq 0} \frac{[\mathbb{CP}^k]}{k+1} \mathrm{S}_k(x,y)),$$

where $\exp_{\mathrm{MU}}(z)$ is the compositional inverse of $\log_{\mathrm{MU}}(z)$, and $\mathrm{S}_k(x, y)$ is the symmetric sum

$$\mathrm{S}_k(x,y) = x^k + x^{k-1}y + \cdots + xy^{k-1} + y^k.$$

Notice that even though the expressions for $\mathscr{F}_{\mathrm{MU}}(x,y)$ *and* $x[-1]y$ *above appear to have denominators, these denominators cancel away in the ring* $\mathrm{MU}^*(pt)$ *once one expands the expression as a power series in* x *and* y.

One may notice that the formal difference $x[-1]y$ in each of the above examples appears to be divisible by the expression $(x-y)$. In fact, this is always true as we now show

CLAIM 3.8. *Given an arbitrary formal group law, there is a unique homogeneous degree 0 element* $q(x,y) \in R[\![x,y]\!]$ *so that*

$$x[-1]y = (x-y) \cdot q(x,y).$$

Furthermore, $q(x,y)$ *is invertible and* $q(x,y) \equiv 1 \mod (y)$.

PROOF. Consider the formal expansion of the expression $(y+z)[-1]y := \mathscr{F}(y+z,[-1]y)$. On setting z as 0, we see that the expression vanishes. Therefore, it is divisible by z. Setting z as $(x-y)$, we conclude that there is a power series $q(x,y)$ that satisfies the relation

$$(x-y)\,q(x,y) := \mathscr{F}(y+x-y,[-1]y) = x[-1]y.$$

Since $(x-y)$ is not a zero divisor in $R[\![x,y]\!]$, we see that $q(x,y)$ is unique. Next, by setting y as 0, we see that $q(x,0) = 1$. In particular, $q(x,y)$ has the form 1 mod (y), and is therefore a unit. □

REMARK 3.9. *It is not hard to show using the definition of the power series* $q(x,y)$ *that the coefficients in its expansion generate the same sub algebra of* R *as the coefficients of the formal group law* $\mathscr{F}(x,y)$. *To see this, first observe that the power series* $[-1]y$ *can be expressed in terms of the coeffiients of* $q(x,y)$ *using the fact that* $[-1]y = -yq(0,y)$. *Next, observe that* $\mathscr{F}(x,y) = x[-1]([-1]y) = (x-[-1]y)q(x,[-1]y)$. *These two observations together establish what we seek to show.*

REMARK 3.10. *In example 3.6, one may verify that* $q(x,y)$ *is the following (symmetric) expression*

$$q(x,y) = \frac{x+y}{x\sqrt{J(y)}+y\sqrt{J(x)}}.$$

In general however, $q(x,y)$ *need not be symmetric in* x,y *as is easily seen from example 3.3.*

REMARK 3.11. *The universal example 3.7 allows us to deduce some interesting properties about* $q(x,y)$. *For instance, we see that* $q(x,y)$ *has the form*

$$q(x,y) = \sum_{k\geq 0} \frac{[\mathbb{CP}^k]}{k+1} \mathrm{S}_k(x,y) + \sum_{n\geq 1} q_n(x,y)(x-y)^n,$$

where $q_n(x,y)$ *are symmetric power series in* x *and* y. *Note that each individual series* $q_n(x,y)$ *involves denominators. However, on setting* $x=y$, *those terms vanish and we obtain the interesting (universal) relation that does not involve denominators*

$$q(x,x) = \frac{d}{dx}\log_{\mathscr{F}}(x).$$

CLAIM 3.12. *Assume that the $\mathbb{Z}$-algebra R is torsion free. Then, given a formal group law $\mathscr{F}(x,y)$ over R, the power series $q(x,y)$ is symmetric if and only if $\log_{\mathscr{F}}(x)$ is an odd power series. Equivalently, $q(x,y)$ is symmetric if and only if $q(x,x)$ is an even power series. Note that these conditions are automatic if R has no nontrivial elements in degrees $2 \mod 4$.*

PROOF. The equivalence of the two conditions follows from remark 3.11 above. It remains to establish the first condition. Now $\log_{\mathscr{F}}(x)$ is an odd power series if and only if its compositional inverse $\exp_{\mathscr{F}}(x)$ is an odd power series. We will now proceed to show that symmetry of $q(x,y)$ is equivalent to $\exp_{\mathscr{F}}(x)$ being an odd power series. Using the universal example 3.7, we see that the formal difference $x[-1]y$ for the formal group law $\mathscr{F}(x,y)$ has the form

$$x[-1]y = \exp_{\mathscr{F}}(z), \qquad z = (x-y)s(x,y),$$

with $s(x,y)$ being a symmetric power series

$$s(x,y) = \sum_{k\geq 0} \frac{l_k}{k+1} \mathrm{S}_k(x,y), \qquad \text{where} \qquad \log_{\mathscr{F}}(x) = \sum_{k\geq 0} \frac{l_k}{k+1} x^{k+1}.$$

Note that $s(x,y)$ is invertible in $(R\otimes\mathbb{Q})[\![x,y]\!]$, and so z can be chosen to be a power series generator. We therefore have an inclusion

$$(R\otimes\mathbb{Q})[\![z]\!] \subset (R\otimes\mathbb{Q})[\![x,y]\!], \qquad z \longmapsto (x-y)s(x,y).$$

It follows that $q(x,y)\,s(x,y)^{-1} = \exp_{\mathscr{F}}(z)/z$ is symmetric if and only if $\exp_{\mathscr{F}}(z)/z$ is even, or that $\exp_{\mathscr{F}}(z)$ is odd. □

Let us now study the divided difference operators in the context of formal group laws.

DEFINITION 3.13. *Consider the formal power series ring $R[\![x_1,x_2,\dots,x_n]\!]$. Let α denote any pair (i,j) for $1 \leq i < j \leq n$. We think of α as a positive root of $\mathrm{U}(n)$ so that the pairs (i,j) are indexed by the set Δ_+ of positive roots of $\mathrm{U}(n)$. Given a formal group law defined over R, we define the generalized divided difference operator A_α as the operator on $R[\![x_1,x_2,\dots,x_n]\!]$*

$$A_\alpha(f) := \frac{f}{x_i[-1]x_j} + \frac{r_\alpha(f)}{x_j[-1]x_i}, \qquad f \in R[\![x_1,x_2,\dots,x_n]\!],$$

where r_α is the reflection on $R[\![x_1,x_2,\dots,x_n]\!]$ given by switching x_i and x_j. Using elementary algebra, one can check that the operator A_α is well defined and does not involve denominators.

REMARK 3.14. *If E is a complex oriented cohomology theory with underlying formal group law $\mathscr{F}_E$ and the coefficients of a point being $E^*(pt) = R$, then the operators A_α have a natural meaning in terms of push-pull oprators on the $\mathrm{U}(n)$-equivariant E-cohomology ring of the flag variety $\mathrm{U}(n)/T$ (see [**BE**]). More precisely, recall that the $\mathrm{U}(n)$-equivariant E-cohomology of a $\mathrm{U}(n)$-space X is defined as the E-cohomology of the space $\mathrm{EU}(n) \times_{\mathrm{U}(n)} X$ with $\mathrm{EU}(n)$ being the principal contractible $\mathrm{U}(n)$-space. For $X = \mathrm{U}(n)/T$, the $\mathrm{U}(n)$-equivariant E-cohomology ring is isomorphic to $R[\![x_1,x_2,\dots,x_n]\!]$, supporting the operator A_α that is defined as the* *pushforward in equivariant E-cohomology followed by the pullback:* $\pi^* \circ \pi_*$, *where* π *denotes the* $\mathrm{U}(n)$*-equivariant fibration*

$$\pi : \mathrm{U}(n)/T \longrightarrow \mathrm{U}(n)/\mathrm{U}_\alpha(n),$$

with $\mathrm{U}_\alpha(n)$ *being the maximal compact subgroup in the parabolic subgroup corresponding to the positive root* α.

CLAIM 3.15. *Given a root* $\alpha \in \Delta_+$ *defined by the pair* $(i < j)$, *let* $q(\alpha)$ *denote the unit* $q(x_i, x_j)$ *as defined in* 3.8. *Then the intersection of the kernels of all the operators* A_{α_i}, *where* $\alpha_i = x_i - x_{i+1}$ *is a simple root, is a rank one free module over the ring of symmetric power series* $R[\![x_1, x_2, \ldots, x_n]\!]^{\Sigma_n}$ *generated by the unit* $q(\Delta_+)$, *where*

$$q(\Delta_+) := \prod_{\alpha \in \Delta_+} q(\alpha).$$

PROOF. Given a simple root $\alpha_i \in \Delta_+$, let us rewrite the action of A_{α_i} on f as

$$A_{\alpha_i}(f) = \frac{f}{x_i[-1]x_{i+1}} + r_i(\frac{f}{x_i[-1]x_{i+1}}).$$

Hence f is in the kernel of A_{α_i} if the expression $\frac{f}{x_i[-1]x_{i+1}}$ switches sign under r_i. Notice that the ratio of any two such elements is invariant under r_i. On the other hand, by claim 3.8, it follows that given an r_i-invariant element g, the expression $q(\alpha_i)g$ is in the kernel of A_{α_i}. In particular, we have shown that for a fixed α_i, the kernel of A_{α_i} is *precisely* the rank one module of r_i invariants generated by the unit $q(\alpha_i)$. Now let us fix $\alpha_i \in \Delta_+$, and consider

$$\check{q}(\alpha_i) := \prod_{\beta \in \Delta_+, \beta \neq \alpha_i} q(\beta), \qquad \text{so that} \quad q(\alpha_i)\,\check{q}(\alpha_i) = q(\Delta_+).$$

Since r_i permutes all positive roots besides α_i, we see that $\check{q}(\alpha_i)$ is an r_i-invariant unit. Hence the kernel of A_{α_i} is a rank one free module of r_i-invariants generated by the element $q(\Delta_+)$. Taking intersection over all simple roots $\alpha_i \in \Delta_+$ we get the required result. □

EXAMPLE 3.16. *For the multiplicative formal group law of example* 3.3, *the intersection of the kernels of all the generalized divided difference operators* A_{α_i} *is a rank one free module over symmetric power series* $R[\![x_1, x_2, \ldots, x_n]\!]^{\Sigma_n}$ *generated by the unit* $q(\Delta_+)$, *where*

$$q(\Delta_+) = \frac{1}{(1-\beta x_2)(1-\beta x_3)^2 \ldots (1-\beta x_n)^{n-1}}.$$

Note that $q(\Delta_+)$ *is not* Σ_n*-invariant for any* $n > 1$.

Let D_α denote the classical divided difference operator (i.e. the divided difference operator for the additive formal group law). Definition 3.13 and claim 3.8 imply that we have

$$A_\alpha(f) = D_\alpha(q(\alpha)^{-1}f), \quad \text{in other words} \quad A_\alpha = D_\alpha \circ Q(\alpha)^{-1}, \tag{3.1}$$

where $Q(\alpha)$ denotes the operator given by multiplication with $q(\alpha)$. In particular, the q-twisted operators $Q(\alpha_i) \circ A_{\alpha_i}$ satisfy the braid relations, and generate an algebra isomorphic to the nilHecke algebra, namely, the algebra generated by the operators D_{α_i}. We also have

THEOREM 3.17. *Given a formal group law defined over R, let $\mathscr{A}(n)$ denote the algebra of operators on $R[\![x_1, x_2, \ldots, x_n]\!]$ generated by multiplication operators, and the generalized divided difference operators A_{α_i} for $1 \leq i < n$. Then $\mathscr{A}(n)$ is identically the same as the (completed) affine nilHecke algebra over the ground ring R. In other words, $\mathscr{A}(n)$ agrees with the algebra generated by the operators D_{α_i} and multiplication operators with respect to $R[\![x_1, x_2, \ldots, x_n]\!]$. In particular, $\mathscr{A}(n)$ is a free (left or right) module of rank $n!$ over the subalgebra $R[\![x_1, x_2, \ldots, x_n]\!]$. Alternatively, $\mathscr{A}(n)$ is a matrix algebra of rank $(n!)^2$ over the subalgebra $R[\![x_1, x_2, \ldots, x_n]\!]^{\Sigma_n}$. It follows that $R[\![x_1, x_2, \ldots, x_n]\!]^{\Sigma_n} \subset \mathscr{A}(n)$ is the center.*

PROOF. By (3.1), we see that the operators A_{α_i} are of the form $D_{\alpha_i} \circ Q(\alpha_i)^{-1}$, where $Q(\alpha_i)$ is the invertible multiplication operator corresponding to $q(\alpha_i)$. It follows that the operators A_{α_i} generate the same algebra as D_{α_i} when extended with multiplication operators, which is the affine nilHecke algebra by definition (once we complete polynomials to power series). The rest of the claim follows from well-known results on the affine nilHecke algebra. □

REMARK 3.18. *The above theorem may come as a surprise to the reader, since it has been known for some time that for an arbitrary compact Lie group G, the push-pull operators A_{α_i} defined as in remark* 3.14 *and acting on the equivariant cohomology $E^*_G(G/T)$, do not satisfy the braid relations, unless the formal group law $\mathscr{F}_E$ underlying the cohomology theory E is highly restrictive (see theorem 3.7 in* [**BE**]*, see also* [**HMSZ**]*). It is possible that theorem* 3.17 *only holds for the compact Lie group $G = \mathsf{U}(n)$, though we have not verified this. It is important to note that the classes $q(\alpha_i)$ that allow for the proof of the above theorem have been studied before (see* [**C**], [**Na**]*), though the main observation of theorem* 3.17 *appears to be new.*

4. Deformed $GL(2)$ foam evaluation

4.1. $GL(2)$ foams and their colorings. The original formulation [**Kh1**] of $SL(2)$ link homology did not use foams. Hints at foams appeared in the work of Clark, Morrison, and Walker [**CMW**] and Caprau [**Ca1, Ca2**], who used disorientation lines on surfaces involved in the construction of $SL(2)$ homology to control minus signs that appear throughout the theory. This allowed them to establish full functoriality of the theory under cobordisms rather than the functoriality up to an overall minus sign, as shown in the earlier work [**J, Kh3, BN**]. One can think of disorientation lines as remnants of the 2-facets of $GL(2)$ foams along which they were attached to the 1-facets.

Earliest constructions of $SL(3)$ and $SL(N)$ link homology for $N > 3$ used foams explicitly [**Kh2**] and implicitly [**KRo2**].

Blanchet [**B**] pioneered the use of foams for the $SL(2)$ (more precisely, $GL(2)$) homology theory. A detailed exploration of various flavours of $GL(2)$ foams and applications can be found in [**EST1, EST2**].

We find its useful to follow the $GL(2)$ foam calculus of Beliakova, Hogancamp, Putyra, Wehrli [**BHPW**]. That is the calculus deformed in this section.

We consider $GL(2)$ foams (or, simply, foams) in this paper. A closed $GL(2)$ foam F is a combinatorial compact two-dimensional CW-complex embedded in $\mathbb{R}^3$

(or $\mathbb{S}^3$). The only allowed singularities of the CW-complex are *singular circles*, such that any point on the circle has a neighbourhood homeomorphic to the product of the tripod and the interval.

The set of points of F on its singular circles is denoted $s(F)$, and connected components of $F \setminus s(F)$ are called *facets*. Facets of F are subdivided into 1-facets and 2-facets. One-facets are also called *thin* facets, two-facets are also called *double* or *thick* facets. We require that along each singular circle two 1-facets and one 2-facet meet, see Figure 4.11.

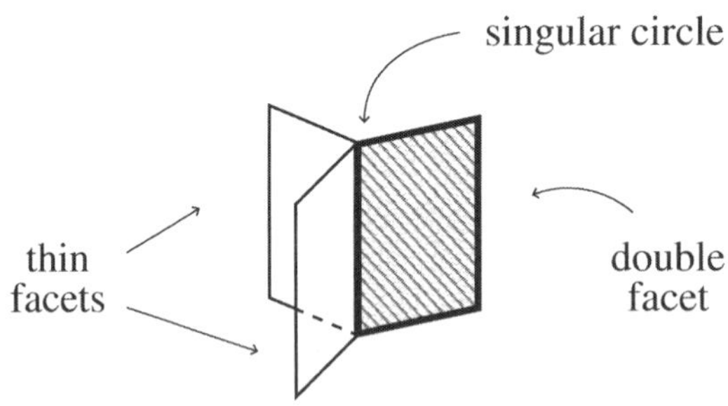

FIGURE 4.11. Part of a singular circle and its neighbourhood, with two thin and one double facet.

This implies, in particular, that no 'monodromy' is possible along any singular circle, so it has a neighbourhood in F homeomorphic to the product of $\mathbb{S}^1$ and a tripod with 'two thin legs and a double leg'.

Each facet is oriented in such a way that all three facets along any singular circle induce a compatible orientation on this circle, see Figure 4.12. In most diagrams that follow, it is clear whether a facet is thin or double, and we usually omit the corresponding label 1 or 2.

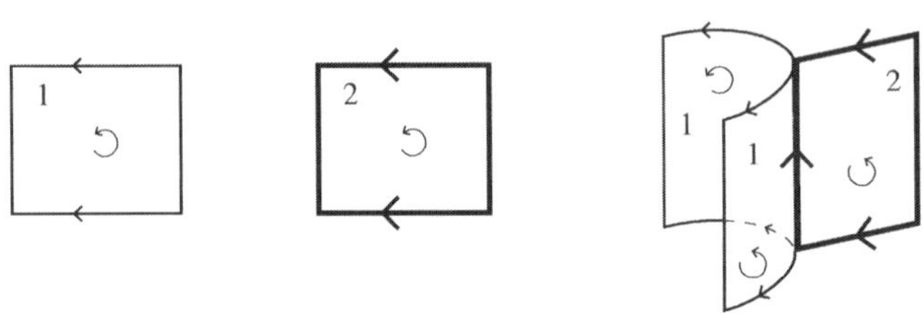

FIGURE 4.12. Left: oriented think and thick facets, with induced orientations on the top and bottom boundary. Right: compatibility between orientations of a singular circle and adjacent thin facets and the double facet. Also shown induced orientations of the top and bottom foam's boundary. They will be needed when we pass from closed foams to foams with boundary in Section 4.6.

Thin facets may carry dots, which can move freely along a facet, but cannot jump to an adjacent facet. If a facet carries n dots, we may record them as a single dot with label n. It is possible to allow similar decorations on 2-facets, namely symmetric polynomials in two variables, but we avoid doing so in the paper, instead moving any such decoration from a 2-facet to the coefficient of the foam.

Remark: Unlike $GL(N)$ foams for $N \geq 3$ and $SL(N)$ foams for $N \geq 4$, $GL(2)$ foams can not have singular vertices.

A *coloring* (or *admissible coloring*) of a foam F is a map c from the set of its thin facets to the set $\{1, 2\}$ such that along any singular circle, the two thin facets are mapped to different numbers. It is convenient to extend c to double facets, coloring each double facet by the set $\{1, 2\}$. This produces the flow condition, that the union of colors of 1-facets along each singular circle is the color of the double facet, that is, the entire set $\{1, 2\}$.

Notice that $F_{12}(c)$ does not depend on the coloring c and is a closed surface which is the union of closures of 1-facets of F. We denote it by F_{12} and call the *thin surface* of F. Likewise, $F_{1\cap 2}(c)$ does not depend on c and is the union of closures of 2-facets of F. We denote it by $F_{1\cap 2}$ and call the *double surface* of F. The boundary of $F_{1\cap 2}$ is exactly the set of singular circles of F.

Often it is convenient to identify a facet f with its closure $\overline{f}$ in F. In particular, the Euler characteristics of f and $\overline{f}$ are equal, since the two spaces differ only by a union of circles, which is the boundary of $\overline{f}$. From now on, unless otherwise specified, by a facet we mean a closed facet.

Surface F_{12} has finitely many connected components $\Sigma_1, \dots, \Sigma_n$. Each component may contain one or more singular circles. The union of these singular circles is zero when viewed as an element of $H_1(\Sigma_k, \mathbb{Z}/2)$, for any k, due to our orientation requirements on F. In particular, each Σ_k admits exactly two checkerboard colorings of its regions, where along each singular circle in Σ_k the coloring is reversed.

A choice of such coloring for each Σ_k is equivalent to a coloring of F. Hence, F has 2^n colorings, where n is the number of connected components of F_{12}.

Quantity $\theta^+(c) = \theta^+_{12}(c)$ counts the number of positive circles for a coloring c, see Figure 4.13.

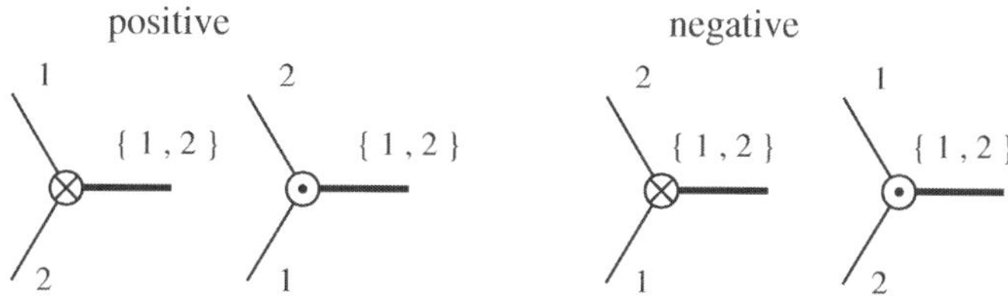

FIGURE 4.13. $\theta^+_{12}(c)$ or simply $\theta^+(c)$ counts the number of positive circles.

4.2. Deformed evaluation for $GL(2)$ foams. Modified Robert-Wagner evaluation formula, in the $GL(2)$ case, specializes to

$$\langle F, c\rangle = (-1)^{\theta^+_{12}(c)} \frac{(x_1 - x_2)^{\chi(F_{1\cap 2}(c))} x_1^{d_1(c)} x_2^{d_2(c)}}{(x_1[-1]x_2)^{\chi_1(c)/2}(x_2[-1]x_1)^{\chi_2(c)/2}} \tag{4.1}$$

Since

$$x_1[-1]x_2 = (x_1 - x_2)p_{12}^{-1}, \quad x_2[-1]x_1 = (x_2 - x_1)p_{21}^{-1}, \tag{4.2}$$

we have

$$\frac{(x_1 - x_2)^{\chi_{1\cap 2}(c)}}{(x_1[-1]x_2)^{\chi_1(c)/2}(x_2[-1]x_1)^{\chi_2(c)/2}} = (-1)^{\chi_2(c)/2} \frac{p_{12}^{\chi_1(c)/2} p_{21}^{\chi_2(c)/2}}{(x_1 - x_2)^{\chi_{12}(c)/2}} \tag{4.3}$$

In the 2-color case, $F_{1\cap 2}(c) = F_{1\cap 2}$ is the union of facets of thickness two and does not depend on c. Likewise, $F_{12}(c) = F_{12}$ does not depend on c either. Its Euler characteristic is denoted $\chi_{12}(F) = \chi(F_{12})$.

Equation (4.1) can be rewritten

$$\begin{aligned}
\langle F, c\rangle &= (-1)^{\theta_{12}^+(c)+\chi_2(c)/2} \frac{x_1^{d_1(c)} x_2^{d_2(c)}}{(x_1 - x_2)^{\chi(F_{12})/2} p_{12}^{-\chi_1(c)/2} p_{21}^{-\chi_2(c)/2}} \\
&= (-1)^{\theta_{12}^+(c)+\chi_2(c)/2} \frac{x_1^{d_1(c)} x_2^{d_2(c)} p_{12}^{\chi_1(c)/2} p_{21}^{\chi_2(c)/2}}{(x_1 - x_2)^{\chi(F_{12})/2}} \\
&= (-1)^{s'(F,c)} \frac{P(F,c)}{Q(F,c)}\, p_{12}^{\chi_1(c)/2} p_{21}^{\chi_2(c)/2} \\
&= (-1)^{\chi(F)/2} \langle F, c\rangle_{RW}\, p_{12}^{\chi_1(c)/2} p_{21}^{\chi_2(c)/2}.
\end{aligned}$$

Above, $s'(F,c) = \theta_{12}^+(c) + \chi_2(c)/2$ and $d_1(c)$, $d_2(c)$ is the number of dots on thin facets colored by 1, resp. 2 by c. We see that the original evaluation $\langle F, c\rangle_{RW}$ is scaled by an invertible element, which is a product of powers of p_{12} and p_{21} and a sign. Also, the power of $x_1 - x_2$ in the denominator depends on F only. Let us write down the formula again.

$$\langle F, c\rangle \;=\; (-1)^{\theta_{12}^+(c)+\chi_2(c)/2} \frac{x_1^{d_1(c)} x_2^{d_2(c)} p_{12}^{\chi_1(c)/2} p_{21}^{\chi_2(c)/2}}{(x_1 - x_2)^{\chi(F_{12})/2}} \tag{4.4}$$

We now define

$$\langle F\rangle \;=\; \sum_c \langle F, c\rangle, \tag{4.5}$$

the sum over all colorings of F. Let $E_1 = x_1 + x_2$, $E_2 = x_1 x_2$. The symmetric group S_2 acts on $\mathbf{k}[\![x_1, x_2]\!]$ by permuting x_1, x_2.

THEOREM 4.1. *$\langle F\rangle \in \mathbf{k}[\![x_1, x_2]\!]^{S_2} \cong \mathbf{k}[\![E_1, E_2]\!]$ for any $GL(2)$ foam F.*

In other words, $\langle F\rangle$ is a power series in x_1, x_2 that is symmetric under the permutation action of S_2 on x_1, x_2. Equivalently, it is a power series in elementary symmetric functions E_1, E_2. Consider the chain of inclusions

$$\mathbf{k} \;\subset\; \mathbf{k}[\![E_1, E_2]\!] \;\subset\; \mathbf{k}[\![x_1, x_2]\!] \;\subset\; \mathbf{k}[\![x_1, x_2]\!]\left[\frac{1}{x_1 - x_2}\right]. \tag{4.6}$$

Denote these rings by

$$\widetilde{R} = \mathbf{k}[\![E_1, E_2]\!], \tag{4.7}$$

$$R' = \mathbf{k}[\![x_1, x_2]\!], \tag{4.8}$$

$$R'' = \mathbf{k}[\![x_1, x_2]\!]\left[\frac{1}{x_1 - x_2}\right], \tag{4.9}$$

resulting in the chain of ring inclusions

$$\mathbf{k} \subset \widetilde{R} \subset R' \subset R''. \tag{4.10}$$

The theorem above has already been proved in Section 2 for general N, see Theorem 2.3. We include a more detailed proof for the special case $N = 2$ to make this section independent from Section 2.

PROOF. The evaluation $\langle F, c\rangle$ can be written, via (4.1), as a power series in x_1, x_2 with coefficients in $\mathbf{k}$ divided by a power of $x_1 - x_2$, either positive or negative, thus it belongs to the ring R'', see above.

Group S_2 acts on 2-colorings of F by transposing the colors 1 and 2. This action is compatible with the evaluation in the sense that $\sigma(\langle F, c\rangle) = \langle F, \sigma(c)\rangle$, where $\sigma = (12)$ is the nontrivial element of S_2. Therefore, $\langle F\rangle$ is in the subring $(R'')^{S_2}$ of S_2-invariants of R''.

$\langle F\rangle$ potentially has a denominator $(x_1 - x_2)^{\chi(F_{12}(c))/2}$. Surface $F_{12}(c) = F_{12}$ is a union of connected components $\Sigma_1, \dots, \Sigma_m$, each one contributing $(x_1 - x_2)^{\chi(\Sigma_k)/2}$ to the product. Only connected components of genus 0 have positive Euler characteristic, $\chi(\Sigma_k) = 2$, and contribute $x_1 - x_2$ to the denominator.

Consider one such component Σ and a coloring c. The Kempe move on Σ replaces c with a coloring $c_1 = (c, \Sigma)$ which is identical to c outside Σ and swaps colors $1, 2$ of c on Σ. We compare $\langle F, c\rangle$ and $\langle F, c_1\rangle$ in formula (4.4).

If there are t_i dots on color i facets of Σ under c, $i = 1, 2$, then

$$x_1^{d_1(c)} x_2^{d_2(c)} = x_1^{t_1} x_2^{t_2} u, \ x_1^{d_1(c_1)} x_2^{d_2(c_1)} = x_1^{t_2} x_2^{t_1} u,$$

for a monomial u in x_1, x_2 counting dots on facets not in Σ.

If Σ has r singular circles, let $\theta^+(c, \Sigma)$ be the number of positive circles on Σ under c and $\theta^-(c, \Sigma)$ be the number of negative circles. Under the swap $c \leftrightarrow c_1$, positive circles on Σ become negative circles on Σ and vice versa, so that $\theta^+(c, \Sigma) + \theta^+(c_1, \Sigma) = r$.

Let $\widetilde{\chi}_i(\Sigma) = \chi(\Sigma \cap F_i(c))$, $i = 1, 2$, be the Euler characteristic of the union of color i facets of Σ, for coloring c. We have

$$\chi_2(c_1) = \chi_2(c) + \widetilde{\chi}_1(\Sigma) - \widetilde{\chi}_2(\Sigma) = \chi_2(c) + \chi(\Sigma) - 2\widetilde{\chi}_2(\Sigma),$$

since $\chi(\Sigma) = \chi_1(\Sigma) + \widetilde{\chi}_2(\Sigma)$. Defining integer ℓ by $2\ell = \widetilde{\chi}_2(\Sigma) - \widetilde{\chi}_1(\Sigma)$, we have

$$\chi_2(c_1) \ = \ \chi_2(c) - 2\ell, \ \chi_1(c_1) \ = \ \chi_1(c) + 2\ell.$$

Consequently, one can write

$$p_{12}^{\chi_1(c_1)/2} p_{21}^{\chi_2(c_1)/2} = p_{12}^{\ell} u', \ \ p_{12}^{\chi_1(c)/2} p_{21}^{\chi_2(c)/2} = p_{21}^{\ell} u',$$

for a suitable monomial u' in p_{12}, p_{21}, possibly with negative exponents.

Also,

$$\frac{(-1)^{s'(F,c_1)}}{(-1)^{s'(F,c)}} = \frac{(-1)^{\theta^+(c_1) + \chi_2(c_1)/2}}{(-1)^{\theta^+(c) + \chi_2(c)/2}} = (-1)^{r + \chi(\Sigma)/2 - \widetilde{\chi}_2(\Sigma)}.$$

When $\Sigma \cong \mathbb{S}^2$,

$$r + \chi(\Sigma)/2 - \widetilde{\chi}_2(\Sigma) \equiv 1 + r - \widetilde{\chi}_2(\Sigma) \equiv 1 (\mathrm{mod}\ 2),$$

since $r \equiv \widetilde{\chi}_2(\Sigma) (\mathrm{mod}\ 2)$. The last comparison modulo 2 can be proved by induction on r, by removing an innermost singular circle of Σ. This operation reduces r by 1 and changes $\widetilde{\chi}_2(\Sigma)$ by ± 1. We see that $(-1)^{s'(F,c_1)} = -(-1)^{s'(F,c)}$.

Putting these relations together,

$$\langle F,c\rangle + \langle F,c_1\rangle = (-1)^{s'(F,c)} \cdot \left(x_1^{t_1}x_2^{t_2}p_{21}^{\ell} - x_1^{t_2}x_2^{t_1}p_{12}^{\ell}\right)uu'(x_1-x_2)^{-\chi_{12}(F)/2}.$$

The expression $(x_1^{t_1}x_2^{t_2}p_{21}^{\ell} - x_1^{t_2}x_2^{t_1}p_{12}^{\ell})$ is divisible by x_1-x_2 and allows to cancel out that term from the denominator.

Repeating this argument simultaneously for all $\mathbb{S}^2$-components of F_{12} shows that $\langle F\rangle \in \mathbf{k}[\![x_1,x_2]\!]$. Permutation action of S_2 on $\mathbf{k}[\![x_1,x_2]\!]$ and on colorings shows that $\langle F\rangle \in \mathbf{k}[\![x_1,x_2]\!]^{S_2} = \mathbf{k}[\![E_1,E_2]\!]$. □

The sum $\chi(F) = \chi_1(c) + \chi_2(c)$ does not depend on the coloring c and is the Euler characteristic of the surface F. In particular, in the symmetric case (when $p_{12}=p_{21}$), one has that

$$\langle F\rangle = \langle F\rangle_{RW} \cdot (-p_{12})^{\chi(F)/2} \tag{4.11}$$

so that the new evaluation is proportional to the original one with the coefficient that depends only on $\chi(F)$. We expect that non-symmetric case will prove more interesting.

4.3. Examples.

Example 1: Let $F=\mathbb{S}^2_{1,n}$ be the two-sphere of thickness one with n dots (or, equivalently, with a single dot labelled n). Here the lower index $(1,n)$ lists thickness followed by the number of dots.

$\mathbb{S}^2_{1,n}$ has two colorings c_1 and c_2, where in the coloring c_i the 2-sphere carries color i. For the coloring c_1

$$F_1(c_1)\cong \mathbb{S}^2,\ F_2(c_1)=\emptyset,\ F_{1\cap 2}(c_1)=\emptyset,$$

and likewise for c_2, so that

$$\langle \mathbb{S}^2_{1,n},c_1\rangle = \frac{x_1^n}{x_1[-1]x_2} = \frac{x_1^n p_{12}}{x_1-x_2},$$
$$\langle \mathbb{S}^2_{1,n},c_2\rangle = \frac{x_2^n}{x_2[-1]x_1} = \frac{x_2^n p_{21}}{x_2-x_1},$$

and

$$\langle \mathbb{S}^2_{1,n}\rangle = \frac{x_1^n p_{12} - x_2^n p_{21}}{x_1-x_2}.$$

To explicitly cancel x_1-x_2 in the denominator, expand p_{12} and p_{21} into power series and then cancel. The result is a power series symmetric in x_1,x_2 with coefficients which are polynomials in $\beta_{i,j}$.

We denote

$$\rho_n = \langle \mathbb{S}^2_{1,n}\rangle = \frac{x_1^n p_{12} - x_2^n p_{21}}{x_1-x_2}. \tag{4.12}$$

Note that the following relation holds:

$$\rho_{n+2} - E_1\rho_{n+1} + E_2\rho_n = 0, \tag{4.13}$$

where, recall, $E_1 = x_1+x_2$, $E_2 = x_1x_2$. It follows from the relation

$$(x_1^{n+2}p_{12} - x_2^{n+2}p_{21}) - (x_1+x_2)(x_1^{n+1}p_{12} - x_2^{n+1}p_{21}) + x_1x_2(x_1^n p_{12} - x_2^n p_{21}) = 0.$$

Relation (4.13) allows to inductively write $\rho_n = \langle \mathbb{S}^2_{1,n}\rangle$ as a linear combination of $\rho_0 = \langle \mathbb{S}^2_{1,0}\rangle$ and $\rho_1 = \langle \mathbb{S}^2_{1,1}\rangle$ with coefficients in $\mathbb{Z}[E_1,E_2]$. The latter are

$$\rho_0 = \langle \mathbb{S}^2_{1,0}\rangle = \frac{p_{12}-p_{21}}{x_1-x_2}, \qquad \rho_1 = \langle \mathbb{S}^2_{1,1}\rangle = \frac{x_1p_{12}-x_2p_{21}}{x_1-x_2}. \tag{4.14}$$

Example 2: Let the foam F be a thin two-torus T^2 with n dots and standardly embedded in $\mathbb{R}^3$ (embedding of a surface does not influence its evaluation). As in the previous example, there are two colorings, c_1 and c_2, with $F_1(c_1) \cong T^2$, $F_2(c_1) = \emptyset$, $F_{1\cap 2}(c_1) = \emptyset$, and $\langle T^2 \rangle = x_1^n + x_2^n$.

Example 3: Closed surface M of genus $g \geq 1$ with n dots.

(4.15)
$$\langle M \rangle = \frac{x_1^n p_{12}^{1-g}}{(x_1 - x_2)^{1-g}} + \frac{x_2^n p_{21}^{1-g}}{(x_2 - x_1)^{1-g}} = (x_1^n p_{12}^{1-g} + (-1)^{g-1} x_2^n p_{21}^{1-g})(x_1 - x_2)^{g-1}.$$

Example 4: F is 2-sphere $\mathbb{S}^2$ of thickness two, also denoted $\mathbb{S}^2_2$. It has a unique coloring c, with the facet labelled by $\{1,2\}$ and $F_1(c) = F_2(c) = F_{1\cap 2}(c) \cong \mathbb{S}^2$ so that

$$\begin{aligned}\langle \mathbb{S}^2_2 \rangle = \langle \mathbb{S}^2_2, c \rangle &= \frac{(x_1 - x_2)^2}{(x_1[-1]x_2)(x_2[-1]x_1)} = \frac{(x_1 - x_2)^2 p_{12} p_{21}}{(x_1 - x_2)(x_2 - x_1)} \\ &= -p_{12} p_{21}\end{aligned}$$

Denote the value of this foam by ρ, so that

$$\rho = \langle \mathbb{S}^2_2 \rangle = -p_{12} p_{21}. \tag{4.16}$$

Note that ρ is an invertible element of the ground ring. In the original case, when $p(x,y) = 1$, the double sphere $\mathbb{S}^2_2$ evaluates to -1.

Example 5: An oriented closed surface M of genus $g \geq 1$ and thickness two:

$$\langle M \rangle = \frac{(x_1 - x_2)^{2-2g} p_{12}^{1-g} q_{21}^{1-g}}{(x_1 - x_2)^{1-g}(x_2 - x_1)^{1-g}} = (-p_{12} p_{21})^{1-g} = \rho^{1-g}. \tag{4.17}$$

In the special case, when $g = 1$ so that $M = T^2$ is a two-torus, $\langle T^2_2 \rangle = 1$.

Example 6: The theta-foam Θ with n_1 and n_2 dots on thin facets, suitably oriented.

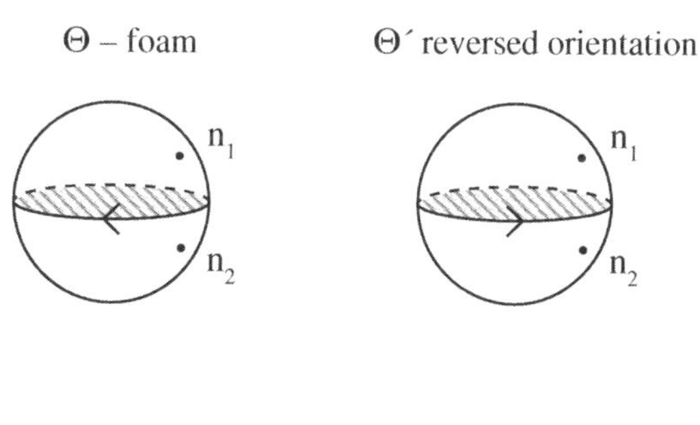

FIGURE 4.31. On the left: Θ-foam. On the right: same foam Θ' with the reversed orientation.

Let c_1 be the coloring of Θ with its top facet colored 1. Then the bottom facet is colored 2. Surfaces $F_1(c_1)$, $F_2(c_1)$, and $F_{12}(c_1) = F_{12}$ are all 2-spheres, with Euler characteristics 2. The sign $\theta_{12}^{+}(c_1) = 1$.

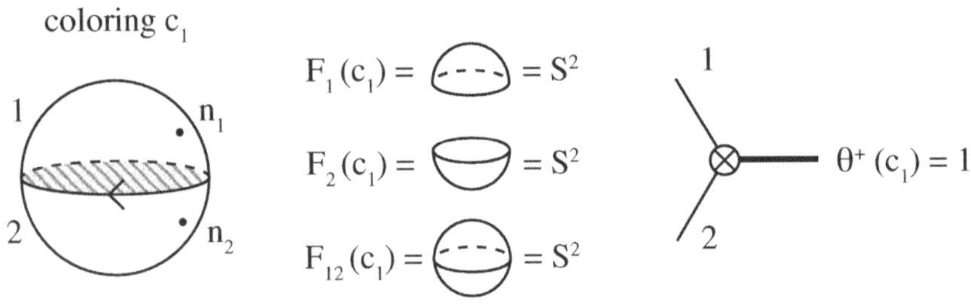

FIGURE 4.32. Computing $\langle \Theta, c_1 \rangle$.

For the sign,

$$s'(\Theta, c_1) = \theta^+(c_1) + \chi_2(c_1)/2 = 1 + 1 = 2, \quad (-1)^{s'(\Theta, c_1)} = 1.$$

We get

$$\langle \Theta, c_1 \rangle = \frac{x_1^{n_1} x_2^{n_2} p_{12} p_{21}}{x_1 - x_2}, \quad \langle \Theta, c_2 \rangle = \frac{-x_1^{n_2} x_2^{n_1} p_{12} p_{21}}{x_1 - x_2},$$

where c_2 is the other coloring (with the opposite sign in the evaluation and transposed exponents of x_1, x_2). Assuming $n_1 \geq n_2$,
(4.18)

$$\langle \Theta \rangle = (x_1 x_2)^{n_2} \frac{x_1^{n_1 - n_2} - x_2^{n_1 - n_2}}{x_1 - x_2} p_{12} p_{21} = (x_1 x_2)^{n_2} h_{n_1 - n_2 - 1}(x_1, x_2) p_{12} p_{21},$$

where $h_k(x_1, x_2) = x_1^k + x_1^{k-1} x_2 + \cdots + x_2^k$ is the k-th complete symmetric function in x_1, x_2. Note that we can write

$$\langle \Theta \rangle = -E_2^{n_2} h_{n_1 - n_2 - 1}(x_1, x_2) \rho, \tag{4.19}$$

and that $h_k(x_1, x_2)$ is a polynomial in E_1, E_2, the latter elementary symmetric functions in x_1, x_2. Also, $\langle \Theta \rangle$ is the product of a Schur function for $GL(2)$ and $-\rho$.

Note that if the two thin facets of the theta-foam carry the same number of dots, $n_1 = n_2$, then it evaluates to zero, $\langle \Theta \rangle = 0$. If we reverse the orientation of Θ to get a foam Θ', then $\langle \Theta' \rangle = -\langle \Theta \rangle$. In general, if foam F contains k singular circles and $\overline{F}$ is given by reversing the orientation of F, then $\langle \overline{F} \rangle = (-1)^k \langle F \rangle$.

Recall that our ground ring $\widetilde{R}$ is the power series $\mathbf{k}[\![E_1, E_2]\!]$, where $\mathbf{k} = \mathbb{Z}[\beta_{i,j}]$ is polynomials in various negative degree generators with integer coefficients, see formulas (2.7), (4.7). Let R be the subring of $\widetilde{R} = \mathbf{k}[\![E_1, E_2]\!]$ generated by $E_1, E_2, \rho_0, \rho_1, \rho^{\pm 1}$ over $\mathbb{Z}$:

$$R = \langle E_1, E_2, \rho_0, \rho_1, \rho^{\pm 1} \rangle \subset \widetilde{R} = \mathbf{k}[\![E_1, E_2]\!]. \tag{4.20}$$

In all the examples above, the foam evaluates to an element of this subring. We will see soon that this is true for any closed foam and that the ground ring of the theory can be reduced from the rather large power series ring $\mathbf{k}[\![E_1, E_2]\!]$ to the subring R, which is finitely generated over the image of $\mathbb{Z}$ in $\mathbf{k}$.

Let us summarize that

$$
\begin{aligned}
\rho_0 &= \langle \mathbb{S}^2_{1,0} \rangle = \frac{p_{12}-p_{21}}{x_1-x_2}, \\
\rho_1 &= \langle \mathbb{S}^2_{1,1} \rangle = \frac{x_1 p_{12}-x_2 p_{21}}{x_1-x_2}, \\
\rho &= \langle \mathbb{S}^2_{2,0} \rangle = -p_{12}p_{21}
\end{aligned} \tag{4.21}
$$

are the evaluations of the thin 2-sphere with zero dots, with one dot, and the double 2-sphere, respectively. The subring R is graded, with homogeneous generators in degrees

generator	ρ_0	ρ_1	ρ	E_1	E_2
degree	-2	0	0	2	4

.

Notice that only ρ_0 has a negative degree. Using that $E_1^2-4E_2=(x_1-x_2)^2$, it is easy to compute

$$
\rho_1^2 - E_1\rho_1\rho_0 + E_2\rho_0^2 = \frac{E_1^2-4E_2}{(x_1-x_2)^2}\, p_{12}p_{21} = -\rho. \tag{4.22}
$$

Define the ring

$$
\overline{R} = \mathbb{Z}[E_1,E_2,\rho_0,\rho_1,(\rho_1^2-E_1\rho_1\rho_0+E_2\rho_0^2)^{-1}] \tag{4.23}
$$

as the localization of the polynomial ring with generators E_1, E_2, ρ_0, ρ_1 at the element

$$
\rho = -(\rho_1^2-E_1\rho_1\rho_0+E_2\rho_0^2). \tag{4.24}
$$

There is an obvious homomorphism $\overline{R} \longrightarrow R \subset \widetilde{R}$, and we now prove that it is an isomorphism between $\overline{R}$ and R.

Consequently, we can think of R as the localization,

$$
R \cong \mathbb{Z}[E_1,E_2,\rho_0,\rho_1,(\rho_1^2-E_1\rho_1\rho_0+E_2\rho_0^2)^{-1}]. \tag{4.25}
$$

We will show that this localization has a basis over $\mathbb{Z}[E_1,E_2]$:

$$
\mathscr{B} := \{\rho_1^{n_1}\rho_0^{n_2}\rho^{n_3}, \quad n_1\in\{0,1\}, n_2\in\mathbb{Z}_+, n_3\in\mathbb{Z}\} \tag{4.26}
$$

To establish isomorphism $\overline{R}\cong R$ of rings, denote by

$$
R^- := \mathbb{Z}[\rho_0,\rho_1^{\pm 1}][\![E_1,E_2]\!] \tag{4.27}
$$

the graded ring of power series in E_1, E_2 with coefficients in the ring $\mathbb{Z}[\rho_0,\rho_1^{\pm1}]$. In this definition, we view ρ_0, ρ_1 as additional generators and not as power series.

LEMMA 4.2. *Ring R^- is naturally a subring of $\widetilde{R}$, via power series expansions* (4.21) *for ρ_0 and ρ_1.*

PROOF. We can write the power series

$$
p_{12} = 1 + Ax_1 + Bx_2 + \sum_{i+j>1} \beta_{ij}x_1^i x_2^j,
$$

where $A=\beta_{1,0}$ and $B=\beta_{0,1}$, also see formula (2.4). The power series for ρ_1 is invertible, since the expansion starts with $1+A(x_1+x_2)=1+AE_1$ followed by higher degree terms in x_1 and x_2 with coefficients in the variables β_{ij} for $i+j>1$,

$$
\rho_1 \longmapsto \frac{x_1p_{12}-x_2p_{21}}{x_1-x_2} = 1 + A(x_1+x_2) + h.o.t.,
$$

where $h.o.t.$ stands for 'higher order terms'. Furthermore, the series expansion for ρ_1 does not involve the coefficient B. The series

$$\rho_0 \longmapsto \frac{p_{12} - p_{21}}{x_1 - x_2} = A - B + h.o.t.$$

begins with the element $A - B$ followed by higher degree terms in x_1 and x_2 with coefficients only involving the parameters β_{ij} for $i+j > 1$. Consider the homomorphism

$$\tau : R^- = \mathbb{Z}[\rho_0, \rho_1^{\pm 1}][\![E_1, E_2]\!] \longrightarrow \widetilde{R},$$

given by expanding ρ_0 and ρ_1 as power series, so that

$$\tau(\rho_0) = A - B + h.o.t., \qquad \tau(\rho_1) = 1 + AE_1 + h.o.t.$$

To show that τ is injective, compose τ with the involution of $\widetilde{R}$ that sends the generator $B = \beta_{0,1}$ to $\tau(\rho_0)$, and fixes all other generators (generators E_1, E_2 and $\beta_{i,j}$ for $(i,j) \neq (0,1)$). So the question reduces to showing injectivity of the map

$$\pi : R^- \longrightarrow \widetilde{R}, \quad \pi(\rho_0) = B, \quad \pi(\rho_1) = \tau(\rho_1) = 1 + AE_1 + h.o.t., \quad \pi(E_i) = E_i, \ i = 1, 2.$$

Now consider any homogeneous element of degree $2n$ in the kernel of π

$$\sum_{i,j,k,\ i+2j=n+k} \rho_0^k f_{ijk}(\rho_1) E_1^i E_2^j,$$

where $f_{ijk}(\rho_1)$ is a Laurent polynomial in ρ_1. Mapping to $\widetilde{R}$ under π and observing that the elements B^k are linearly independent over the subring of $\tilde{R}$ given by power series in E_1 and E_2 with values in the polynomial algebra $\mathbb{Z}[\beta_{i,j}, (i,j) \neq (0,1)]$, we deduce that for any fixed $k \geq 0$ one has relations

$$0 = \sum_{i+2j=2n+k} f_{ijk}(\tau(\rho_1)) E_1^i E_2^j.$$

Notice that for any k, the above expression is a finite sum. So by multiplying by a suitable power of $\tau(\rho_1)$, we may assume that each Laurent polynomial $f_{ijk}(\rho_1)$ is in fact a polynomial in ρ_1. The algebraic independence of the classes $\tau(\rho_1), E_1, E_2$ easily implies that each $f_{ijk}(\rho_1)$ must be trivial. In other words, the map π is injective, which is what we wanted to prove. □

COROLLARY 4.3. *The power series homomorphism $\overline{R} \longrightarrow \widetilde{R}$ takes $\overline{R}$ isomorphically onto the subring R of $\widetilde{R}$. Moreover, the ring R has a basis over $\mathbb{Z}[E_1, E_2]$ given by*

$$\mathscr{B} := \{\rho_1^{n_1} \rho_0^{n_2} \rho^{n_3}, \quad n_1 \in \{0,1\}, n_2 \in \mathbb{Z}_+, n_3 \in \mathbb{Z}\}.$$

PROOF. By definition, the image of $\overline{R}$ in $\widetilde{R}$ is equal to the ring R. Now both rings $\overline{R}$ and R are generated as modules over $\mathbb{Z}[E_1, E_2]$ by the set of elements of $\mathscr{B}$. To be more precise, both rings $\overline{R}$ and R have a collection of generators $\mathscr{B}(\overline{R})$ and $\mathscr{B}(R)$, respectively, as defined above that are compatible under the map from $\overline{R}$ to R. Hence, to demonstrate the isomorphism between $\overline{R}$ and R, it is sufficient to show that the elements $\mathscr{B}(R)$ are linearly independent over $\mathbb{Z}[E_1, E_2]$ when seen as elements in R, thereby showing that the elements $\mathscr{B}(R)$ form a $\mathbb{Z}[E_1, E_2]$-module basis of R. It follows from this that the collection $\mathscr{B}(\overline{R})$ also forms a $\mathbb{Z}[E_1, E_2]$-module basis of $\overline{R}$, and consequently, that the map from $\overline{R}$ to R is an isomorphism.

In what follows, we will actually show that the elements $\mathscr{B}(R)$ are linearly independent over $\mathbb{Z}[\![E_1, E_2]\!]$ in the larger ring R^-, once we observe that the ring R

is contained in the image of $R^- \subset \widetilde{R}$. For this it suffices to show that ρ^{-1} is in R^-, which follows from formula (4.22) that expresses ρ^{-1} as a power series in E_1 and E_2 with polynomial coefficients in $\rho_0, \rho_1^{\pm 1}$:

$$\rho^{-1} = -(\rho_1^2 - E_1\rho_1\rho_0 + E_2\rho_0^2)^{-1} = -\rho_1^{-2}(1 - E_1\rho_0\rho_1^{-1} + E_2\rho_0^2\rho_1^{-2})^{-1},$$

and then formally expanding the inverse as a power series. This shows that the inclusion $R \subset \widetilde{R}$ factors through the subring R^-.

It remains to show linear independence of the elements $\mathscr{B}(R)$ over $\mathbb{Z}[\![E_1, E_2]\!]$ inside R^-. Since the set of elements $\{\rho_0^{n_2}\}$ are linearly independent over $\mathbb{Z}[\![E_1, E_2]\!]$, it is sufficient to show that the sub-collection of $\mathscr{B}(R)$ given by the elements $\{\rho_1^{n_1}\rho^{n_3}\}$ is linearly independent over $\mathbb{Z}[\rho_0][\![E_1, E_2]\!]$. Let us consider a homogeneous relation

$$0 = \sum_{n:=(n_1,n_3)} A_n(\rho_0, E_1, E_2)\rho_1^{n_1}\rho^{n_3}, \tag{4.28}$$

where the indexing set is some finite subset of distinct pairs $n := (n_1, n_3)$ as above with $A_n(\rho_0, E_1, E_2)$ being a homogeneous element of $\mathbb{Z}[\rho_0][\![E_1, E_2]\!]$. Reducing relation (4.28) mod ρ_0 and using equation (4.22), we obtain the relation in $\mathbb{Z}[\rho_1^{\pm 1}][\![E_1, E_2]\!]$

$$0 = \sum_{n:=(n_1,n_3)} (-1)^{n_3} A_n(0, E_1, E_2)\rho_1^{n_1+2n_3},$$

which is clearly true only if $A_n(0, E_1, E_2) = 0$ for all n. This condition implies that each $A_n(\rho_0, E_1, E_2)$ is divisible by ρ_0. We may therefore factor ρ_0 out of the entire relation (4.28), and repeat the argument (note that ρ_0 is not a zero divisor). This shows that $A_n(\rho_0, E_1, E_2)$ must be trivial for all n, which is what we needed to establish.

□

REMARK 4.4. *The inclusion $R \subset R^-$ is dense in the power series ring topology. In order to show this, it is sufficient to show that ρ_1^{-1} can be described in terms of a power series in E_1 and E_2, with coefficients that are polynomials in $\rho_0, \rho_1, \rho^{\pm 1}$. This follows from formula* (4.22) *which implies that*

$$\rho_1^{-1} = -\rho^{-1}(1 + \rho_0^2\rho^{-1}E_2)^{-1}(\rho_1 - \rho_0 E_1).$$

Notice that in addition to the chain of ring inclusions in formulas (4.6)-(4.10), there is also a chain of inclusions

$$R \subset \widetilde{R} \subset R' \subset R''. \tag{4.29}$$

The example 6 above for the evaluation of the Θ-foam is straightforward to generalize to $GL(N)$, where Θ-foam has a disk of thickness N with N disks of thickness one attached to it, carrying $n_1, \ldots, n_N$ dots, respectively, where we can assume $n_1 \geq n_2 \geq \cdots \geq n_N$, see Figure 4.33.

Let $\lambda_i = n_i - N + i$, so that $\lambda = (\lambda_1, \ldots, \lambda_N)$ is a partition iff $n_i > n_{i+1}$ for all i. Denote this foam by Θ_λ. One can compute the foam evaluation

$$\langle \Theta_\lambda \rangle = \pm \sum_{\sigma \in S_N} \frac{(-1)^{\ell(\sigma)} \prod_{i=1}^N x_{\sigma(i)}^{n_i}}{\prod_{i<j}(x_i - x_j)} \prod_{i \neq j} p_{ij} = \pm s_\lambda \prod_{i \neq j} p_{ij}, \tag{4.30}$$

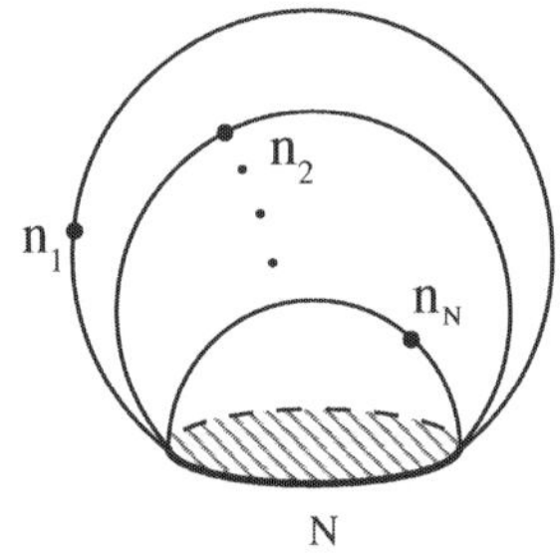

FIGURE 4.33. Θ-foam for $GL(N)$.

where s_λ is the $GL(N)$ Schur function for the partition λ. The last equality holds if λ is a partition, otherwise $\langle \Theta_\lambda \rangle = 0$. One can argue that our deformation does not go far enough, since it does not deform Schur functions in an interesting way and only scales them by the product of p_{ij}'s. At least it does deform the value of the thin 2-sphere with dots and other closed surfaces in a non-trivial way.

4.4. Skein relations.

PROPOSITION 4.5. *The skein relation (*singular neck-cutting relation*) in Figure* 4.41 *holds.*

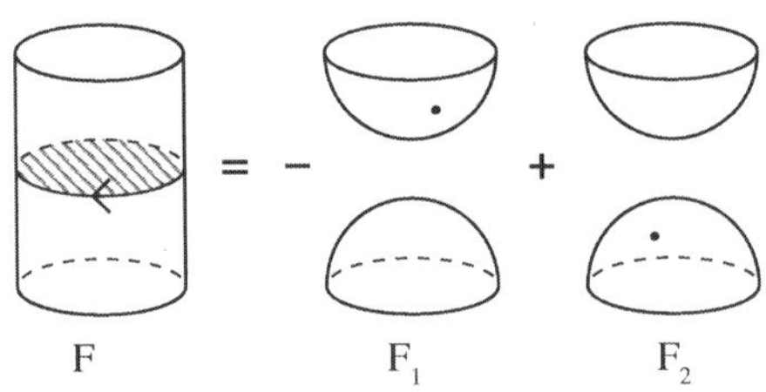

FIGURE 4.41. Singular neck-cutting relation.

Proof: Coloring c of F induces a coloring c' of F_1, F_2 (the latter two foams differ only by dot placement, and we use c' to denote corresponding coloring of both foams). Coloring c' has opposite colors on the two disks of F_1 (and F_2). If a coloring c_1 of F_1 and F_2 has the same color on the two disks, $\langle F_1, c_1 \rangle = \langle F_2, c_1 \rangle$, since dots will contribute with the same x_i, $i \in \{1,2\}$, to the evaluations, and this coloring will not contribute to the difference $\langle F_2 \rangle - \langle F_1 \rangle$. Thus, we can restrict to colorings c' as above, in bijection with colorings c of F.

If the top facet of c is colored 1, see Figure 4.42, then the circle of F in the figure is positive and $\theta^+(c) = \theta^+(c') + 1$. Also, $\chi_2(F, c) = \chi_2(F_1, c') = \chi_2(F_2, c')$, so that $-(-1)^{s'(F,c)} = (-1)^{'(F_1,c')} = (-1)^{'(F_2,c')}$.

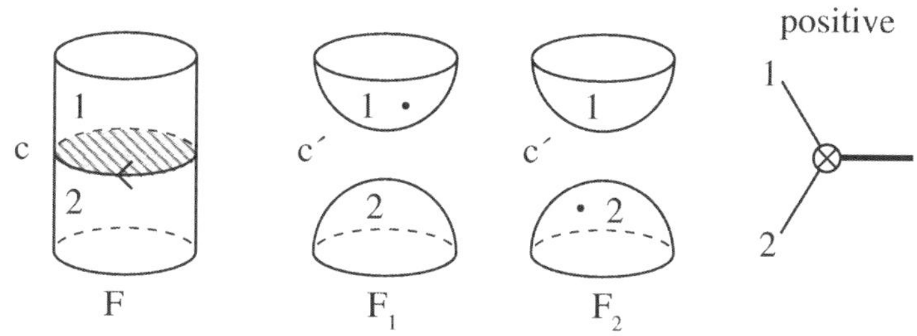

FIGURE 4.42. When top facet is colored 1.

We have $\chi_{12}(F,c) = \chi_{12}(F_1,c') - 2$, so that $\langle F,c\rangle$ has an additional $(x_1 - x_2)$ in the numerator, compared to $\langle F_1,c'\rangle$ and $\langle F_2,c'\rangle$. Due to a dot on facet colored 1 there is an extra x_1 in $\langle F_1,c'\rangle$ and an extra x_2 in $\langle F_2,c'\rangle$. More accurately, we can write

$$\langle F,c\rangle = -(x_1 - x_2)y, \ \langle F_1,c'\rangle = x_1 y, \ \langle F_2,c'\rangle = x_2 y$$

for some y, so that $\langle F,c\rangle = -\langle F_1,c'\rangle + \langle F_2,c'\rangle$.

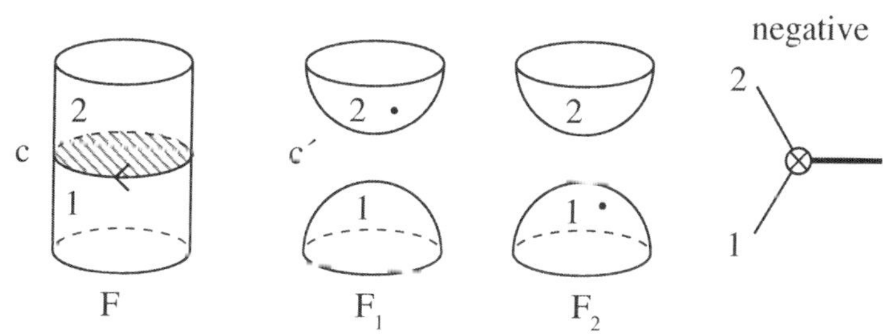

FIGURE 4.43. When top facet is colored 2.

The other case is when the top facet of F is colored 2 by c, see Figure 4.43. In this case the singular circle of F in the figure is negative for the coloring c, so that $\theta^+(c) = \theta^+(c')$ and $s'(F,c) = s'(F_1,c') = s'(F_2,c')$. This change of sign is balanced by the opposite coloring of the two disks in F_1, F_2, so that

$$\langle F,c\rangle = (x_1 - x_2)y, \ \langle F_1,c'\rangle = x_2 y, \ \langle F_2,c'\rangle = x_1 y$$

for some y, and we still have $\langle F,c\rangle = -\langle F_1,c'\rangle + \langle F_2,c'\rangle$. Summing over all c implies the proposition. $\square$

Reversing the orientation of the singular circle (and hence of the entire connected component of F) changes the signs in the relation, see Figure 4.44 and Proposition 4.9 below.

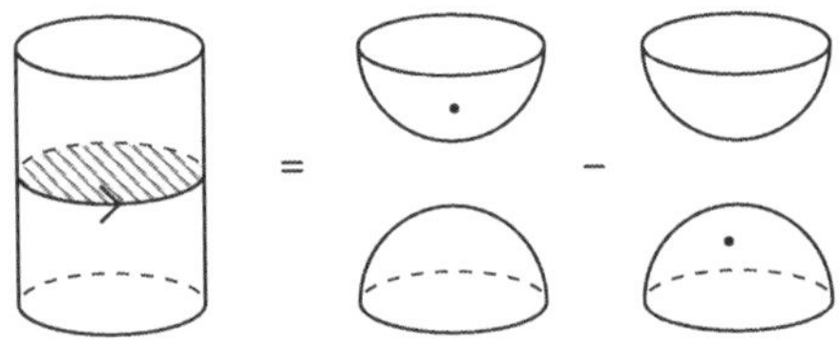

FIGURE 4.44. Relation for the other orientation.

PROPOSITION 4.6. *The skein relation (*canceling double disks*) in Figure* 4.45 *holds.*

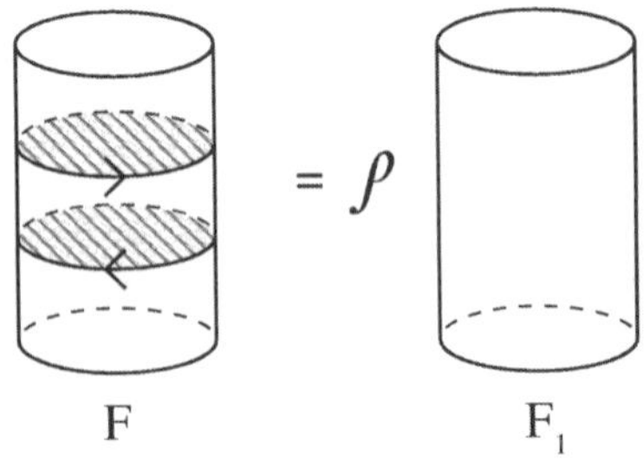

FIGURE 4.45. Canceling parallel double disks.

Proof: There is a bijection between colorings c of F and colorings c_1 of F_1, see Figure 4.46.

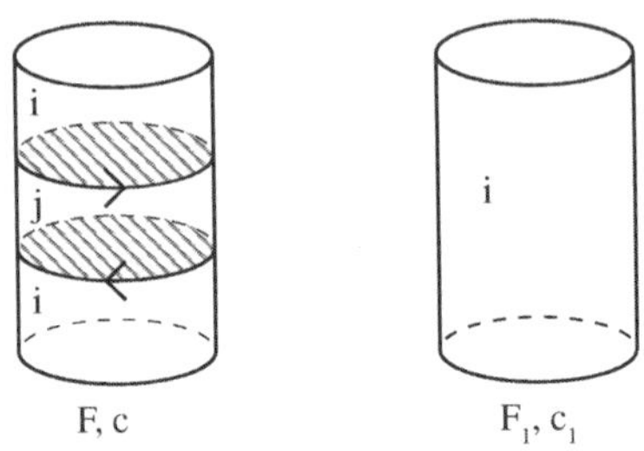

FIGURE 4.46. A coloring c of F and the corresponding coloring c_1 of F_1.

One checks that $\chi_k(F_1, c_1) = \chi_k(F, c) - 2$, $k = 1, 2$, and $\chi_{12}(F, c) = \chi_{12}(F_1, c_1)$. For any coloring, $\theta^+(c') \equiv \theta^+(c) (\text{mod } 2)$, since the two singular circles in F_1 have the same parity, and $(-1)^{s'(F,c)} = -(-1)^{s'(F_1,c_1)}$. Comparing the contributions,

$$\langle F, c\rangle = -p_{12}p_{21}\langle F_1, c_1\rangle = \rho\langle F_1, c_1\rangle.$$

Summing over all c, the result follows. $\square$

Since ρ is invertible, this relation shows that either of the two foams in Figure 4.45 can be written as the other foam times $\rho^{\pm 1}$.

Reversing orientation of the two singular circles on the left hand side of Figure 4.45 gives a similar skein relation, with no sign added since the parity of the number of singular circles is the same on both sides of the relation.

PROPOSITION 4.7. *The skein relation* (neck-cutting relation) *in Figure* 4.47 *holds.*

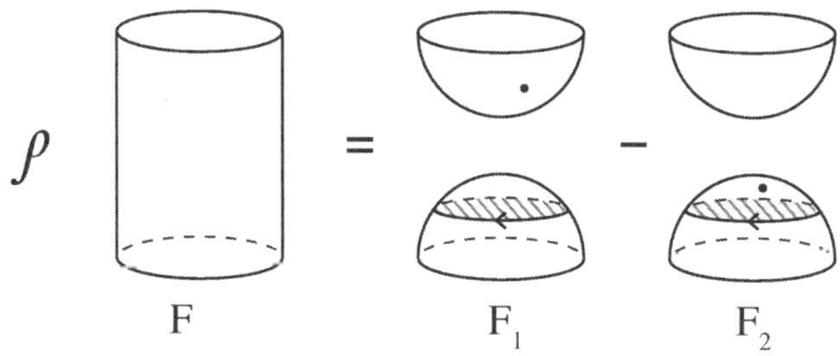

FIGURE 4.47. Neck-cutting relation.

Again, that ρ is invertible, and the relation allows us to do a surgery on an annulus which is part of a thin facet of F.

Proof: Apply Figure 4.45 relation to pass to a tube with two double disks and then use Figure 4.44 relation to do surgery on the top double disk. □

Doing the surgery on the bottom double disk using Figure 4.41 results in a similar relation, depicted in Figure 4.48 where singular disks now appear at the top rather than the bottom on the right hand side.

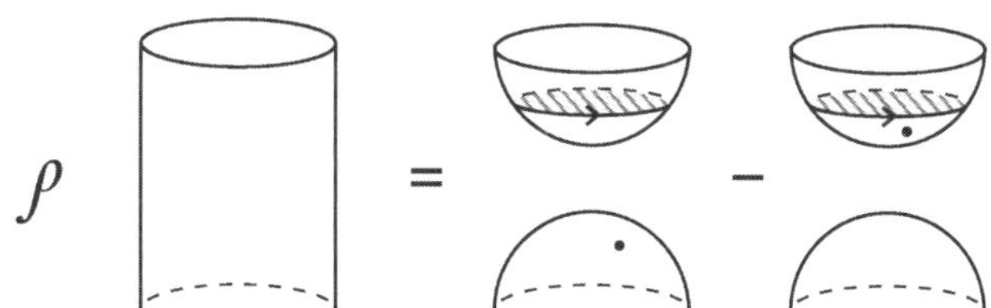

FIGURE 4.48. Neck-cutting relation with double disks at the top.

PROPOSITION 4.8. *If a double disk D^2 bounding a singular circle in a foam F can be completed to a 2-sphere without additional intersections with F, denote by*

F_1 the foam given by removing the 2-disk from F and adding its complement in $\mathbb{S}^2$, see Figure 4.49. Then $\langle F\rangle = -\langle F_1\rangle$.

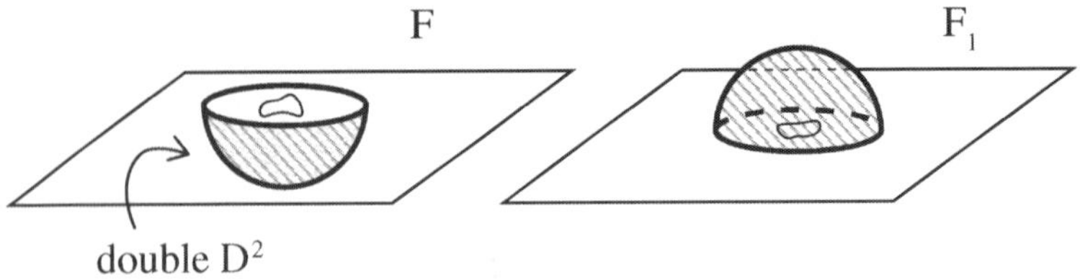

FIGURE 4.49. Double disk flipping.

Proof: There is a bijection between colorings c of F and colorings c_1 of F_1, with the only difference in evaluations coming from the type of the singular circle, so that $s'(F,c) = s'(F_1,c_1) \pm 1$ and $\langle F,c\rangle = -\langle F_1,c_1\rangle$. □

PROPOSITION 4.9. *If $\overline{F}$ is a foam F with the reversed orientation of all facets, then $\langle\overline{F}\rangle = (-1)^k\langle F\rangle$, where k is the number of singular circles of F.*

Proof: Each coloring c of F is a coloring of $\overline{F}$ as well, and $\langle\overline{F},c\rangle = (-1)^k\langle F,c\rangle$, since $s'(\overline{F},c) = (-1)^k s'(F,c)$ as the type (positive or negative) of each singular circle of F is reversed in $\overline{F}$. Summing over c implies the proposition. □

This proposition can be applied, for instance, to the neck-cutting relation in Figure 4.47. Reversing the orientation of singular circles in F_1, F_2 reverses the orientation of all facets as well. Since F has one less singular circle than F_1, F_2, there'll be an additional overall minus sign, which can go either to the left or right hand side.

PROPOSITION 4.10. *For a foam F_2 with a facet with two dots, the relation*

$$\langle F_2\rangle = E_1\langle F_1\rangle - E_2\langle F_0\rangle$$

holds, where F_1 and F_0 are the foams with one fewer and two fewer dots on the same facet, see Figure 4.410.

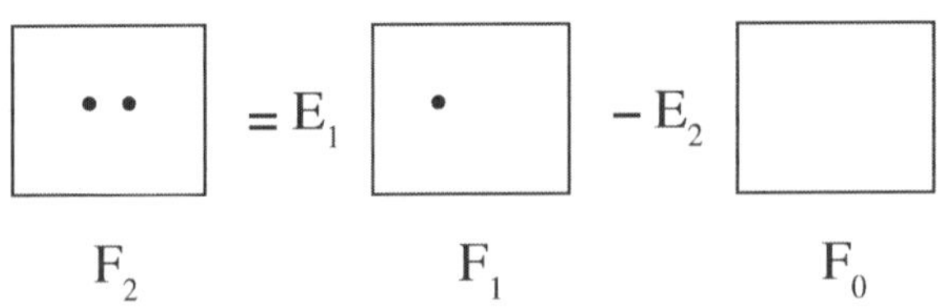

FIGURE 4.410. Dot reduction relation.

Proof: Follows, since $x_i^2 = E_1x_i - E_2$ for $i = 1, 2$. □

PROPOSITION 4.11. *(Double facet neck-cutting relation) Evaluations of foams F and F_1 in Figure* 4.411 *satisfy* $\langle F\rangle = \rho\langle F_1\rangle$.

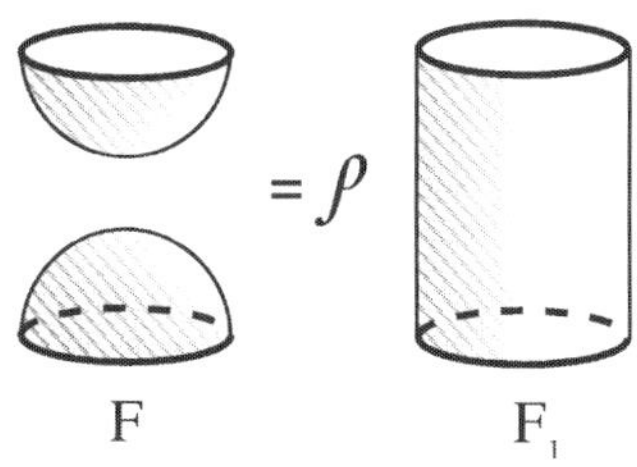

FIGURE 4.411. Neck cutting on a double facet.

Proof: Again, there is a bijection between colorings c of F and colorings c_1 of F_1. Difference in the Euler characteristics $\chi_i(F,c) = \chi_i(F_1,c_1) + 2$, for $i = 1,2$, contributes the term $-p_{12}p_{21} = \rho$ to the evaluation of $\langle F, c\rangle$ compared to that of $\langle F_1, c_1\rangle$. Summing over c implies the proposition. □

PROPOSITION 4.12. *(Dot migration relations)*

(1) *Evaluations of foams F_1, F_2 and F_1 in Figure* 4.412 *satisfy* $\langle F_1\rangle + \langle F_2\rangle = E_1\langle F\rangle$.
(2) *Figure* 4.413 *relation holds.*

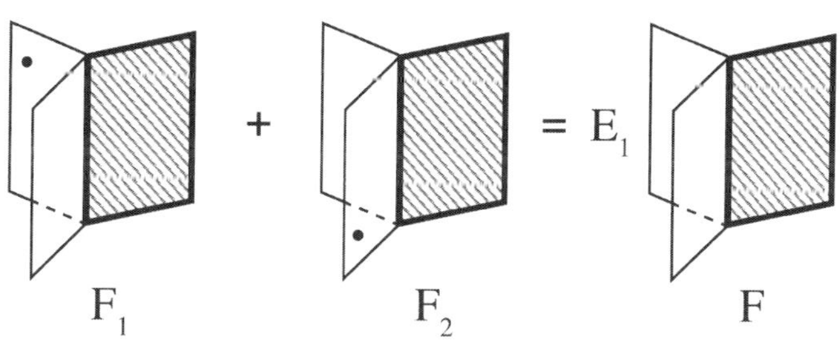

FIGURE 4.412. Dot migration relation.

Proof: Follows, since these foams differ only by dot placement, any for any coloring c the two facets with dots carry opposite colors. These dots contribute x_1 and x_2 to the evaluation. Consequently,

$$\langle F_1, c\rangle + \langle F_2, c\rangle = (x_1 + x_2)\langle F_3, c\rangle = E_1\langle F_3, c\rangle$$

The same argument implies the second relation. □

PROPOSITION 4.13. *Skein relation in Figure* 4.414 *holds.*

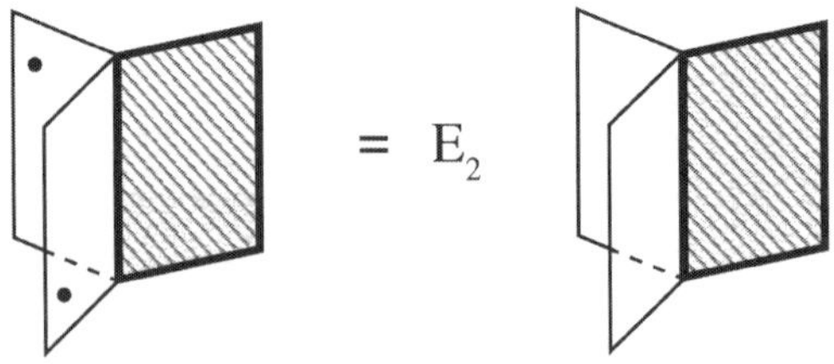

FIGURE 4.413. Second dot migration relation.

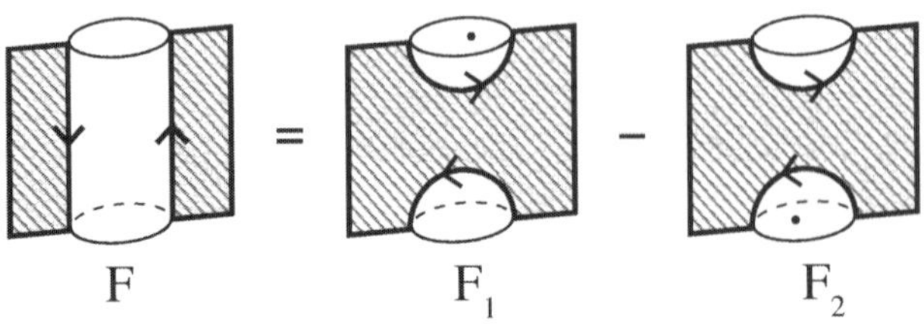

FIGURE 4.414. Cutting a tube with two singular edges.

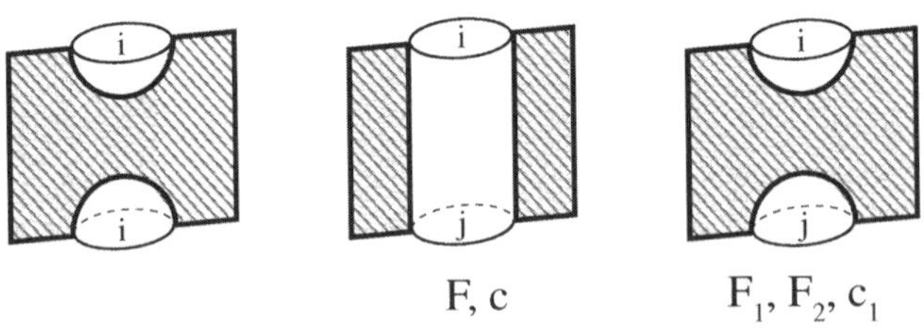

FIGURE 4.415. Left: a coloring of F_1, F_2 that does not come from a coloring of F. Right: a coloring c' of F and induced coloring c_1 of F_1, F_2.

Proof: Colorings c' of F_1, F_2 that do not come from colorings of F have the property that the front thin bottom and back thin top facets are colored by the same color, see Figure 4.415 left.

The dots on F_1, F_2 will have the same color and these terms will cancel out from the difference, with $\langle F_1, c'\rangle - \langle F_2, c'\rangle = 0$.

The remaining colorings c_1 of F_1, F_2 are in bijection with colorings c of F, see Figure 4.415 right. For these colorings we have

$$\chi_{12}(F,c) = \chi_{12}(F_k, c_1) - 2, \ \chi_\ell(F,c) = \chi_\ell(F_k, c_1), \ \ell, k = 1, 2.$$

The rest of the computation is similar to that in the proof of Proposition 4.7. If $i = 1, j = 2$, one checks that $\theta^+(c_1) = \theta^+(c)$ and the signs $(-1)^{s'}$ are the same in the three evaluations. Due to difference in χ_{12}, the evaluation $\langle F, c \rangle$ will acquire $(x_1 - x_2)$ in the numerator compared to the other two foams. This will be matched by the dots, contributing x_1 to $\langle F_1, c_1 \rangle$ and x_2 to $\langle F_2, c_1 \rangle$, correspondingly.

If $i = 2, j = 1$, there will be sign difference $(-1)^{s'(F,c)} = -(-1)^{s'(F_k,c_1)}$, $k = 1, 2$. Dots will now contribute x's with the opposite indices to the evaluations of F_1, F_2, canceling the sign difference, so that again $\langle F, c \rangle = \langle F_1, c_1 \rangle - \langle F_2, c_1 \rangle$. □

This relation with the opposite singular circles orientation, see Figure 4.416, can be obtained from that in Figure 4.414 by looking at foams there from the opposite side of the plane. Furthermore, rotating foams in Figure 4.414 by 180° (and using dot migration relation in Figure 4.412 twice) yields a similar to Figure 4.414 relation but with a different distribution of dots across thin facets.

FIGURE 4.416. Tube-cutting with the other orientation.

In a foam F, let γ be a curve that connects two points on singular lines and lies in a single facet of F, see Figure 4.417 left. Let us call such a curve a *proper* curve. The foam $F_1 = m(\gamma, F)$ on that figure on the right is called the modification of F along γ. In the undeformed case, when $p(x, y) = 1$, $GL(2)$ foam evaluation satisfies $\langle F \rangle = \pm \langle F_1 \rangle$, with the sign depending on orientation of singular edges of F, see [**BHPW**, Equation (2.10)].

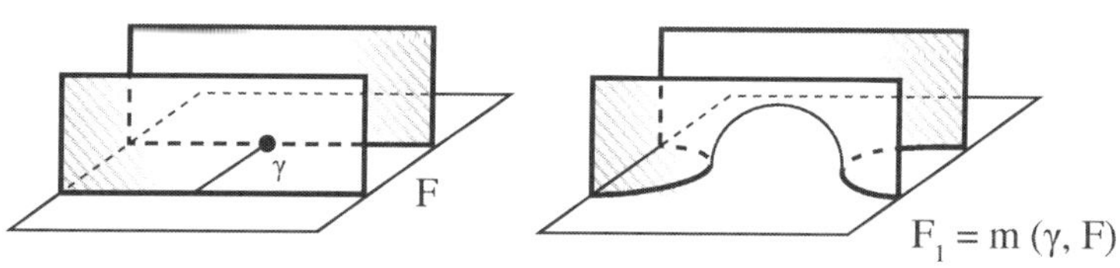

FIGURE 4.417. Modifying foam F along curve γ in a thin facet. γ connects two points on the singular set of F.

The relation is more subtle in our case. We start with orientations of singular edges as shown on Figure 4.418; note that choosing orientation of one edge forces the orientation of the other edge of F shown. Choose a coloring c of F and denote by c' the corresponding coloring of F_1 (there is a bijection between colorings of F and F_1), see Figure 4.418.

We have

$$\chi_{12}(F_1, c_1) = \chi_{12}(F, c), \ \chi_i(c_1) = \chi_i(c) - 2, \ \chi_j(c_1) = \chi_j(c).$$

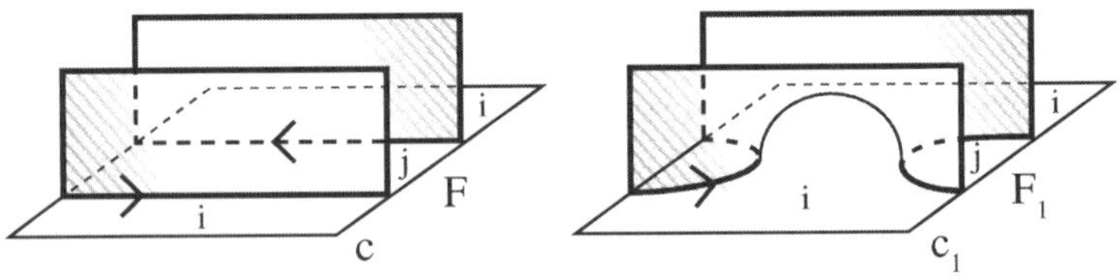

FIGURE 4.418. Notice orientation of singular edges.

The number of singular circles of F_1 is one less or more of that of F, depending on whether the two singular edges shown in F belong to different or the same singular circle.

If $i = 1, j = 2$ then these circles are negative, they make no contribution to $\theta^+(c)$ and $\theta^+(c_1)$, and $s'(F, c) = s'(F_1, c_1)$, since $\chi_2(c) = \chi_2(c_1)$ and $\theta^+(c) = \theta^+(c_1)$. If $i = 2, j = 1$, the circles are positive and $(-1)^{\theta^+(c)} = -(-1)^{\theta^+(c_1)}$. Also, the Euler characteristics $\chi_2(c), \chi_2(c_1)$ differ by two and contribute a sign to the difference as well. We again get $s'(F, c) = s'(F_1, c_1)$, so that for any coloring $s'(F, c) = s'(F_1, c_1)$.

Combining these computations,

$$\langle F, c\rangle = p_{ij}\langle F', c'\rangle. \tag{4.31}$$

Note that p_{ij} is not an element of our ground ring $\mathbf{k}[\![E_1, E_2]\!]$ and summing this equality over all colorings c of F will not get an immediate relation between evaluations of F and F'.

We now look at the oppositely oriented case, see Figure 4.419. Circles now carry opposite signs from that of the previous case, and one can check that $s'(F, c) = -s'(F_1, c_1)$ in each of the cases $(i, j) = (1, 2)$ and $(i, j) = (2, 1)$. A similar computation now gives

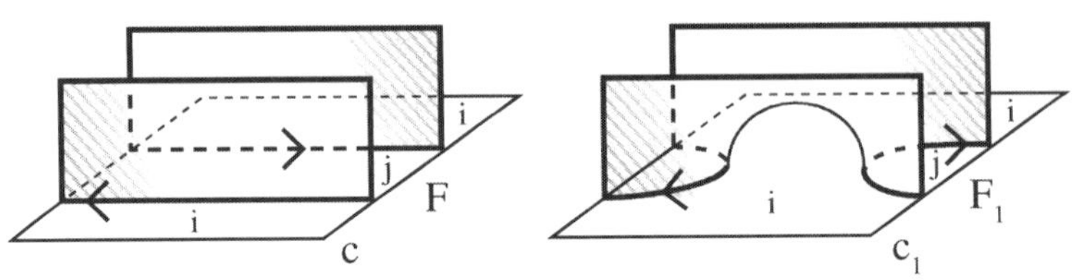

FIGURE 4.419. Singular edges have the opposite orientation from that in the previous figure.

$$\langle F, c\rangle = -p_{ij}\langle F_1, c_1\rangle, \tag{4.32}$$

which is similar to (4.31) but with an additional sign.

Consider the thin surface F_{12} of F and choose a connected component Σ in it. Recall that we are looking at modifications of F along proper curves γ and now restrict to γ on a component Σ. Notice that the double facets at the endpoints of γ are pointing in the same direction relative to Σ, either both outward or both inward. Also, if we were to redraw F in Figure 4.418 keeping orientations of the singular edges but drawing double facets on the opposite side of Σ ('below', rather than 'above'), the type of the diagram would change to the one in Figure 4.419, and vice versa.

Proper disjoint curves or arcs $\gamma_1, \gamma_2 \in \Sigma$ are called *complementary* if for a coloring c of F they lie in differently colored regions. This property does not depend on the choice of c.

To a pair (γ_1, γ_2) of complementary arcs we assign a sign $s(\gamma_1, \gamma_2)$. Namely, consider the four double facets of F at the endpoints of γ_1 and γ_2. If these four facets all point into the same connected component of $\mathbb{R}^3 \setminus \Sigma$, we set $s(\gamma_1, \gamma_2) = 1$. Otherwise we define $s(\gamma_1, \gamma_2) = -1$.

An example when $s(\gamma_1, \gamma_2) = 1$ is shown in Figure 4.420. In general, γ_1, γ_2 don't have to have an endpoint on the same singular circle.

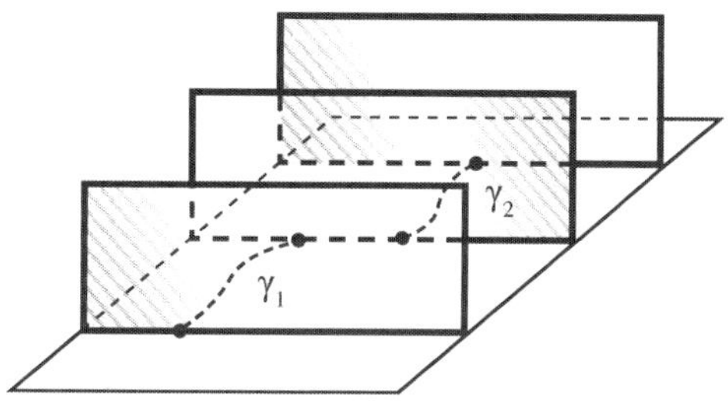

FIGURE 4.420. Complementary proper arcs γ_1, γ_2 with $s(\gamma_1, \gamma_2) = 1$.

Given complementary proper arcs γ_1, γ_2 in Σ, we can do commuting modifications along γ_1, γ_2 to get from F to the foam

$$F_2 = m(\gamma_1, m(\gamma_2, F)) = m(\gamma_2, m(\gamma_1, F)).$$

PROPOSITION 4.14. *For F and F_2 as above, $\langle F \rangle = s(\gamma_1, \gamma_2)\rho \cdot \langle F_2 \rangle$.*

Proof: For a coloring c of F curves γ_1 and γ_2 lie in differently colored regions of Σ, say i and j-colored regions, $\{i, j\} = \{1, 2\}$. When $s(\gamma_1, \gamma_2) = 1$, orientations on singular edges will make one of the curves γ the type in Figure 4.418 and the other in Figure 4.419, with (i, j) replaced by (j, i) in one of these two cases. Using equations (4.31) and (4.32), we obtain $\langle F, c \rangle = -p_{12}p_{21}\langle F_2, c_2 \rangle = \rho \langle F_2, c_2 \rangle$ for the corresponding coloring c_2 of F_2.

When $s(\gamma_1, \gamma_2) = -1$, orientations on singular edges will make both γ_1, γ_2 either the type in Figure 4.418 or the type in Figure 4.419, with (j, i) in place for (i, j) for one of γ_1, γ_2. This will introduce a minus sign, with $\langle F, c \rangle = -\rho \langle F_2, c_2 \rangle$. □

This proposition may be generalized in some cases when one of γ_1, γ_2 is not a proper arc. One would need γ_2 to be a proper arc in $m(\gamma_1, F)$, in the region of color opposite to that of γ_1, with a coloring of F naturally converted to a coloring of $m(\gamma_1, F)$. We provide an example of such pair of arcs in Figure 4.421 and leave the details to the reader.

COROLLARY 4.15. *$\langle F \rangle = \rho \langle F_1 \rangle$ for foams F, F_1 in Figure 4.422.*

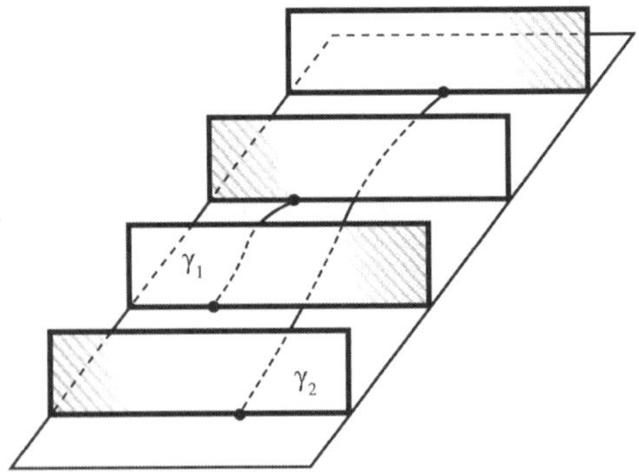

FIGURE 4.421. Arc γ_2 is not proper but becomes proper in $m(\gamma_1, F)$.

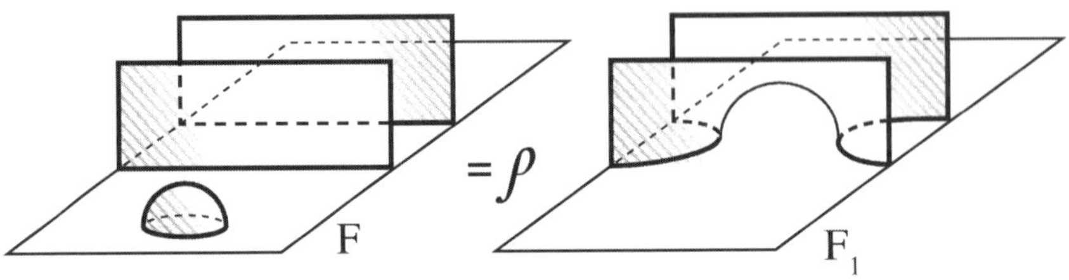

FIGURE 4.422. Notice additional double cap on the foam F, used to create a pair of complementary proper arcs on it.

This follows from Proposition 4.14 using the pair of arcs in Figure 4.423 with $s(\gamma_1, \gamma_2) = 1$. □

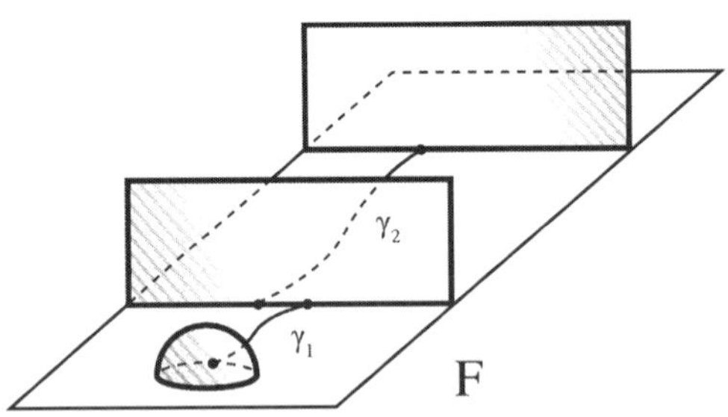

FIGURE 4.423. Arcs γ_1,γ_2 are complementary proper with $s(\gamma_1, \gamma_2) = 1$.

COROLLARY 4.16. *Figure* 4.424 *relation on foam evaluations holds.*

The corollary follows at one from the previous one. □

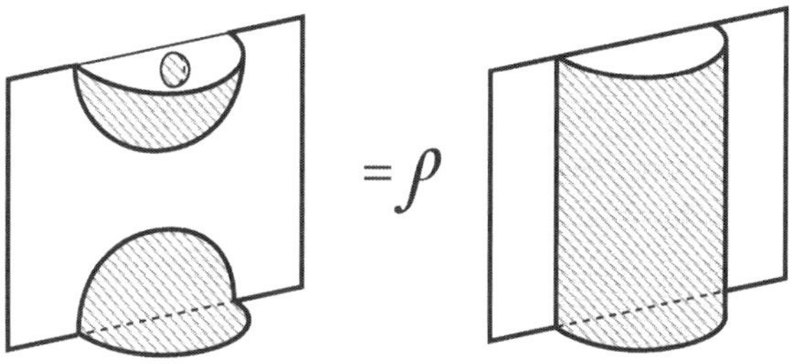

FIGURE 4.424. The bubble on the thin plane on the LHS foam points its double facet toward us, being on the same side of the thin plane as portions of the other double facets shown on the LHS.

4.5. Pre-foams and ground ring reduction.

To prove Proposition 4.19 below, it is convenient to introduce the notion of a $GL(2)$ pre-foam and its evaluation. An (oriented) $GL(2)$ pre-foam (or prefoam) F has the same local structure as a $GL(2)$ foam, but without an embedding into $\mathbb{R}^3$. It has oriented thin and double facets, with the orientation of facets being compatible along singular edges as in Figure 4.12. In particular, orientations of facets induce orientations of singular circles. Vice versa, an orientation of a singular circle in a connected component of a pre-foam will induce orientations on all facets of that component.

Along each singular edge a preferred facet out of two adjacent thin facets is specified. One can encode this choice by an arrow (a normal direction) out of the singular edge and into the thin surface of the pre-foam (the union of its thin facets). A pre-foam may carry dots on its thin facets.

A $GL(2)$ foam F gives rise to a $GL(2)$ pre-foam, also denoted F. Embedding of foam F in $\mathbb{R}^3$ together with orientation of singular circles induces an order on the two thin facets attached to a given singular circle. Namely, look in the direction of the orientation on the circle and choose the first thin facet counterclockwise starting from the double facet attached to the circle. This is then the preferred facet for the singular circle in the underlying pre-foam F.

Coloring c of a pre-foam is defined in the same way as for foams. For each coloring c, surfaces $F_1(c)$ and $F_2(c)$ inherit orientations from the facets of F they contain. Surface $F_{12}(c)$ is orientable as well, say with orientation matching that of thin facets of $F_{12}(c)$ colored 1 and opposite to that of thin facets colored 2.

Orientation requirements for facets ensure that each connected component of the thin surface F_{12} of a pre-foam F will admit two checkerboard colorings, so that a pre-foam F will admit 2^k colorings, where k is the number of connected components of F.

Given a coloring c of F, the preferred thin facet at a singular circle u allows to label the circle *positive* or *negative*, as in Figure 4.13. Namely, if the preferred facet is colored 1, the circle is *positive*. If the preferred facet is colored 2, the circle is negative.

Define $\theta^+(c) = \theta^+_{12}(c)$ as the number of positive singular circles for the coloring c.

Thus, in a pre-foam F, each singular circle u comes with both an orientation (induced from the orientation of attached facets and, vice versa, determining them) and a choice of preferred thin facet (normal direction to the thin surface F_{12}) along u. The evaluation of F, though, will only depend on the choice of preferred facet at each singular circle, not on its orientation.

Unlike the foam case, in a $GL(2)$ pre-foam we can reverse the thin normal direction (reverse the choice of preferred thin facet) at any subset of its singular circles without making any other changes, such as reversing orientations of facets or singular circles, changing the embedding into $\mathbb{R}^3$, etc. In a foam, the analogous operation of reversing the cyclic order of facets at a single circle via a simple modification of the embedding is possible only sometimes, see Figure 4.49 for an example.

Recall the chain of inclusions of rings $R \subset \widetilde{R} \subset R' \subset R''$ defined in formulas (4.6)-(4.10) and (4.20).

Now, to a coloring c of a pre-foam F we assign an element $\langle F, c\rangle \in R''$ using the formula (4.4). Furthermore, define $\langle F\rangle$ via the formula (4.5).

PROPOSITION 4.17. *Evaluation $\langle F\rangle$ of any $GL(2)$ prefoam F belongs to the subring $\mathbf{k}[\![E_1, E_2]\!]$ of R''.*

Proof: Our proof of this result for foams, Theorem 4.1, extends to prefoams without change. □

PROPOSITION 4.18. *Evaluation $\langle F\rangle$ of any $GL(2)$ prefoam F belongs to the subring R of $\widetilde{R} = \mathbf{k}[\![E_1, E_2]\!]$.*

Proof: Evaluations of surfaces and theta-foams, with dots, in Section 4.3 depend only on the pre-foam structure, not on an embedding in $\mathbb{R}^3$. Skein relations described in Section 4.4 extend, with suitable care, to pre-foams. In Figure 4.41 relation, a pre-foam on the LHS induces pre-foam structures on terms on the right, with orientations of facets in the RHS coming from those of the LHS. With this convention, Figure 4.41 relation holds for pre-foams, where in the pre-foam F on the LHS one also remembers the cyclic order of the facets

Relations in Figures 4.44, 4.45 extend likewise. In Figure 4.47 choice of orientations of all foams (respectively, pre-foams) is encoded in the orientation of the singular circle on the RHS (equivalently, of the cyclic order of the 3 facets at the circle).

Analogue of Proposition 4.9 for pre-foams is that $\langle F'\rangle = (-1)^k\langle F\rangle$, where F' is obtained from F by reversing the cyclic order of facets at some k singular circles of F.

Figure 4.410 relation obviously extends to pre-foams. In the double facet neck-cutting relation in Figure 4.411 relation pre-foam F_1 on the right induces an orientation on the pre-foam F on the left. With this convention, Figure 4.411 relation extends to pre-foams. Dot migration relations in Figures 4.412 and 4.413 as well as the tube-cutting relation in Figure 4.414 extend to pre-foams.

Modification $m(\gamma, F)$ in Figure 4.418 can be done to a pre-foam F, assuming compatible orientations and cyclic orders along the two singular edges of F. Proposition 4.14 will hold for pre-foams as well, again assuming compatibility of the orientations and cyclic orders along the three singular edges shown in Figure 4.420.

Starting with a pre-foam F, look at the thin surface F_{12}. It may have several connected components, some of which are connected in F by double facets. Applying the double neck-cutting relation in Figure 4.411, using multiplicativity of $\langle F\rangle$ on the disjoint union of pre-foams, and the evaluation of closed double surfaces (Examples 4, 5 in Section 4.3), we can reduce the evaluation to the case when F_{12} is connected and each double facet is a disk. Applying the singular neck-cutting relation in Figure 4.41 along each singular circle of F, the evaluation reduces to that of a closed thin surface, possibly with dots, see Examples 1-3 in Section 4.3. All coefficients in the skein relations and in the evaluation of closed surfaces belong to the ring R, implying the proposition. □

The proposition implies the next result.

PROPOSITION 4.19. *Evaluation $\langle F\rangle$ of any closed foam F coincides with the evaluation of the associated pre-foam. In particular, it belongs to the subring R of $\widetilde{R}=\mathbf{k}[\![E_1,E_2]\!]$.*

Proof: Foam F lives in $\mathbb{R}^3$, but to evaluate it using the formulas (4.4) and (4.5) we can pass to the associated pre-foam and evaluate it instead. □

Consequently, evaluations of all closed foams belong to the subring R of $\widetilde{R}$. It can then be chosen as the ground ring of the theory instead of $\widetilde{R}$, in the $GL(2)$ case.

4.6. $GL(2)$ webs, their state spaces, and direct sum decompositions.

We define $GL(2)$ closed webs Γ as generic intersections of $GL(2)$ foams with planes $\mathbb{R}^2$ in $\mathbb{R}^3$. A $GL(2)$ web Γ is a plane trivalent oriented graph with *thin* and *thick* (or *double*) edges and vertices as in Figure 4.61.

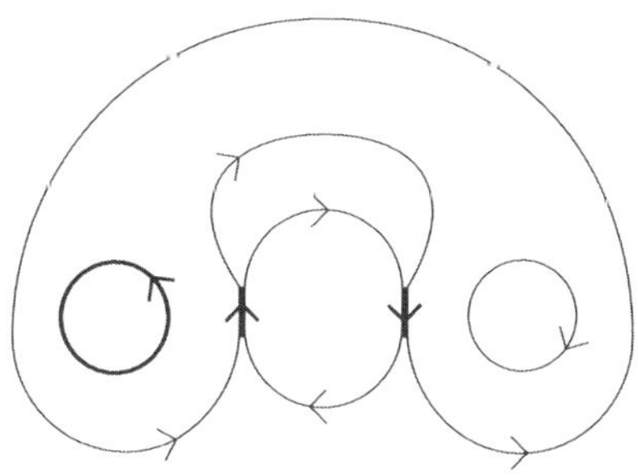

FIGURE 4.61. A $GL(2)$ web Γ with two thick edges, four thin edges, one thin and one thick circle.

Vertices of $GL(2)$ foams may be of two types. In one type, a pair of oriented thin edges flows into the vertex and a double edge flows out. In the other type, a double edge flows in and a pair of oriented thin edges flows out of the vertex. The web in Figure 4.61 has two vertices of each type.

Single and double closed loops are allowed, as well as the empty web. The union of thin edges of Γ is called the *thin one-manifold* of Γ, or the *thin cycles* of Γ and denoted $\Gamma_{(1)}$. For Γ in Figure 4.61, the thin one-manifold $\Gamma_{(1)}$ has three connected components.

One defines $GL(2)$ foams with boundary given by a $GL(2)$ web Γ in the usual way. We use Figure 4.12 as the convention for the induced orientation of the web that is the boundary of a $GL(2)$ foam. Note that $GL(2)$ foams F in $\mathbb{R}^2 \times [0,1]$ with the boundary $(-\Gamma_0) \sqcup \Gamma_1$, where $\Gamma_i = F \cap (\mathbb{R}^2 \times \{i\})$, $i = 0, 1$, may be viewed as cobordisms between Γ_0 and Γ_1.

Define Foam_2 as the category where objects are $GL(2)$ webs Γ and morphisms from Γ_0 to Γ_1 are isotopy classes (rel boundary) of $GL(2)$ foams with the boundary $(-\Gamma_0) \sqcup \Gamma_1$. Composition is the concatenation of foams.

Define the degree of a foam F, not necessarily closed, as

$$\deg(F) = -\chi(F_{12}) + 2|d(F)|, \tag{4.33}$$

where $d(F)$ is the number of dots of F. Thin surface F_{12} of F is well-defined for foams with boundary. The boundary of F_{12} is the union of thin circles on the boundary of F.

For closed foams F, $\deg(F)$ equals the degree of $\langle F \rangle$, viewed as a homogeneous element of either $\widetilde{R}$ or its subring R. Degree of a foam is additive under composition of foams.

We define the state space $\langle \Gamma \rangle$ of a $GL(2)$ web Γ using the universal construction as in [**BHMV, Kh2**].

First, let $\mathrm{Fr}(\Gamma)$ be the free graded R-module with a basis $\{[F]\}_F$, over all foams F from the empty web to Γ. The degree of the generator $[F]$ is defined to be $\deg(F)$. Define a bilinear form on $\mathrm{Fr}(\Gamma)$ by

$$([F],[G]) = \langle w(G)F \rangle, \tag{4.34}$$

where $w(G)$ is the reflection of G in the horizontal plane together with the orientation reversal of all facets of G to make F and $w(G)$ composable along F. The foam $w(G)F$ is closed and can be evaluated to an element of R. Given a closed foam H, reflecting it about a plane into a foam H' may add sign to the evaluation, $\langle H' \rangle = (-1)^k \langle H \rangle$, where k is the number of singular circles of H. To get rid of the sign, reverse orientation of all facets of H' to get a foam $w(H)$ with $\langle w(H) \rangle = \langle H \rangle$. A similar argument works for non-closed foams. Consequently, the bilinear form (4.34) is symmetric.

Define the state space $\langle \Gamma \rangle$ as the quotient of $\mathrm{Fr}(\Gamma)$ by the kernel of the bilinear form $(,)$. The state space $\langle \Gamma \rangle$ is a graded R-module, via the degree formula (4.33). As usual in the universal construction, a foam F with boundary $(-\Gamma_0) \sqcup \Gamma_1$ induces a homogeneous R-module map

$$\langle F \rangle \ : \ \langle \Gamma_0 \rangle \longrightarrow \langle \Gamma_1 \rangle$$

of degree $\deg(F)$ taking an element $\langle G \rangle \in \langle \Gamma_0 \rangle$ associated to a foam G with boundary Γ_0 to the element $\langle FG \rangle$ associated to the foam FG with boundary Γ_1. These maps assemble into a functor from the category of $GL(2)$ foams to the category of graded R-modules and homogeneous R-module maps. The results below imply that the functor is monoidal.

The state space of the empty web is naturally isomorphic to the free rank one module over R with a generator in degree zero, $\langle \emptyset \rangle \cong R$.

Let Γ' denote the web Γ with an innermost thin circle (with one of the two orientations) added in a region of Γ. Thus, Γ' depends on the choice of a region of Γ and the orientation of the circle.

PROPOSITION 4.20. *There are natural isomorphisms of graded R-modules*

$$\langle\Gamma'\rangle \;\cong\; \langle\Gamma\rangle\{1\}\oplus\langle\Gamma\rangle\{-1\},$$

for Γ, Γ' as above and $\{m\}$ the grading shift up by m.

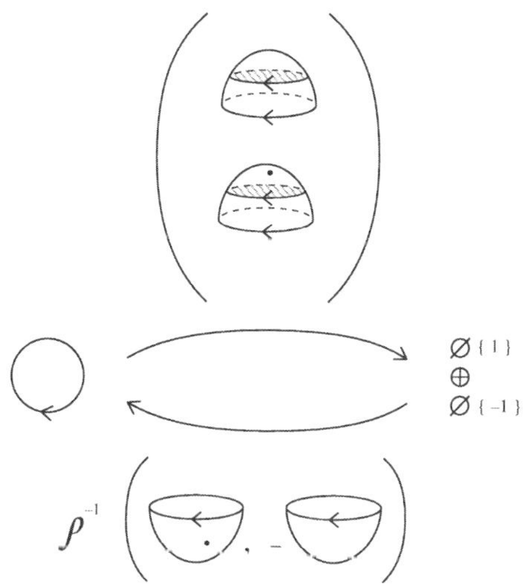

FIGURE 4.62. Direct sum decomposition for an innermost thin circle with the clockwise orientation. For the opposite orientation of the circle, reverse the orientation of the singular circle as well and add an overall minus sign to one of the two maps.

Proof: Foam cobordisms that deliver this direct sum decomposition are shown in Figure 4.62. The composition of the maps in either order is the identity, as follows from Θ-foam evaluations in section 4.3 and neck-cutting relation in Figure 4.47. □

PROPOSITION 4.21. *The saddle cobordism in a thick facet induces a grading-preserving isomorphism between the state spaces of its two boundary webs, see Figure 4.63. The inverse isomorphism is given by the adjoint saddle cobordism scaled by ρ^{-1}.*

Proof: This follows from the thick neck-cutting relation in Figure 4.411. □

PROPOSITION 4.22. *Let Γ be a web and Γ' be Γ with added innermost thick circle. There is a canonical degree zero isomorphism of state spaces*

$$\langle\Gamma'\rangle \;\cong\; \langle\Gamma\rangle$$

given by the cobordisms in Figure 4.64.

PROPOSITION 4.23. *Let web Γ have a thin edge and denote by Γ' the web Γ with an attached double edge along the thin edge. The state spaces of Γ and Γ' are naturally isomorphic as graded R-modules via the maps given in Figure 4.65.*

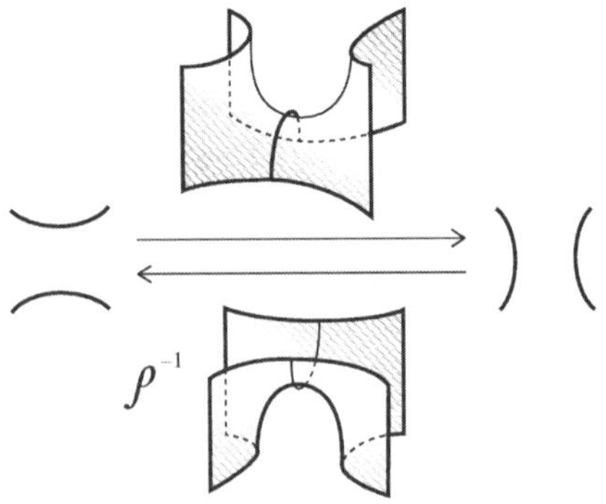

FIGURE 4.63. Saddle isomorphism on a double facet. Double edges on the left and right must carry compatible orientation (that is, extendable to the orientation of the surface).

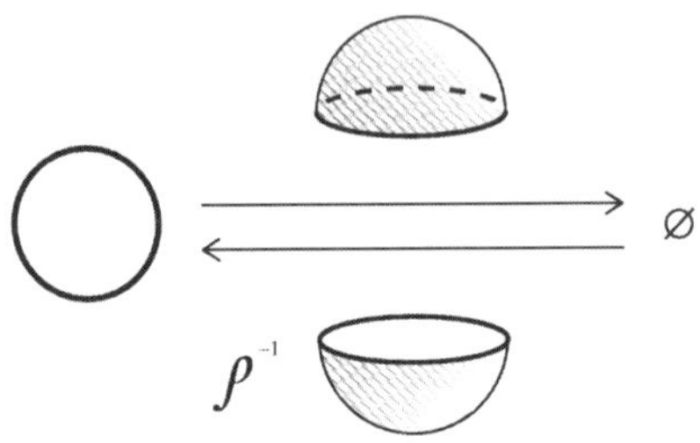

FIGURE 4.64. An isomorphism between a diagram with an innermost double circle and the diagram without it, via double cup and cap cobordisms.

Proof: This follows from relations in Figures 4.424 and 4.45. □

THEOREM 4.24. *$\langle\Gamma\rangle$ is a free graded R-module of graded rank $[2]^m$, where m is the number of components (circles) of the thin one-manifold $\Gamma_{(1)}$ and $[2]=q+q^{-1}$.*

Proof: This can be proved by induction on m. An innermost thin or double circle of Γ, see Figure 4.66, can be removed using isomorphisms in Figures 4.62 and 4.64, respectively.

Now look at $\Gamma_{(1)}$ and choose an innermost circle α in it. We distinguish between innermost circles of Γ and those of $\Gamma_{(1)}$. The latter correspond to thin circles in Γ which may contain vertices and thus have attached double edges. α bounds a disk D^2 in $\mathbb{R}^2$. Double edges emanating out of α split into those inside and outside of D^2. Repeatedly applying the double saddle isomorphism in Figure 4.63, we can reduce to the case when each of these double edges has both endpoints on α. Going along α one encounters $2n$ vertices (an even number due to orientation reversal along α at each vertex). If at two consecutive vertices double edges both point in or out of D^2, one can apply an isomorphism in Figure 4.63 followed by an isomorphism in Figure 4.65 to reduce from $2n$ to $2n-2$ vertices along α. A configuration where such a pair of vertices does not exist is impossible for $n>0$, for then the n ends

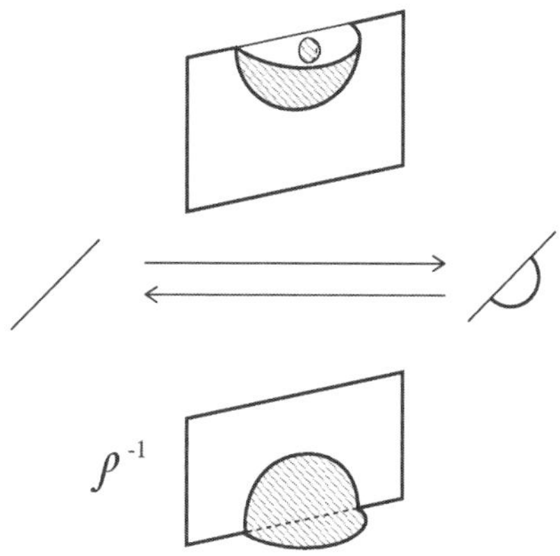

FIGURE 4.65. Mutually-inverse isomorphisms between a thin edge and a thin edge with an attached double edge.

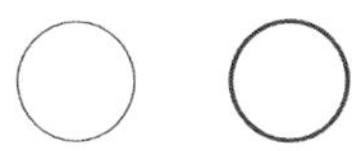

FIGURE 4.66. Thin and double innermost circles.

of double edges pointing into D^2 from α would all have the same orientations and there would be no room for the other n ends of these edges to land. This concludes the inductive argument. □

COROLLARY 4.25. *Associating the state space $\langle\Gamma\rangle$ to a $GL(2)$ web Γ and the map $\langle F\rangle$ of state spaces to a foam F with boundary is a monoidal functor from the category of $GL(2)$ foams to the category of free graded R-modules of finite rank.*

5. Reidemeister moves invariance and link homology

With the state spaces $\langle\Gamma\rangle$ of $GL(2)$ webs Γ defined, we can associate homology groups to a generic projection D of an oriented link $L \subset \mathbb{R}^3$, as follows. Let D has n crossings. We resolve each crossing into two resolutions, 0- and 1-resolutions, as in Figure 5.01.

One of the resolutions consists of two disjoint thin edges, the other contains a double edge and four adjoint thin edges. All the edges are oriented. Choose a total order on crossings of D. Doing this procedure over all crossings results in 2^n resolutions of D into $GL(2)$ webs $D(\overline{\mu})$, for $\overline{\mu} = (\mu_1, \ldots, \mu_n)$, with $\mu_i \in \{0, 1\}$. In a web $D(\overline{\mu})$ the i-th crossing is resolved according to μ_i.

To a crossing now associate a complex of two webs with boundaries and the differential induced by the "singular saddle" cobordism between them, see Figure 5.02 which sets us the terms in the complex, and Figure 5.03 which depicts "singular saddle" foams inducing the differential. These complexes make sense whenever the two webs are closed on the outside into two closed $GL(2)$ webs. Grading shifts

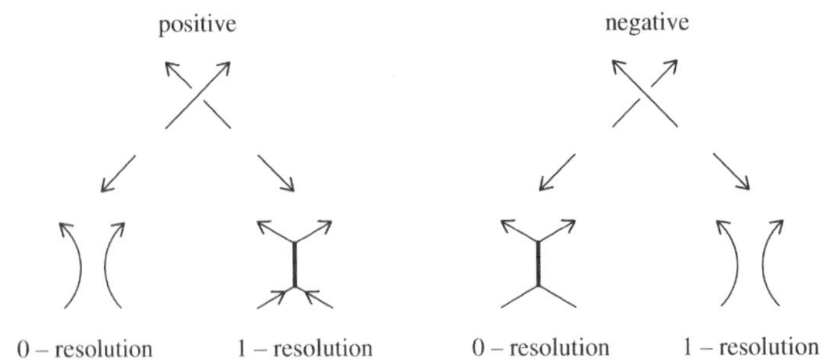

FIGURE 5.01. Resolutions of positive and negative crossings.

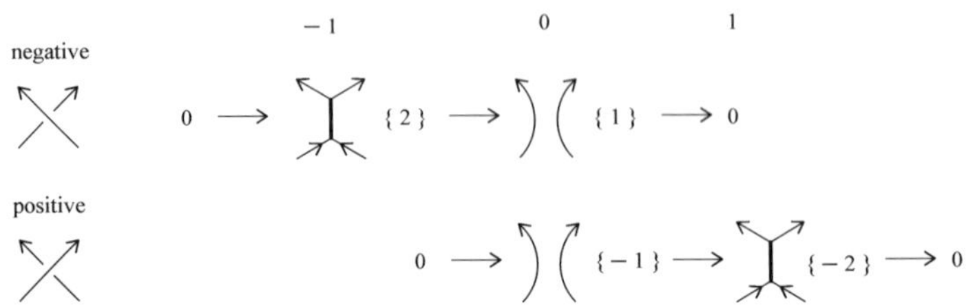

FIGURE 5.02. Complexes associated to positive and negative crossings. Numbers at the top show homological gradings of the terms. Resolution into two edges is always in homological degree 0.

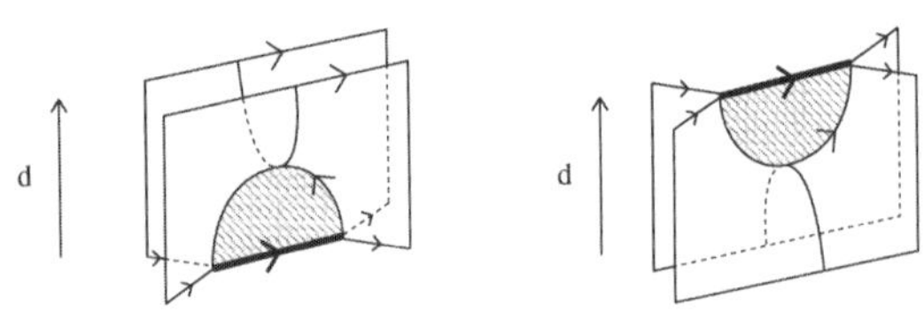

FIGURE 5.03. Foams that induce the differential in the complexes for positive and negative crossings. Upward-pointing arrows next to the foams indicate the 'direction' of the differential.

are inserted to make the map induced by the "singular saddle" cobordism grading-preserving (and, later, to have full invariance under the Reidemeister I move, rather than an invariance up to an overall grading shift).

In this way, one can form a commutative n-dimensional cube which has the graded R-module $\langle D(\overline{\mu})\rangle$ in its vertex labelled by the sequence $\overline{\mu}$ and maps induced by "singular saddle" foams associated to oriented edges of the cube. The maps commute for every square of the cube.

This setup with "singular saddle" cobordisms goes back to Blanchet [**B**], and is also visible in the earlier papers of Clark-Morrison-Walker [**CMW**] and Caprau [**Ca1**,

Ca2], where the double facet is not there, but its boundary, a singular edge along the foam, together with a choice of normal direction, is present.

The commuting cube of graded R-modules $\langle\Gamma(\overline{\mu})\rangle$ and grading-preserving homomorphisms between them collapses, in the standard way upon adding minus signs, to a complex of graded R-modules with a degree-preserving differential. This complex starts in the homological degree – minus the number of negative crossings of D and ends in the homological degree which is the number of positive crossings of D.

Denote this complex by $F(D)$.

THEOREM 5.1. *For two diagrams D_1 and D_2 of an oriented link L, complexes $F(D_1)$ and $F(D_2)$ are chain homotopy equivalent as complexes of graded R-modules.*

Proof: Consider the Reidemeister move R1, undoing a positive curl in Figure 5.04.

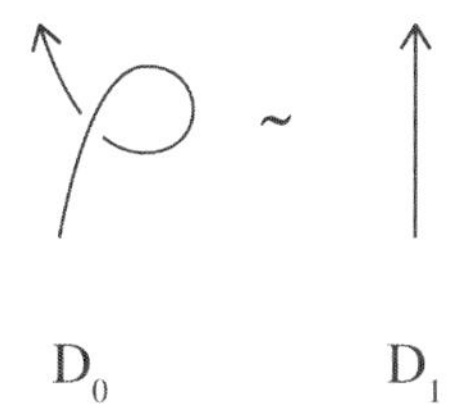

FIGURE 5.04. Reidemeister move R1, for a positive twist.

PROPOSITION 5.2. *The following relations hold on maps f_0, g_0, h and d in Figures 5.05, 5.06:*

$$dh = \mathrm{id}, \tag{5.1}$$

$$df_0 = 0, \tag{5.2}$$

$$g_0 f_0 = \mathrm{id}_{F(D_1)}, \tag{5.3}$$

$$\mathrm{id} = f_0 g_0 + hd. \tag{5.4}$$

The map id *in the first equation is the identity of the complex $F(D_0(1))$, associated to the diagram in the top right corner of Figure* 5.05, *while* id *in the last equation is the identity of the complex $F(D_0(0))$ associated to the diagram the top left corner of the figure.*

Proof: is a direct computation using skein relations derived in Section 4.4. This proof is very similar to the proof of the invariance under the Reidemeister move in [**MSV**], that does it in the non-equivariant $GL(N)$ case, in particular see Figure 8 there. □

COROLLARY 5.3. *Complexes $F(D_0)$ and $F(D_1)$, for diagrams in Figure* 5.04, *are chain homotopy equivalent as complexes of graded R-modules.*

PROPOSITION 5.4. *For each pair of the diagrams D_0, D_1 in Figure* 5.07, *which shows Reidemeister moves R2 and R3, complexes $F(D_0)$ and $F(D_1)$ are chain homotopy equivalent as complexes of graded R-modules.*

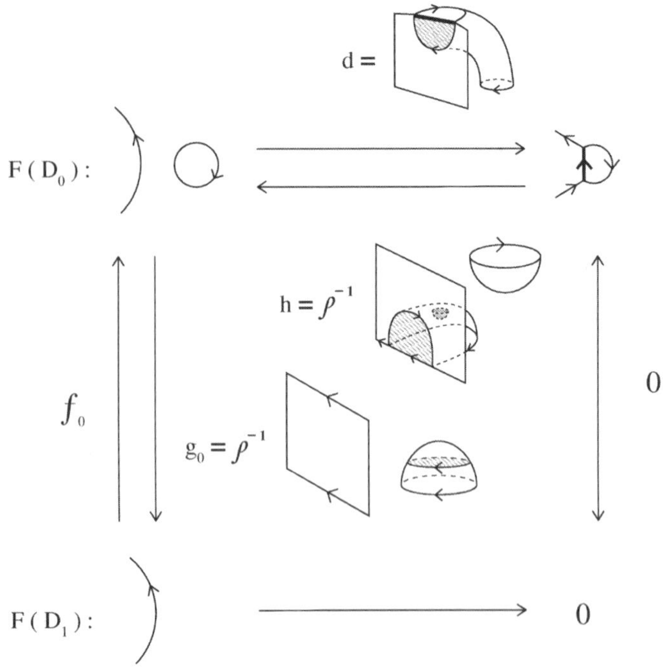

FIGURE 5.05. Top row, together with the right-pointing arrow d, encodes the complex $F(D_0)$. Top left-pointing arrow h is a self-homotopy of $F(D_0)$. Down and up arrows h_0 and g_0 are maps of complexes $F(D_0)$ and $F(D_1)$. Map f_0 is given in the next figure.

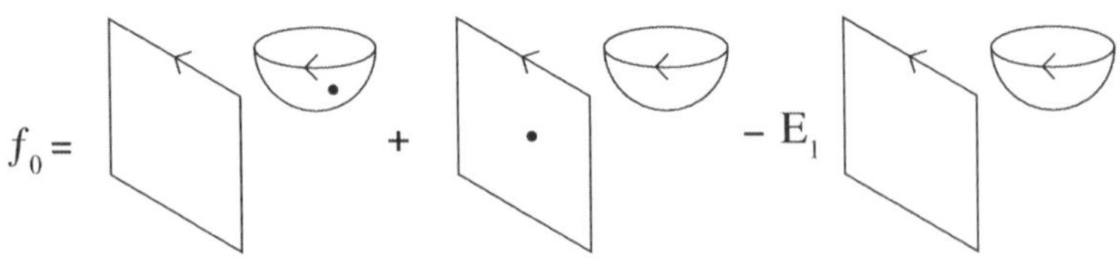

FIGURE 5.06. Map $f_0 : F(D_1) \longrightarrow F(D_0)$ of complexes.

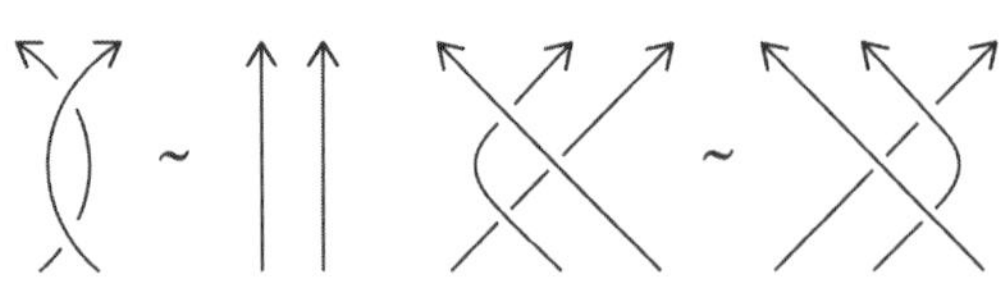

FIGURE 5.07. Reidemeister moves R2 and R3.

Proof: For the Reidemeister R2 move, relation (4.416) used in the direct sum of decomposition of a web Γ_1 with a digon facet into the sum of two copies of the simpler web Γ_0 is no different from the corresponding decomposition in the usual

$SL(N)$ graphical calculus, for an arbitrary N (see [**Kh2**, Proposition 8] for the analogous decomposition in the non-equivariant $SL(3)$ case). As one of the relations for this decomposition, the relation of removing a bubble on a double facet with at most one dot on one of the two thin facets is identical with the corresponding relation in the usual $SL(N)$ foam calculus, whether for the standard calculus or the equivariant one. Bubble removal relation follows from the combination of theta foam evaluation in Example 6 in Section 4.3 for $n_1, n_2 \leq 1$ and Proposition 4.11. For essentially the same relations in the $SL(3)$ case see, for instance, the top two relations in [**Kh2**, Figure 18].

For this reason, the usual proof of the Reidemester R2 relation, when both strands are oriented in the same direction, as in Figure 5.07 left, repeats without any changes in our case, see for instance [**Kh2**, Section 5.2], [**MSV**, Theorem 7.1], and many other sources.

Reideimeister move R2 for the oppositely oriented strands can be shown by simplifying the complex for the two crossings by splitting off contractible summands using the maps in Figure 5.05. The remaining complex has a single term that differs from two parallel lines by a pair of double lines connecting them. The standard cobordisms between these two diagrams need to be twisted by a pair of double caps on two thin facets to make them the inverses of each other (a double cap is shown, for instance, at the top of Figure 4.65 and in Figure 4.422). We leave the details to an interested reader.

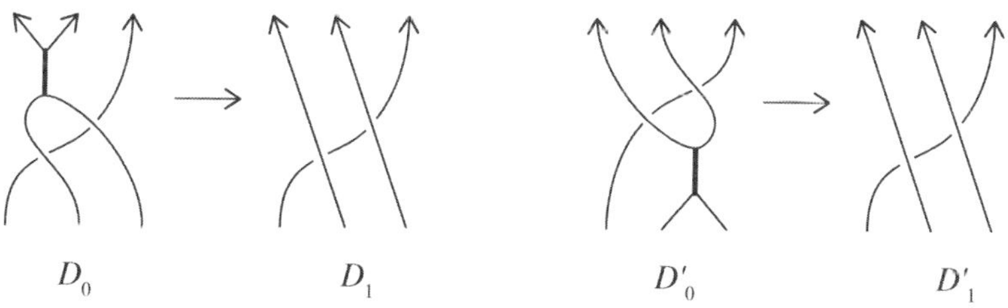

FIGURE 5.08. Two partial resolutions of each of D and D'. Note that D_1 and D'_1 are identical diagrams.

Consider the Reidemeister R3 move in Figure 5.07. Denote by D and D' the diagrams on the left and right of this move.

We start by resolving a single crossing in each of D and D', see Figure 5.08. Complexes $C(D)$ and $C(D')$ are isomorphic to cones of maps $C(D_0) \longrightarrow C(D_1)$ and $C(D'_0) \longrightarrow C(D'_1)$ built out of foams between complete resolutions of these diagrams.

Tangle diagrams D_1 and D'_1 are canonically isomorphic, and their resolutions result in the total complex of the square shown in Figure 5.09 with the differential coming from the four foams associated to the arrows of the diagram, with each foam a standard singular saddle in the appropriate position.

Consider now the diagram D_0 and its resolution in Figure 5.010. Maps ψ_k, $k = 1, \ldots, 4$ are homomorphisms between state spaces of web induced by appropriate foams (singular saddle foams). Summing over all possible resolutions

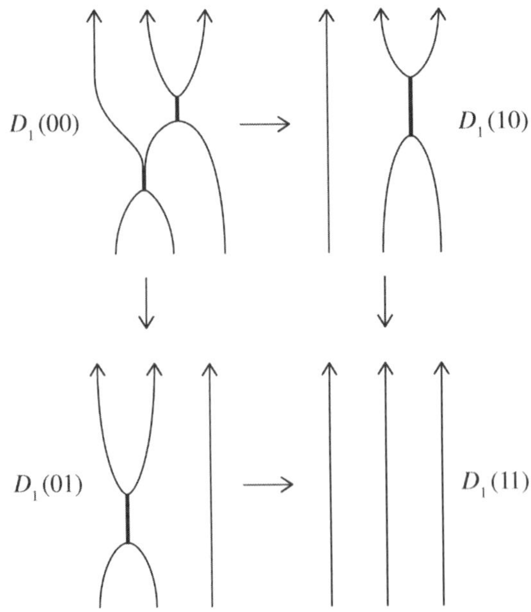

FIGURE 5.09. Resolution of the diagram $D_1 \cong D_1'$.

of crossings of D_0 not shown in the diagram gives homomorphisms, also denoted $\psi_1, \dots, \psi_4$, of corresponding complexes.

The four terms $C(D_0(k\ell))$, $k, \ell \in \{0,1\}$, will also map to the corresponding four terms $C(D_1(k\ell))$ in $C(D_1)$ in Figure 5.09 to constitute a 3-dimensional cube diagram (not shown).

The complex $C(D_0(00))$ of the diagram in the upper left of Figure 5.010 is isomorphic (and not just homotopy equivalent) to the complex $C(D_2)$ of the diagram D_2 shown in Figure 5.011 left.

Foam F_0 going from $D_0(00)$ to D_2 'straightens out' the long thin arc u_0 of $D_0(00)$ by canceling in pairs the four vertices on this arc where double edges meet u_0. Arc u_0 becomes the rightmost arc u_2 of D_2. Seam edges that cancel the four vertices in pairs are shown in Figure 5.012 as two arcs in the upper half of the diagram. The upper half shows the thin facet of F_0 where singular vertices along u_0 are cancelled in pairs. These cancellations are done via singular arcs, shown in Figure 5.012 top, along which double facets are attached to the thin facet.

Foam F_1 goes back from D_2 to $D_0(00)$ and is given by reflecting F_0 in the horizontal plane. The thin facet of F_1 is shown as the lower half of Figure 5.012. Semicircles depict singular edges along the thin facet.

Denote the maps F_0, F_1 induce on state spaces and on complexes built out of the state spaces of all resolutions of $D_0(00)$ and D_2 by

$$\tau_0 : C(D_0(00)) \longrightarrow C(D_2), \quad \tau_1 : C(D_2) \longrightarrow C(D_0(00)). \tag{5.5}$$

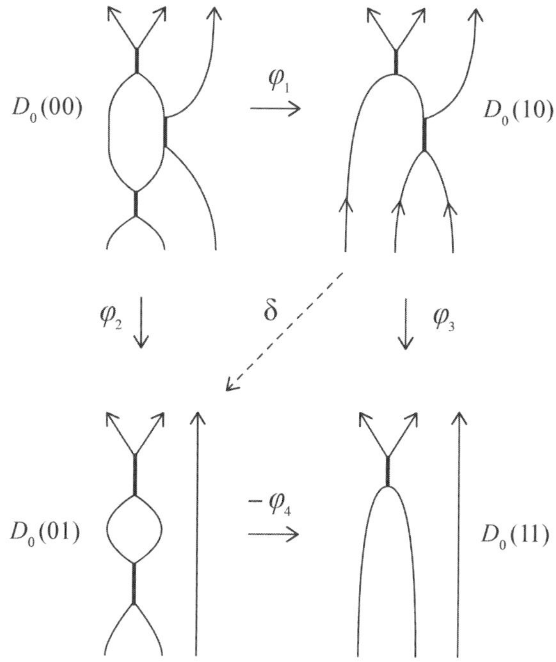

FIGURE 5.010. Resolution of the diagram D_0.

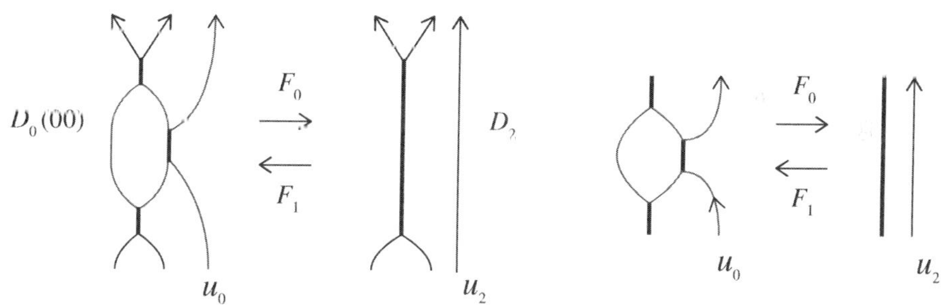

FIGURE 5.011. Diagrams $D_0(00)$ and D_2 have isomorphic state spaces for any resolution of these diagrams. Complexes $C(D_0(00))$ and $C(D_2)$, with the differentials induced by various foams between their resolutions, are canonically isomorphic, $C(D_0(00)) \cong C(D_2)$ in the abelian category of complexes (before factoring by homotopies).

We know that both τ_0 and τ_1 are isomorphisms of the state spaces and corresponding complexes, since annihilating a digon facet with a thick edge is an isomorphism, see Proposition 4.23.

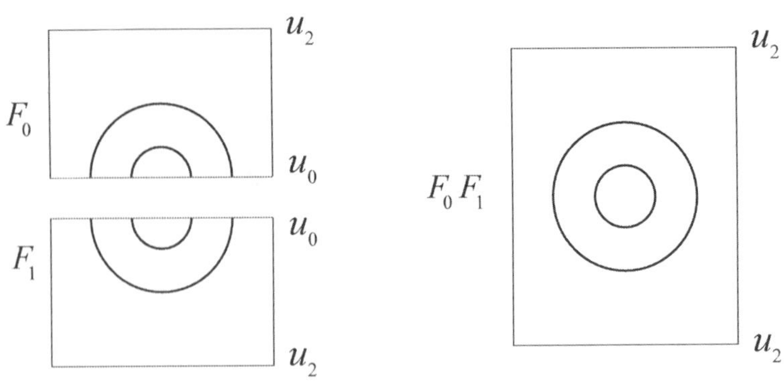

FIGURE 5.012. Flattened thin facets of F_0 and F_1 containing arcs u_0, u_2. Composition F_0F_1 contains thin surface S (shown on the right) given by gluing the two thin surfaces along the common arc u_0. This surface has two singular circles where double facets attach.

More precisely, $\tau_0\tau_1 = -\rho^{-1}\mathrm{Id}$. Indeed, the composition $\tau_0\tau_1$ is an endomorphism of the state space $\langle\Gamma\rangle$ for each web resolution Γ of D_2 and the induced endomorphism of the complex $C(D_2)$. The map $\tau_0\tau_1 : \langle\Gamma\rangle \longrightarrow \langle\Gamma\rangle$ transforms arc u_2 of the diagram D_2 to the arc u_0 of $D_0(00)$ and back, via the composition of foams F_0F_1.

Consider the thin surface S bounded by u_2 at the top and bottom of the cobordism F_0F_1. It can be visualized by gluing the two thin surfaces for F_0 and F_1 shown in Figure 5.012 along the common arc u_0, shown in red. Surface S contains two nested singular circles, where double facets of F_0F_1 meet S. Double facets at these two circles attach to S from opposite sides, as one can glean from Figure 5.011. This corresponds to having two double edges attached to arc u_0 on one side and the other double edge attached to u_0 on the other side of the plane, at both endpoints, see the leftmost diagram in Figure 5.011.

Apply Proposition 4.11 at each of these attached double facets to simplify the non-trivial part of the foam F_0F_1 to the surface S with two double disks attached to it from the opposite sides along the two singular circles, with an additional factor ρ^{-2}. We then apply Proposition 4.8 to flip one of the disks to the opposite side, gaining a minus sign, and then use Proposition 4.11 to reduce to the identity foam times $-\rho^{-1}$.

Consequently, maps τ_0 and $-\rho\tau_1$ are mutually-inverse isomorphisms.

Note that diagrams $D_0(11)$ and D_2 are isotopic and their complexes are canonically isomorphic. Complex $C(D_0(01))$ decomposes into direct sum of two copies of $D_0(11)$ in the usual way. The composition $\psi_2\tau_1 : C(D_2) \longrightarrow C(D_0(01))$ is a split inclusion into one of these copies. Since τ_1 is an isomorphism, this composition allows to split off contractible summand

$$0 \longrightarrow C(D_0(00)) \xrightarrow{\cong} \mathrm{im}(\psi_2) \longrightarrow 0$$

from the total complex of D_0, also see Figure 5.010. The map ψ_4 induces an isomorphism from the complementary direct summand of $C(D_0(01))$, also isomorphic to $C(D_2)$, to $C(D_0(11))$, allowing to split the second contractible summand from $C(D_0)$. After removing these contractible summands, the entire complex $C(D_0)$ in Figure 5.010 is downsized to $C(D_0(10))$. The inclusion $C(D_0(10)) \subset C(D_0)$ realizing this chain homotopy equivalence is given in coordinates by (id, δ), see Figure 5.010 with δ the diagonal map induced by the simplest cobordism from $D_0(10)$ to $D_0(01)$, with the property $\psi_3 = \psi_4\delta$.

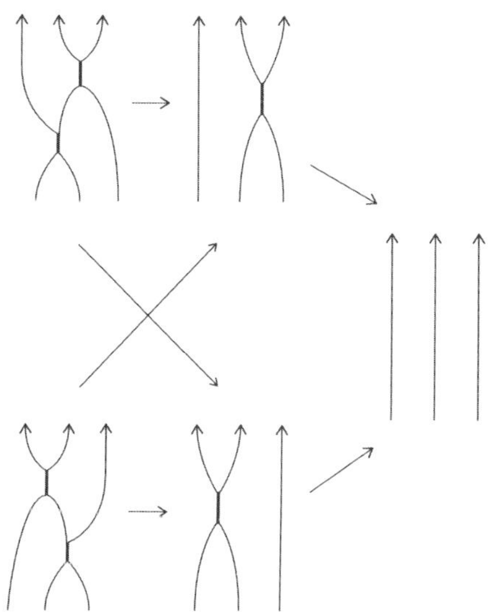

FIGURE 5.013. Common reduction of $C(D)$ and $C(D')$.

Reducing the map of complexes $C(D_0) \longrightarrow C(D_1)$ to the map $C(D_0(10)) \longrightarrow C(D_1)$ via the above inclusion of complexes results in the complex shown in Figure 5.013, with all arrows given by maps induced by the elementary foams between these webs. Signs need to be added to make each square anticommute, but the isomorphism class of the complex does not depend on the distribution of signs. This complex has an obvious symmetry given by reflecting all diagrams and foams about the vertical axis (or plane, in case of foams) and permuting top and bottom terms in the complex.

The cone of the map $C(D_0') \longrightarrow C(D_1')$ in Figure 5.08 right reduces to the isomorphic complex, by removing contractible summands of $C(D_0')$ in the same fashion as for $C(D_0)$. □

This completes the proof of Theorem 5.1. □

Our proofs of the Reidemeister R2 and R3 relations, for upwards orientations and in $N = 2$ case, are essentially identical to those in the usual equivariant case, when $p_{ij} = 1$. This observation mirrors our earlier Theorem 3.17 and Remark 3.18 that our deformation does not change the nilHecke algebra relation. This makes it likely that our $p(x, y)$ deformation does not modify the Soergel category and that the Soergel category will act in the deformed situation as well, with the proofs of Reidemeister R2 and R3 moves for upward orientations identical to that in the $p(x,y)=1$ case. Then $p(x,y)$ deformation would only modify the first Reidemeister move and Reidemeister moves R2 and R3 for non-braid orientations of strands. This expectation mirrors our observation that $p(x,y)$ may only contribute to the deformation of the Frobenius structure, not of multiplication. In the $N=2$ case, similar deformations can be hidden at the level of link homology, see [**V**, **Kh4**].

Acknowledgments. M.K. was partially supported by NSF grants DMS-1664240 and DMS-1807425 while working on this paper. The authors are grateful to Yakov Kononov, Louis-Hadrien Robert and Lev Rozansky for valuable discussions and would like to thank Elizaveta Babaeva[1] for help with producing figures for the paper. The authors would also like to thank Joshua Sussan and the anonymous referee for many valuable corrections to the earlier version of the paper.

References

[Ba] J. C. Baez, *An introduction to spin foam models of BF theory and quantum gravity*, Geometry and quantum physics (Schladming, 1999), Lecture Notes in Phys., vol. 543, Springer, Berlin, 2000, pp. 25–93, DOI 10.1007/3-540-46552-9_2. MR1770708

[BE] P. Bressler and S. Evens, *The Schubert calculus, braid relations, and generalized cohomology*, Trans. Amer. Math. Soc. **317** (1990), no. 2, 799–811, DOI 10.2307/2001488. MR968883

[BN] D. Bar-Natan, *On Khovanov's categorification of the Jones polynomial*, Algebr. Geom. Topol. **2** (2002), 337–370, DOI 10.2140/agt.2002.2.337. MR1917056

[BHPW] A. Beliakova, M. Hogancamp, K. Putyra, and S. Wehrli, *On the functoriality of sl(2) tangle homology*, Algebraic & Geometric Topology **23** (2023) 1303–1361, DOI: 10.2140/agt.2023.23.1303.

[B] C. Blanchet, *An oriented model for Khovanov homology*, J. Knot Theory Ramifications **19** (2010), no. 2, 291–312, DOI 10.1142/S0218216510007863. MR2647055

[BHMV] C. Blanchet, N. Habegger, G. Masbaum, and P. Vogel, *Topological quantum field theories derived from the Kauffman bracket*, Topology **34** (1995), no. 4, 883–927, DOI 10.1016/0040-9383(94)00051-4. MR1362791

[Bo] D. Boozer, (2021) *Computer Bounds for Kronheimer-Mrowka Foam Evaluation*, Experimental Mathematics, DOI: 10.1080/10586458.2021.1982078

[Ca1] C. L. Caprau, sl(2) *tangle homology with a parameter and singular cobordisms*, Algebr. Geom. Topol. **8** (2008), no. 2, 729–756, DOI 10.2140/agt.2008.8.729. MR2443094

[Ca2] C. L. Caprau, *The universal* $\mathfrak{sl}(2)$ *cohomology via webs and foams*, Topology Appl. **156** (2009), no. 9, 1684–1702, DOI 10.1016/j.topol.2009.02.001. MR2521705

[CMW] D. Clark, S. Morrison, and K. Walker, *Fixing the functoriality of Khovanov homology*, Geom. Topol. **13** (2009), no. 3, 1499–1582, DOI 10.2140/gt.2009.13.1499. MR2496052

[C] B. Cooper, *Deformations of nilHecke algebras*, Acta Math. Vietnam. **39** (2014), no. 4, 515–527, DOI 10.1007/s40306-014-0085-9. MR3292580

[ETW] M. Ehrig, D. Tubbenhauer, and P. Wedrich, *Functoriality of colored link homologies*, Proc. Lond. Math. Soc. (3) **117** (2018), no. 5, 996–1040, DOI 10.1112/plms.12154. MR3877770

[EST1] M. Ehrig, C. Stroppel and D. Tubbenhauer, *Generic* $\mathfrak{gl}_2$*-foams, webs, and arc algebras*, arXiv:1601.08010.

[EST2] M. Ehrig, C. Stroppel, and D. Tubbenhauer, *The Blanchet-Khovanov algebras*, Categorification and higher representation theory, Contemp. Math., vol. 683, Amer. Math. Soc., Providence, RI, 2017, pp. 183–226, DOI 10.1090/conm/683. MR3611714

[Ha] M. Hazewinkel, *Formal groups and applications*, Pure and Applied Mathematics, vol. 78, Academic Press, Inc. [Harcourt Brace Jovanovich, Publishers], New York-London, 1978. MR506881

[1] Elizaveta Babaeva, `https://www.behance.net/lizababaiva`

[HBJ] F. Hirzebruch, T. Berger, and R. Jung, *Manifolds and modular forms*, Aspects of Mathematics, E20, Friedr. Vieweg & Sohn, Braunschweig, 1992, DOI 10.1007/978-3-663-14045-0. MR1189136

[HMSZ] A. Hoffnung, J. Malagón-López, A. Savage, and K. Zainoulline, *Formal Hecke algebras and algebraic oriented cohomology theories*, Selecta Math. (N.S.) **20** (2014), no. 4, 1213–1245, DOI 10.1007/s00029-013-0132-8. MR3273635

[HKK] P. Hu, D. Kriz, and I. Kriz, *Field theories, stable homotopy theory, and Khovanov homology*, Topology Proc. **48** (2016), 327–360. MR3465966

[J] M. Jacobsson, *An invariant of link cobordisms from Khovanov homology*, Algebr. Geom. Topol. **4** (2004), 1211–1251, DOI 10.2140/agt.2004.4.1211. MR2113903

[Kh1] M. Khovanov, *A categorification of the Jones polynomial*, Duke Math. J. **101** (2000), no. 3, 359–426, DOI 10.1215/S0012-7094-00-10131-7. MR1740682

[Kh2] M. Khovanov, *sl(3) link homology*, Algebr. Geom. Topol. **4** (2004), 1045–1081, DOI 10.2140/agt.2004.4.1045. MR2100691

[Kh3] M. Khovanov, *An invariant of tangle cobordisms*, Trans. Amer. Math. Soc. **358** (2006), no. 1, 315–327, DOI 10.1090/S0002-9947-05-03665-2. MR2171235

[Kh4] M. Khovanov, *Link homology and Frobenius extensions*, Fund. Math. **190** (2006), 179–190, DOI 10.4064/fm190-0-6. MR2232858

[KR] M. Khovanov and L.-H. Robert, *Foam evaluation and Kronheimer-Mrowka theories*, Adv. Math. **376** (2021), Paper No. 107433, 59 pp., DOI 10.1016/j.aim.2020.107433. MR4178907

[KRW] M. Khovanov, L.-H. Robert, and E. Wagner, *In preparation.*

[KRo1] M. Khovanov and L. Rozansky, *Matrix factorizations and link homology*, Fund. Math. **199** (2008), no. 1, 1–91, DOI 10.4064/fm199-1-1. MR2391017

[KRo2] M. Khovanov and L. Rozansky, *Topological Landau-Ginzburg models on the world-sheet foam*, Adv. Theor. Math. Phys. **11** (2007), no. 2, 233–259. MR2322554

[K1] N. Kitchloo, *Symmetry Breaking and Link Homologies I*, preprint, `arXiv:1910.07443`.

[K2] N. Kitchloo, *Symmetry Breaking and Link Homologies II*, preprint, `arXiv:1910.07444`.

[MSV] M. Mackaay, M. Stošić, and P. Vaz, $\mathfrak{sl}(N)$*-link homology* ($N \geq 4$) *using foams and the Kapustin-Li formula*, Geom. Topol. **13** (2009), no. 2, 1075–1128, DOI 10.2140/gt.2009.13.1075. MR2491657

[MV1] M. Mackaay and P. Vaz, *The universal* sl_3*-link homology*, Algebr. Geom. Topol. **7** (2007), 1135–1169, DOI 10.2140/agt.2007.7.1135. MR2336253

[MV2] M. Mackaay and P. Vaz, *The diagrammatic Soergel category and* $\mathrm{sl}(N)$*-foams, for* $N \geq 4$, Int. J. Math. Math. Sci., posted on 2010, Art. ID 468968, 20, DOI 10.1155/2010/468968. MR2671770

[MOY] H. Murakami, T. Ohtsuki, and S. Yamada, *Homfly polynomial via an invariant of colored plane graphs*, Enseign. Math. (2) **44** (1998), no. 3-4, 325–360. MR1659228

[LS1] R. Lipshitz and S. Sarkar, *A Khovanov stable homotopy type*, J. Amer. Math. Soc. **27** (2014), no. 4, 983–1042, DOI 10.1090/S0894-0347-2014-00785-2. MR3230817

[LS2] R. Lipshitz and S. Sarkar, *A Steenrod square on Khovanov homology*, J. Topol. **7** (2014), no. 3, 817–848, DOI 10.1112/jtopol/jtu005. MR3252965

[M] J. Morava, *On the complex cobordism ring as a Fock representation*, Homotopy theory and related topics (Kinosaki, 1988), Lecture Notes in Math., vol. 1418, Springer, Berlin, 1990, pp. 184–204, DOI 10.1007/BFb0083703. MR1048186

[NN1] M. Nakagawa and H. Naruse, *Universal Gysin formulas for the universal Hall-Littlewood functions*, An alpine bouquet of algebraic topology, Contemp. Math., vol. 708, Amer. Math. Soc., [Providence], RI, 2018, pp. 201–244, DOI 10.1090/conm/708/14267. MR3807757

[NN2] M. Nakagawa and H. Naruse, *Darondeau-Pragacz formulas in complex cobordism*, Math. Ann. **381** (2021), no. 1-2, 335–361, DOI 10.1007/s00208-021-02196-5. MR4322614

[Na] H. Naruse, *Elementary proof and application of the generating functions for generalized Hall-Littlewood functions*, J. Algebra **516** (2018), 197–209, DOI 10.1016/j.jalgebra.2018.09.010. MR3863476

[Nt] S. M. Natanzon, *Cyclic foam topological field theories*, J. Geom. Phys. **60** (2010), no. 6-8, 874–883, DOI 10.1016/j.geomphys.2010.02.004. MR2647289

[QR] H. Queffelec and D. E. V. Rose, *The $\mathfrak{sl}_n$ foam 2-category: a combinatorial formulation of Khovanov-Rozansky homology via categorical skew Howe duality*, Adv. Math. **302** (2016), 1251–1339, DOI 10.1016/j.aim.2016.07.027. MR3545951

[RT] N. Yu. Reshetikhin and V. G. Turaev, *Ribbon graphs and their invariants derived from quantum groups*, Comm. Math. Phys. **127** (1990), no. 1, 1–26. MR1036112

[RW1] L.-H. Robert and E. Wagner, *A closed formula for the evaluation of foams*, Quantum Topol. **11** (2020), no. 3, 411–487, DOI 10.4171/qt/139. MR4164001

[RW2] L.-H. Robert and E. Wagner, *Symmetric Khovanov-Rozansky link homologies* (English, with English and French summaries), J. Éc. polytech. Math. **7** (2020), 573–651, DOI 10.5802/jep.124. MR4086581

[RW3] L.-H. Robert and E. Wagner, *A quantum categorification of the Alexander polynomial*, Geom. Topol. **26** (2022), no. 5, 1985–2064, DOI 10.2140/gt.2022.26.1985. MR4520301

[RWd] D. E. V. Rose and P. Wedrich, *Deformations of colored $\mathfrak{sl}_N$ link homologies via foams*, Geom. Topol. **20** (2016), no. 6, 3431–3517, DOI 10.2140/gt.2016.20.3431. MR3590355

[St] N. P. Strickland, *Formal schemes and formal groups*, Homotopy invariant algebraic structures (Baltimore, MD, 1998), Contemp. Math., vol. 239, Amer. Math. Soc., Providence, RI, 1999, pp. 263–352, DOI 10.1090/conm/239/03608. MR1718087

[TT] V. Turaev and P. Turner, *Unoriented topological quantum field theory and link homology*, Algebr. Geom. Topol. **6** (2006), 1069–1093, DOI 10.2140/agt.2006.6.1069. MR2253441

[Vz] P. Vaz, *The diagrammatic Soergel category and* sl(2) *and* sl(3) *foams*, Int. J. Math. Math. Sci., posted on 2010, Art. ID 612360, 23 pp., DOI 10.1155/2010/612360. MR2652384

[V] P. Vogel, *Functoriality of Khovanov homology*, J. Knot Theory Ramifications **29** (2020), no. 4, 2050020, 66 pp., DOI 10.1142/S0218216520500200. MR4096813

[We] B. Webster, *Knot invariants and higher representation theory*, Mem. Amer. Math. Soc. **250** (2017), no. 1191, v+141, DOI 10.1090/memo/1191. MR3709726

[Wd] P. Wedrich, *Exponential growth of colored HOMFLY-PT homology*, Adv. Math. **353** (2019), 471–525, DOI 10.1016/j.aim.2019.06.023. MR3982970

[W] E. Witten, *Quantum field theory and the Jones polynomial*, Comm. Math. Phys. **121** (1989), no. 3, 351–399. MR990772

[Wu1] H. Wu, *Colored $\mathfrak{sl}_N$ link homology via matrix factorizations*, `arXiv:1110.2076`.

[Wu2] H. Wu, *Equivariant colored $\mathfrak{sl}(N)$-homology for links*, J. Knot Theory Ramifications **21** (2012), no. 2, 1250012, 104 pp., DOI 10.1142/S0218216511009558. MR2885476

[Y] Y. Yonezawa, *Quantum $(\mathfrak{sl}_n, \wedge V_n)$ link invariant and matrix factorizations*, Nagoya Math. J. **204** (2011), 69–123, DOI 10.1215/00277630-1431840. MR2863366

Department of Mathematics, Columbia University, New York, New York 10027
Email address: `khovanov@math.columbia.edu`

Department of Mathematics, Johns Hopkins University, Baltimore, Maryland 21218
Email address: `nitu@math.jhu.edu`

Contemporary Mathematics
Volume **791**, 2024
https://doi.org/10.1090/conm/791/15876

Odd two-variable Soergel bimodules and Rouquier complexes

Mikhail Khovanov, Krzysztof Putyra, and Pedro Vaz

Abstract. We consider the odd analogue of the category of Soergel bimodules. In the odd case and already for two variables, the transposition bimodule cannot be merged into the generating Soergel bimodule, forcing one into a monoidal category with a larger Grothendieck ring compared to the even case. We establish biadjointness of suitable functors and develop graphical calculi in the 2-variable case for the odd Soergel category and the related singular Soergel 2-category. We describe the odd analogue of the Rouquier complexes and establish their invertibility in the homotopy category. For three variables, the absence of a direct sum decomposition of the tensor product of generating Soergel bimodules presents an obstacle for the Reidemeister III relation to hold in the homotopy category.

1. Introduction

In this note we propose an odd analogue of Soergel bimodules for Coxeter type A_1. Soergel bimodules for A_1 are certain bimodules for the algebra of polynomials in two variables. In the odd case it's role is played by the algebra of skew-symmetric polynomials $R = \mathbf{k}\langle x_1, x_2\rangle/(x_1x_2+x_2x_1)$. The substitute for the generating Soergel bimodule B over the polynomial algebra consists of two R-bimodules B and $\underline{B}$ that constitute a biadjoint pair (that is, the functor of tensoring with $\underline{B}$ is both left and right adjoint to tensoring with B).

Starting in Section 2.2 we develop a graphical calculus for the category of odd Soergel bimodules in two variables and define a pair of mutually-inverse functors $\mathcal{R}, \mathcal{R}'$ on the homotopy category of graded R-modules given by the tensor product with complexes of bimodules formed from suitable bimodule maps $B \xrightarrow{m} R\{-1\}$ and $R\{1\} \longrightarrow \underline{B}$, where $\{\pm 1\}$ is a grading shift.

These complexes of bimodules and corresponding functors $\mathcal{R}, \mathcal{R}'$ are odd analogues of the Rouquier complexes that in the even case give rise to a braid group action on the homotopy category of modules over the n-variable polynomial algebra. In Section 6 we explain an obstacle that exists in the odd case to having the braid relation $\mathcal{R}_i\mathcal{R}_{i+1}\mathcal{R}_i \cong \mathcal{R}_{i+1}\mathcal{R}_i\mathcal{R}_{i+1}$. The lack of this braid relation blocks

---

2020 *Mathematics Subject Classification.* Primary 18M30, 18N25, 16D20.

The authors are grateful to the anonymous referee for a thorough and valuable work on the earlier version of the paper and to Cailan Li for additional corrections. M.K. was partially supported by NSF grant DMS-1807425 and Simons Foundation sabbatical award #817792 (Simons Fellows program). P.V. was supported by the Fonds de la Recherche Scientifique - FNRS under Grant no. MIS-F.4536.19.

an attempt, from which this note originated, to define odd HOMFLYPT link homology via braid closures and odd Soergel bimodules, analogous to the original construction of HOMFLYPT link homology via Hochschild homology of Soergel bimodules [**8**]. It's not known either whether the odd counterpart of bigraded $SL(N)$ link homology [**9**] exists for $N > 2$. For the definition and structure of odd $SL(2)$ link homology see [**12–16**].

In Section 5 we identify the Grothendieck ring of the category of odd Soergel bimodules for A_1 and compute a natural semilinear form and trace on that ring.

Rouquier functors on the even Soergel category are closely related to the invertible functors of twisting by a relative spherical object in the Fukaya–Floer categories and in the derived categories of coherent sheaves [**11**]. It should be interesting to explore odd counterparts of such functors; one can, for instance, ask whether there exists an odd counterpart of quiver varieties and associated derived categories of coherent sheaves on them and, more generally, an odd counterpart of algebraic geometry. A simpler problem is to understand the relation between functors $\mathcal{R}, \mathcal{R}'$ and recently constructed odd counterpart of Chuang–Rouquier symmetries and Rickard complexes [**1, 2**].

2. Bimodules for two strands

2.1. Anticommuting polynomials and odd Demazure operators. Let $\mathbf{k}$ be a commutative ring and denote by $R = \mathbf{k}\langle x_1, x_2\rangle/(x_1x_2 + x_2x_1)$ the algebra of anticommuting polynomials in two variables. Let S_2 be the symmetric group on two letters with generator s, acting on R by $s(x_i) = -x_{s(i)}$ and $s(fg) = s(f)s(g)$ for $f, g \in R$.

Define the *odd Demazure operator* $\partial : R \longrightarrow R$ (see [**5–7**]) as follows:

- $\partial(1) = 0$, $\partial(x_1) = \partial(x_2) = 1$.
- The twisted Leibniz rule holds

$$\partial(fg) = (\partial f)g + s(f)\partial g.$$

Note that ring R does not have unique factorizations, for instance $(x_1+x_2)^2 = (x_1 - x_2)^2$ in R.

The equation

$$\partial \circ s = -s \circ \partial \tag{2.1}$$

follows via the Leibniz rule above and checking it on generators of R.

Let $R^s = \ker(\partial) = \mathrm{im}(\partial) \subset R$. Equality $\ker(\partial) = \mathrm{im}(\partial)$ is straighforward to check. The twisted Leibniz rule then implies that R^s is a subring of R. The ring R^s has generators $E_1 = x_1 - x_2, E_2 = x_1x_2$ and the defining relation is that these generators anticommute,

$$R^s = \mathbf{k}\langle E_1, E_2\rangle/(E_1E_2 + E_2E_1), \quad E_1 = x_1 - x_2, \ E_2 = x_1x_2.$$

The action of S_2 on R restricts to an action on R^s, with

$$s(E_1) = E_1, \quad s(E_2) = -E_2.$$

Define the *transposition bimodule* $\underline{R}$ to be free rank one as a left and as a right R-module, with the generator $\underline{1}$ and relations $f\underline{1} = \underline{1}s(f)$. It has a subbimodule

$\underline{R}^s := R^s\underline{1} = \underline{1}R^s \subset \underline{R}$. Denote $\underline{1} \in \underline{R}$ by $\underline{1}^s$ when viewed as an element of $\underline{R}^s \subset \underline{R}$. We fix bimodule isomorphisms

$$\underline{R} \otimes_R \underline{R} \cong R, \quad \underline{1} \otimes \underline{1} \longmapsto 1, \tag{2.2}$$

$$\underline{R}^s \otimes_{R^s} \underline{R}^s \cong R^s, \quad \underline{1}^s \otimes \underline{1}^s \longmapsto 1^s, \tag{2.3}$$

$$\underline{R}^s \otimes_{R^s} \underline{R} \cong {}_{R^s}R_R, \quad \underline{1}^s \otimes \underline{1} \longmapsto 1, \tag{2.4}$$

$$\underline{R} \otimes_{R^s} \underline{R}^s \cong {}_R R_{R^s}, \quad \underline{1} \otimes \underline{1}^s \longmapsto 1. \tag{2.5}$$

The first map is an isomorphism of R-bimodules, the second – that of R^s-bimodules.

The odd Demazure operator can be written as a map

$$\partial' : R \longrightarrow \underline{R}^s, \quad \partial'(f) = \underline{1}\partial(f). \tag{2.6}$$

It is then naturally a map of R^s-bimodules. We can also write it as a bimodule map

$$\partial'' : \underline{R} \longrightarrow R^s, \quad \partial''(f\underline{1}g) = \partial(s(f)g). \tag{2.7}$$

The ring R is a free rank two left and right module over R^s,

$${}_{R^s}R \cong R^s \cdot 1 \oplus R^s \cdot x_i, \quad R_{R^s} \cong 1 \cdot R^s \oplus x_i \cdot R^s, \quad i \in \{1,2\}.$$

Note that in the bimodule $R \otimes_{R^s} R$ we have

$$x_1 \otimes 1 - 1 \otimes x_1 \;=\; x_2 \otimes 1 - 1 \otimes x_2, \tag{2.8}$$

since $x_1 - x_2 \in R^s$.

We make R into a graded ring, with $\deg(x_1) = \deg(x_2) = 2$. Then R^s has an induced grading, and ∂ is a degree -2 map. Bimodules $\underline{R}$ and $\underline{R}^s$ are naturally graded, with $\deg(\underline{1}) = \deg(\underline{1}^s) = 0$.

2.2. Biadjointness. We consider the graded bimodules ${}_R R_{R^s}$, ${}_{R^s}R_R$ and ${}_R\underline{R}_R$ and introduce the following four functors, where gmod stands for the category of graded modules and degree zero maps:

- $F_\uparrow : R^s\text{–gmod} \longrightarrow R\text{–gmod}$ is the induction functor of tensoring with the graded bimodule ${}_R R_{R^s}$.
- $F_\downarrow : R\text{–gmod} \longrightarrow R^s\text{–gmod}$ is the restriction functor; it is isomorphic to tensoring with the bimodule ${}_{R^s}R_R$.
- $F_- : R\text{–gmod} \longrightarrow R\text{–gmod}$ is the functor of tensoring with $\underline{R}$.
- $F_{\underline{s}} : R^s\text{–gmod} \longrightarrow R^s\text{–gmod}$ is the functor of tensoring with $\underline{R}^s$.

The endofunctors F_- and $F_{\underline{s}}$ are involutive. Fix a functor isomorphism

$$F_{\underline{s}} \circ F_\downarrow \cong F_\downarrow \circ F_- \tag{2.9}$$

given by the bimodule isomorphism

$$\underline{R}^s \otimes_{R^s} R \cong {}_{R_s}R \otimes_R \underline{R} \cong {}_{R^s}\underline{R}$$

which takes $\underline{1}^s \otimes f$ to $s(f) \otimes \underline{1}$ to $\underline{1}f$. The last term is $\underline{R}$ viewed as (R^s, R)-bimodule with the standard left action of R_s and right action of R. Likewise, there's an isomorphism

$$F_\uparrow \circ F_{\underline{s}} \cong F_- \circ F_\uparrow \tag{2.10}$$

via the corresponding bimodule isomorphisms

$$R \otimes_{R^s} \underline{R}^s \cong \underline{R} \otimes_R R_{R_s} \cong \underline{R}_{R^s}, \quad f \otimes \underline{1}^s \longmapsto \underline{1} \otimes s(f) \longmapsto \underline{1}s(f).$$

These isomorphisms can be thought of as "sliding" involutive functors F_- and $F_{\underline{s}}$ through the induction and restriction functors $F_\uparrow$ and $F_\downarrow$.

We depict natural transformations between compositions of these functors by drawing planar diagrams, with regions labelled by categories R−gmod (white regions) and R^s−gmod (shaded regions), following the usual string diagram notation. Identity natural transformation of $F_\uparrow$ (respectively $F_\downarrow$) is denoted by a vertical line, with the shaded region to the left (respectively, to the right), see equation (2.11) below. We denote the identity functor on a category $\mathcal{C}$ by $\mathbb{1}_{\mathcal{C}}$ or just by $\mathbb{1}$.

(2.11)

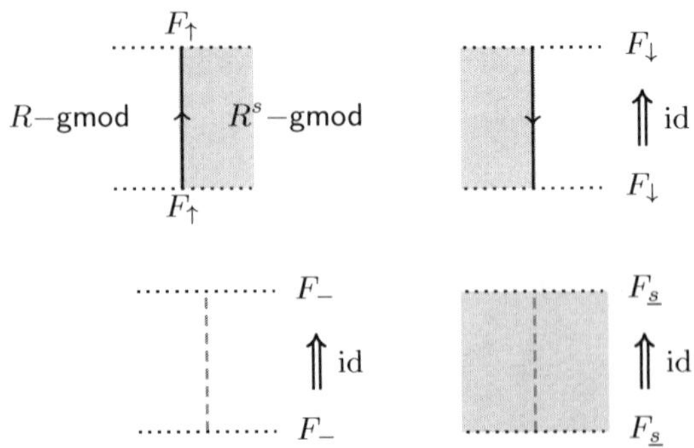

The identity natural transformation of F_- and $F_{\underline{s}}$ is denoted by a vertical dashed orange line, in a white or shaded region, respectively, see (2.11) above.

Sliding isomorphisms above are shown as crossings of strands, which are mutually-inverse isomorphisms, see equations (2.12)-2.12 below.

(2.12)

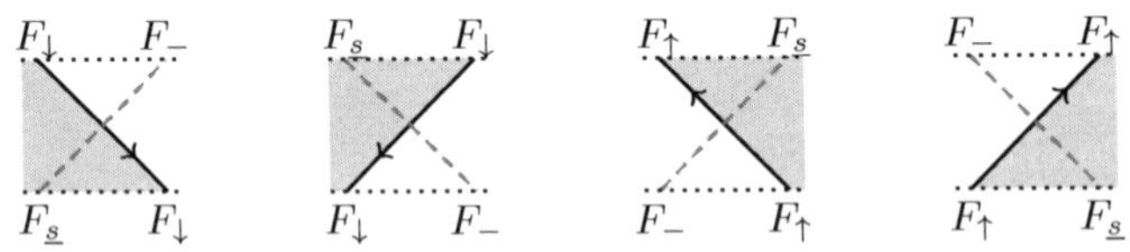

(2.13)

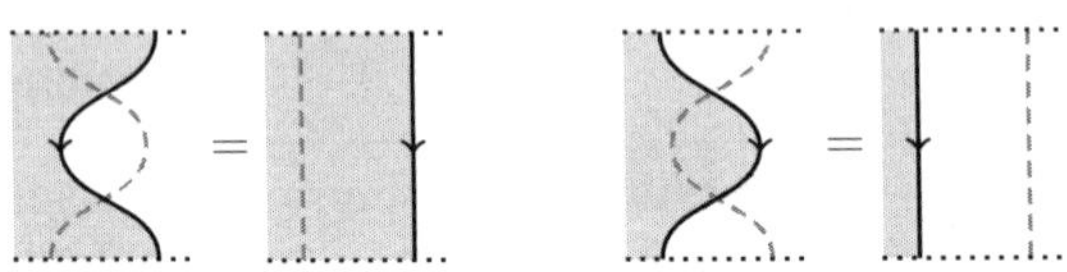

(2.14)

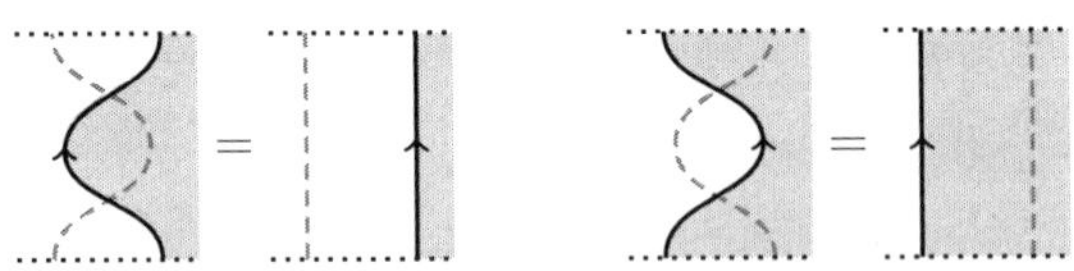

PROPOSITION 2.1. *The following pairs of functors are adjoint pairs:* $(F_\uparrow, F_\downarrow)$ *and* $(F_\downarrow, F_- \circ F_\uparrow)$.

PROOF. Induction functor $F_\uparrow$ is left adjoint to the restriction functor $F_\downarrow$. Adjointness natural transformations come from standard bimodule homomorphisms

(2.15) $$\alpha_0 \ : \ R \otimes_{R^s} R \longrightarrow R, \quad f \otimes g \longmapsto fg, \ f, g \in R,$$

and

$$\beta_0 \ : \ R^s \longrightarrow {}_{R^s}R_{R^s}, \quad f \longmapsto f, \ f \in R^s.$$

These adjointness maps and the corresponding adjointness isotopy relations are shown in equation (2.16). We use α_0, β_0, etc. to denote both bimodule maps and the corresponding natural transformations of functors (id_R stands for the identity functor in the category $R-\mathsf{gmod}$). The composition $F_\downarrow \circ F_\uparrow$ is written $F_{\downarrow\uparrow}$, for brevity.

(2.16)

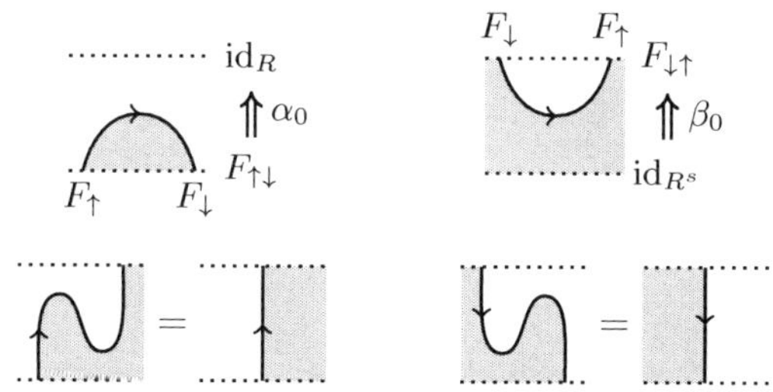

For the second adjoint pair of functors, we consider corresponding bimodules and define bimodule homomorphisms

$$\alpha_1 \ : \ {}_{R^s}R \otimes_R \underline{R} \otimes_R R_{R^s} \longrightarrow R^s, \quad \alpha_1(f \otimes \underline{1} \otimes g) = \partial(s(f)g).$$

This is just the map ∂'' in (2.7), under the bimodule isomorphism ${}_{R^s}R \otimes_R \underline{R} \otimes_R R_{R^s} \cong {}_{R^s}\underline{R}_{R^s}$.

$$\beta_1 \ : \ R \longrightarrow \underline{R} \otimes_R R \otimes_{R^s} R, \quad \beta_1(f) = \underline{1} \otimes x_1 \otimes f - \underline{1} \otimes 1 \otimes x_1 f, \ f \in R.$$

Due to (2.8),

$$\beta_1(f) = \underline{1} \otimes (x_1 \otimes 1 - 1 \otimes x_1)f = \underline{1} \otimes (x_2 \otimes 1 - 1 \otimes x_2)f.$$

To prove that β_1 is a well-defined bimodule map we check that $\ell_{x_i}(\beta_1(1)) = r_{x_i}(\beta_1(1))$, where ℓ_x, respectively r_x, denotes the left, respectively right, multiplication by x in an R bimodule:

$$\begin{aligned}
\ell_{x_1}(\beta(1)) &= x_1\underline{1} \otimes x_1 \otimes 1 - x_1\underline{1} \otimes 1 \otimes x_1 = \underline{1} \otimes x_2x_1 \otimes 1 - \underline{1} \otimes x_2 \otimes x_1 \\
&= \underline{1} \otimes (1 \otimes x_2x_1 - x_2 \otimes x_1) = \underline{1} \otimes (1 \otimes x_2 - x_2 \otimes 1)x_1 \\
&= \underline{1} \otimes (1 \otimes x_1 - x_1 \otimes 1)x_1 = r_{x_1}(\beta(1)), \\
\ell_{x_2}(\beta(1)) &= x_2\underline{1} \otimes (x_2 \otimes 1 - 1 \otimes x_2) = \underline{1} \otimes (x_1x_2 \otimes 1 - x_1 \otimes x_2) \\
&= \underline{1} \otimes (1 \otimes x_1x_2 - x_1 \otimes x_2) = \underline{1} \otimes (1 \otimes x_1 - x_1 \otimes 1)x_2 = r_{x_2}(\beta(1)).
\end{aligned}$$

We check the adjointness relation $(\alpha_1 \otimes \mathrm{id}) \circ (\mathrm{id} \otimes \beta_1) = \mathrm{id}$, where id stands for the identity homomorphism of suitable bimodules:

$$1 \overset{\mathrm{id} \otimes \beta_1}{\longmapsto} 1 \otimes \underline{1} \otimes (x_1 \otimes 1 - 1 \otimes x_1) \overset{\alpha_1 \otimes \mathrm{id}}{\longmapsto} \partial(x_1)1 - \partial(1)x_1 = 1.$$

To check the other adjointness relation $(\mathrm{id} \otimes \alpha_1) \circ (\beta_1 \otimes \mathrm{id}) = \mathrm{id}$ we compute the corresponding endomorphism of the (R, R^s)-bimodule $\underline{R} \otimes_R R_{R^s}$ (functor $F_- \circ F_\uparrow$ is given by tensoring with this bimodule):

$$\begin{aligned}
\underline{1} \otimes 1 &\overset{\beta_1 \otimes \mathrm{id}}{\longmapsto} \underline{1} \otimes (x_1 \otimes 1 - 1 \otimes x_1) \otimes \underline{1} \otimes 1 \\
&\overset{\mathrm{id} \otimes \alpha_1}{\longmapsto} \underline{1} \otimes x_1\alpha_1(1 \otimes \underline{1} \otimes 1) - \underline{1} \otimes 1\alpha_1(x_1 \otimes \underline{1} \otimes 1) \\
&= \ \underline{1} \otimes x_1\partial(1) - \underline{1} \otimes 1\partial(s(x_1)) = 0 - \underline{1} \otimes 1(-1) = \underline{1} \otimes 1.
\end{aligned}$$

□

Diagrams for maps α_1, β_1 and their adjointness relations are depicted below. To shorten notations, we write $F_{\downarrow-\uparrow}$ for $F_\downarrow \circ F_- \circ F_\uparrow$, etc.

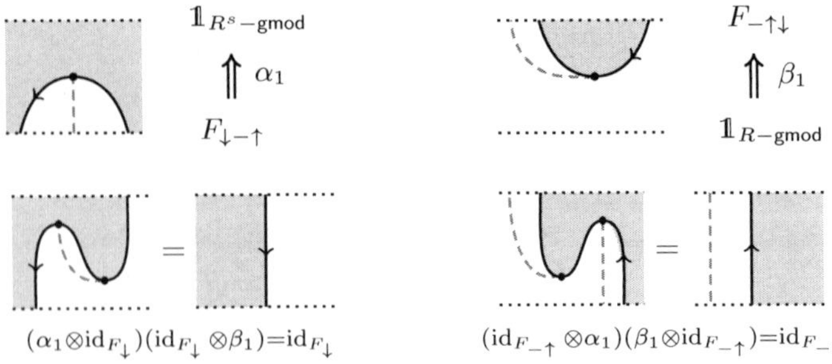

It is also natural to define another "cup" morphism, with the dotted line entering the local minimum in the middle, see below, as the composition of β_1 and the isomorphism (2.9). We call this cup *balanced* and denote the morphism by $\widetilde{\beta}_1$. Then the diagram representing β_1 can be rewritten as the composition of a balanced cup and a crossing, see below (the map $\psi\colon F_{-\uparrow} \to F_{\uparrow\underline{s}}$ is the crossing isomorphism in (2.12)).

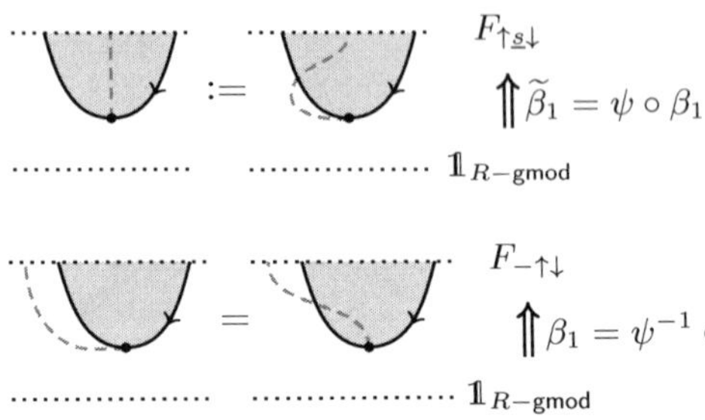

The balanced cup morphism is

$$\widetilde{\beta}_1\colon R \longrightarrow R\otimes\underline{R}^s\otimes R,\ \widetilde{\beta}_1(f) = -x_2\otimes\underline{1}^s\otimes f - 1\otimes\underline{1}^s\otimes x_1 f = -x_1\otimes\underline{1}^s\otimes f - 1\otimes\underline{1}^s\otimes x_2 f.$$

As $\widetilde{\beta}_1$ is a bimodule map, $f \in R$ can also be placed on the far left in the formula.

PROPOSITION 2.2. *Functors $F_{\uparrow\downarrow} = F_\uparrow \circ F_\downarrow$ and $F_{\uparrow\underline{s}\downarrow} = F_\uparrow \circ F_{\underline{s}} \circ F_\downarrow$ are biadjoint.*

PROOF. This follows from Proposition 2.1 and functor isomorphisms (2.9) and (2.10). □

The corresponding biadjointness maps are shown below. We fix these biadjointness maps $\alpha_2, \beta_2, \alpha_3, \beta_3$.

(2.17)

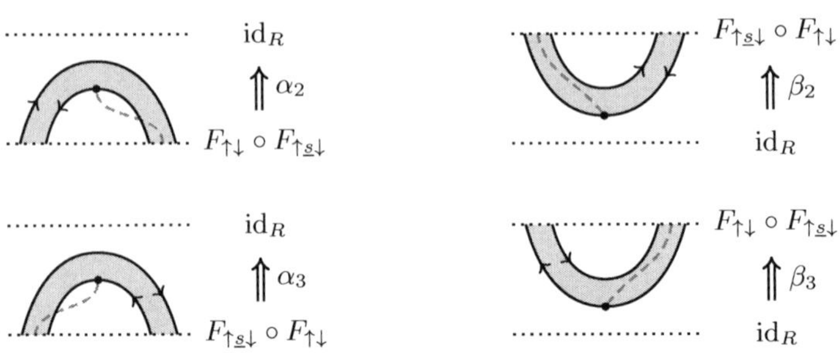

A quick computation shows that the two bimodule maps on the left of the diagram (2.19) below differ by a minus sign. We can then define a trivalent vertex, with a orange dashed line entering it from below, as in (2.19) on the right. This bimodule map is given by

$$\underline{R} \longrightarrow R \otimes_{R^s} R, \quad \underline{1} \longrightarrow 1 \otimes x_2 - x_2 \otimes 1 = x_1 \otimes 1 - 1 \otimes x_1. \tag{2.18}$$

It may be interesting to compare our dashed line and vertex with dashed lines in Ellis-Lauda [**6**].

(2.19)

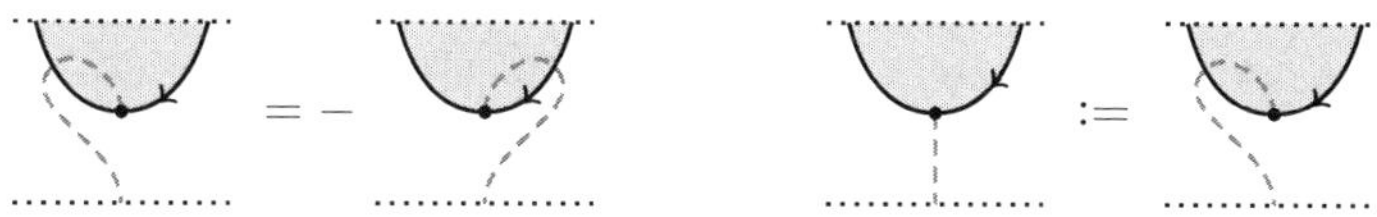

Likewise, there's a sign in a similar relation given by reflecting these diagrams about a horizontal axis and reversing the shading of regions, see (2.20) below left. One can then define the reflected trivalent vertex as in the figure in (2.20) below, on the right. This bimodule map is ∂', see formula (2.6),

$$R \longrightarrow \underline{R}^s, \quad f \longmapsto \underline{1}^s \partial(f), \quad f \in R.$$

(2.20)

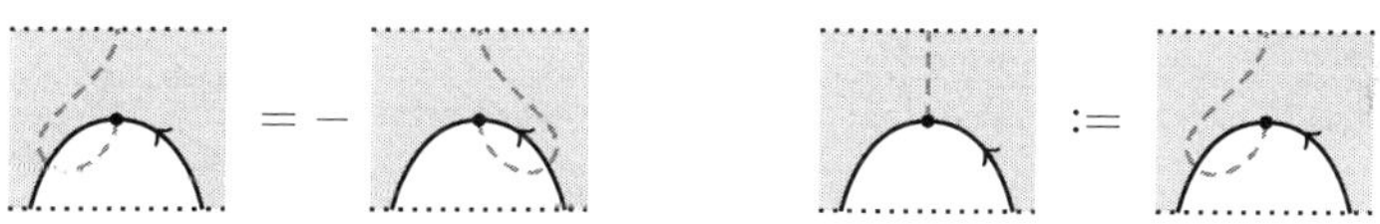

Some other relations in this graphical calculus are shown below.

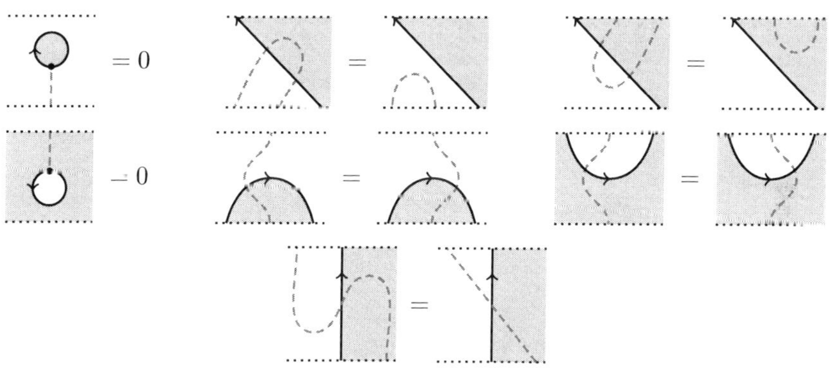

Recall that a vertical dotted orange line on a white (respectively, blue or shaded) background denotes the identity map of R-bimodule $\underline{R}$ (respectively, of $\underline{R}^s$-bimodule $\underline{R}^s$), see below.

(2.21)

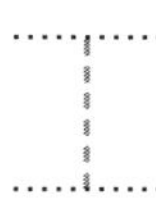 $= \mathrm{id}_{\underline{R}}$, 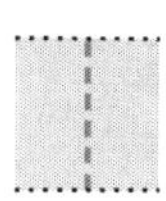 $= \mathrm{id}_{\underline{R}^s}$.

The isomorphism (2.2) between $\underline{R} \otimes_R \underline{R}$ and R can be represented by orange dashed "cup" and "cap" maps, see below. These maps satisfy the following relations

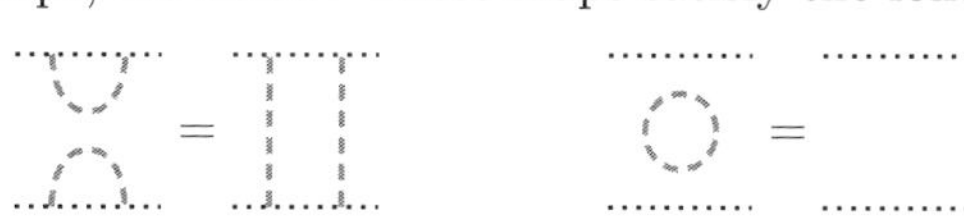

as well as the isotopy relations on the cup and the cap. Orange dashed cup and cap maps have degree 0. Isomorphism (2.3) is represented by oranged dashed cup and cap maps on a blue (shaded) background, with the following relations:

and the isotopy relations.

Likewise, isomorphisms (2.4) and (2.5) are represented by the orange dashed cup and cap diagrams in white-blue (or white-shaded) regions, as shown below for the isomorphisms (2.4), together with suitable relations on them, including isotopies.

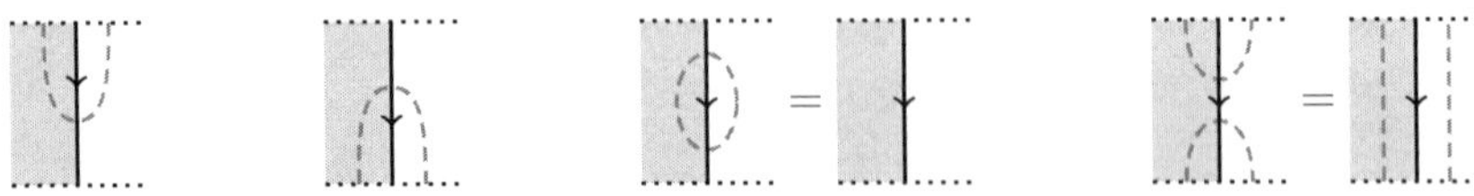

All eight possible orange dashed cup and cap maps have degree 0.

2.3. Bimodules B and $\underline{B}$ and their tensor products.

Define graded R-bimodules

$$B := R\otimes_{R^s} R\{-1\}, \quad \underline{B} := R\otimes_{R^s} \underline{R}^s \otimes_{R^s} R\{-1\} = R\underline{\otimes}_s R\{-1\}. \tag{2.22}$$

We use a shorthand and denote $M\otimes_{R^s}\underline{R}^s\otimes_{R^s}N$ (respectively, its element $m\otimes\underline{1}^s\otimes n$) by $M\underline{\otimes}_s N$ (respectively, by $m\underline{\otimes}_s n$). Likewise, $M\otimes_R \underline{R}\otimes_R N$ (and its element $m\otimes\underline{1}\otimes n$) can be denoted $M\underline{\otimes}N$ (and by $m\underline{\otimes}n$).

Endofunctors $F_{\uparrow\downarrow}$ and $F_{\uparrow-\downarrow}$ of the category $R-\mathsf{gmod}$ of graded R-modules are given by tensoring with bimodules $B\{1\}$ and $\underline{B}\{1\}$, respectivley. Natural transformations $\alpha_2,\beta_2,\alpha_3,\beta_3$ can then be rewritten as bimodule maps, denoted the same (the tensor products are over R and $f,g\in R$):

$$\begin{aligned}
&\alpha_2 : B\otimes\underline{B}\longrightarrow R,\ \alpha_2(1\otimes_s f\otimes g\underline{\otimes}_s 1)=\partial(s(fg)),\\
&\beta_2 : R\longrightarrow \underline{B}\otimes B,\ \beta_2(f)=-(x_1\underline{\otimes}_s 1)\otimes(1\otimes_s f)-(1\underline{\otimes}_s 1)\otimes(1\otimes_s x_2 f),\\
&\alpha_3 : \underline{B}\otimes B\longrightarrow R,\ \alpha_3(1\underline{\otimes}_s 1\otimes f\otimes_s g)=\partial(f)g,\\
&\beta_3 : R\longrightarrow B\otimes\underline{B},\ \beta_3(f)=-(x_1\otimes_s 1)\otimes(1\underline{\otimes}_s f)-(1\otimes_s 1)\otimes(1\underline{\otimes}_s x_2 f).
\end{aligned}$$

All four maps have zero degree: $\deg(\alpha_2)=\deg(\beta_2)=\deg(\alpha_3)=\deg(\beta_3)=0$.

We fix graded bimodule isomorphisms

$$\underline{R}\otimes_R B \;\cong\; \underline{B} \;\cong\; B\otimes_R\underline{R} \tag{2.23}$$

given by

$$\underline{1}\otimes f\otimes g \longmapsto s(f)\otimes\underline{1}^s\otimes g,\quad f\otimes\underline{1}^s\otimes g\longmapsto f\otimes s(g)\otimes\underline{1},\quad f,g\in R.$$

We depict the identity maps of B by a blue line

(2.24) 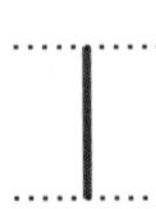 $= \mathrm{id}_B$

Bimodule maps (2.15) and (2.18) are given by the following diagrams (we write $f \otimes_s g$ for $f \otimes_{R^s} g$ and $f \otimes g = 1 \otimes fg = fg \otimes 1$ for $f \otimes_R g$):

(2.25)  $: B \xrightarrow{m} R, \quad f \otimes_s g \mapsto fg,$

(2.26)  $: \underline{R} \xrightarrow{\Delta} B, \quad \underline{1} \mapsto 1 \otimes_s x_2 - x_2 \otimes_s 1 = 1 \otimes_s x_1 - x_1 \otimes_s 1.$

Due to our definition (2.22) of the graded bimodule B, both of these maps have degree 1. The maps in equations (2.25) and (2.26) fit into a short exact sequence

(2.27) $0 \longrightarrow \underline{R}\{1\} \longrightarrow B \longrightarrow R\{-1\} \longrightarrow 0,$ 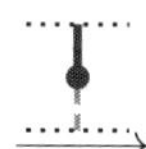 

where we shifted the gradings of the left and right terms to make the differential grading-preserving.

Tensoring this sequence with $\underline{R}$ gives another exact sequence, since $\underline{R}$ is a free left and right R-module:

(2.28) $0 \longrightarrow R\{1\} \longrightarrow B \otimes_R \underline{R} \longrightarrow \underline{R}\{-1\} \longrightarrow 0.$  

The middle term in the second sequence is isomorphic to $\underline{B}$. Here and later we fix the isomorphism $B \otimes_R \underline{R} \cong \underline{B}$ given by

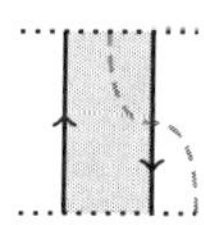

Exactness of sequences (2.27) and (2.28) implies relations

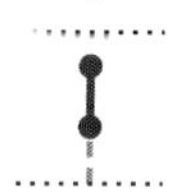 $= 0$  $= 0.$

Note that the two relations are equivalent, due to the isotopy relation on red cups and caps.

LEMMA 2.3. *The following are (R, R)-bimodule maps:*

(2.29)  $: 1 \otimes_s 1 \otimes \underline{1} \mapsto \underline{1} \otimes 1 \otimes_s 1,$

(2.30) 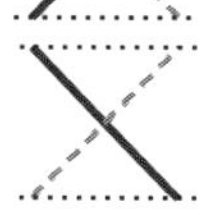 $: \underline{1} \otimes 1 \otimes_s 1 \mapsto 1 \otimes_s 1 \otimes \underline{1},$

(2.31) 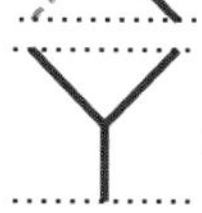 $: 1 \otimes_s 1 \mapsto 1 \otimes_s 1 \otimes_s 1,$

(2.32) 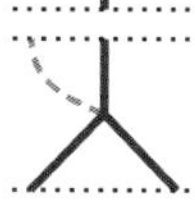 $: 1 \otimes_s f \otimes_s 1 \mapsto \underline{1} \otimes \partial(f) \otimes_s 1.$

Proof is straightforward. □

These maps have degrees $0, 0, -1, -1$, respectively.

The shortcuts below will be useful in the sequel.

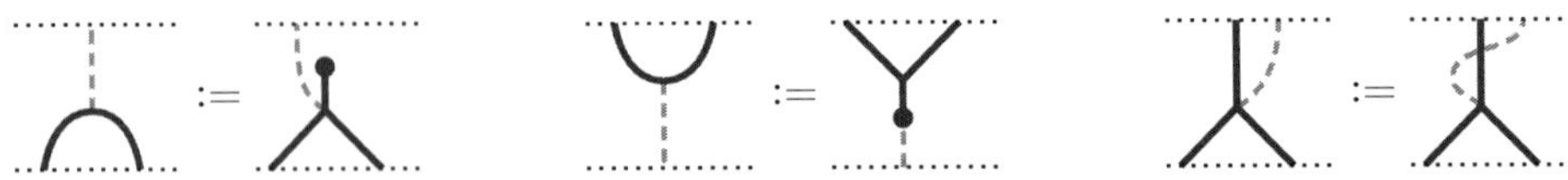

LEMMA 2.4. *The bimodule maps in equations* (2.24) *to* (2.32) *satisfy the relations below.*

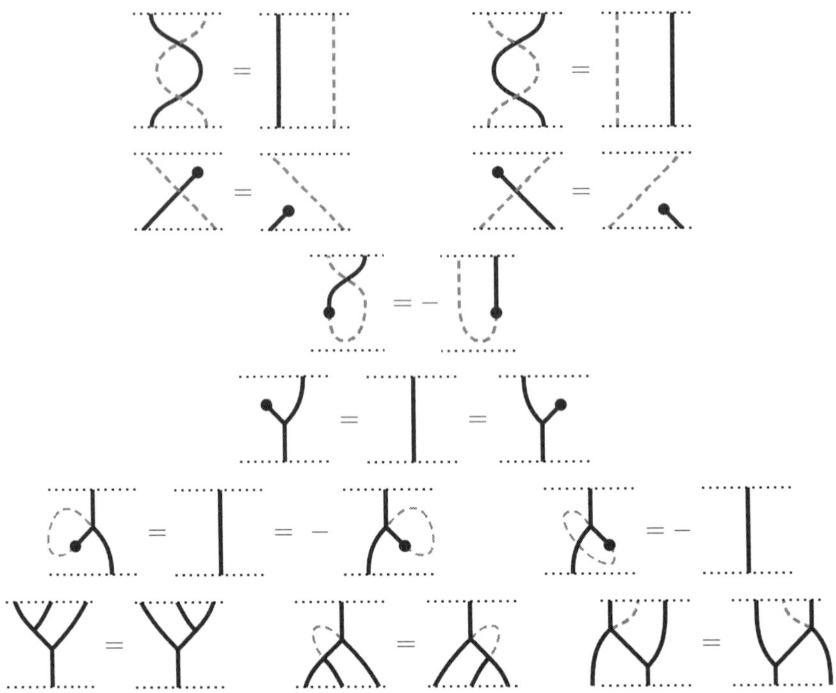

A proof is given by a straighforward computation. □

Dashed orange lines in the present paper are similar to dashed blue lines in Ellis-Lauda's categorification of odd quantum $sl(2)$, see [**6**]. (Compare adjointness relations (3.13), (3.14) in that paper with the adjointness in Proposition 2.2.)

The difference of the present diagrammatical calculus of blue lines (for B) and dashed red lines (for $\underline{R}$) from the earlier calculus in Section 2.2 is that blue regions (for the category $R^s-\mathsf{gmod}$) are now hidden inside blue lines and graphs. Thickening these lines and graphs recovers the earlier diagrammatics, see equation 2.33 below.

(2.33)

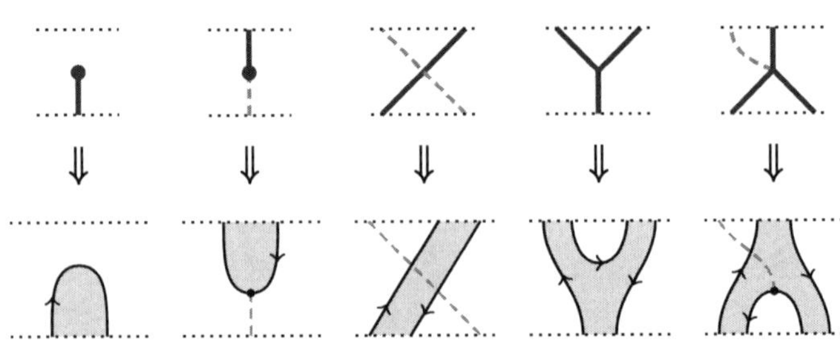

PROPOSITION 2.5. *The following equality holds*

(2.34) 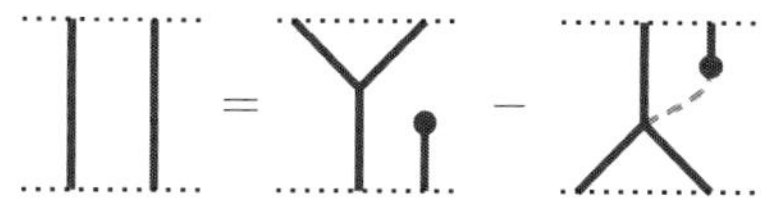

Moreover, both terms on the right hand side are orthogonal idempotents.

PROOF. A direct computation shows that the two terms on the right hand side are orthogonal idempotents. To show that their sum is the identity of $B \otimes_s B$ note that for $f \in R$ the element $P(f) := f - x_2 \partial f$ is in R^s since $\partial P(f) = 0$. This allows writing $f = P(f) + x_2\partial(f)$ (a similar argument appears in [**3**, §2.2]). We then compute

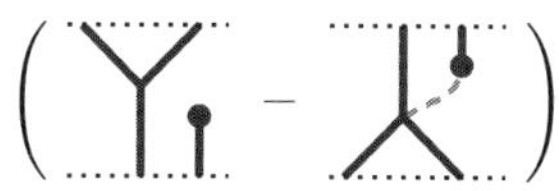 $(1 \otimes_s f \otimes_s 1)$

$$
\begin{aligned}
&= 1 \otimes_s 1 \otimes_s f - 1 \otimes_s 1 \otimes_s x_2\partial f + 1 \otimes_s x_2 \partial f \otimes_s f \\
&= 1 \otimes_s 1 \otimes_s P(f) + 1 \otimes_s x_2 \partial f \otimes_s 1 \\
&= 1 \otimes_s P(f) \otimes_s 1 + 1 \otimes_s x_2\partial f \otimes_s 1 = 1 \otimes_s f \otimes_s 1,
\end{aligned}
$$

as claimed. □

COROLLARY 2.6. *There are direct sum decompositions*

$$B \otimes B \cong B\{-1\} \oplus \underline{B}\{1\} \cong \underline{B} \otimes \underline{B}, \tag{2.35}$$

$$B \otimes \underline{B} \cong B\{1\} \oplus \underline{B}\{-1\} \cong \underline{B} \otimes B. \tag{2.36}$$

PROOF. The first idempotent on the right hand side of (2.34) is a composition of degree 0 maps

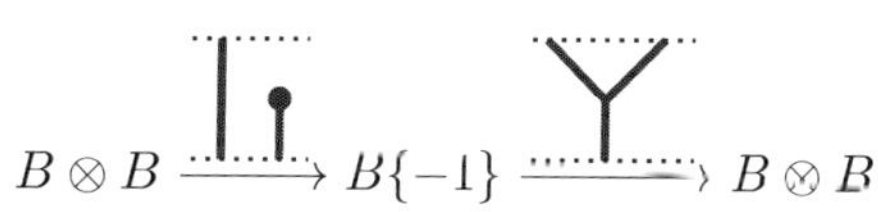

Composing in the opposite direction gives the identity map of $B\{-1\}$, so that this idempotent is a projection onto a copy of $B\{-1\}$. Grading shift is present due to the degree of (2.25) being one.

Likewise, the second idempotent is a composition

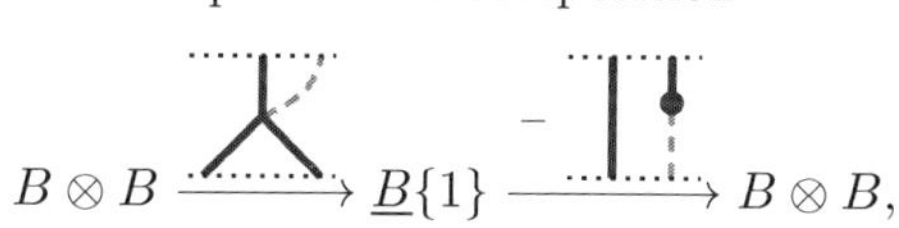

with the composition in the opposite direction equal $\mathrm{id}_{\underline{B}\{1\}}$. Thus, it's a projection onto a graded bimodule isomorphic to $\underline{B}\{1\}$. We obtain a direct sum decomposition $B \otimes B \cong \underline{B}\{-1\} \oplus B\{1\}$ in (2.35) Tensoring with $\underline{R}$ on the right and on the left gives the remaining direct sum decompositions. □

REMARK 2.7. The identity in (2.34) can be expressed in equivalent ways, which result in different presentations of the maps realising the isomorphisms in (2.35) and (2.36). For example, it equals its reflection around a vertical axis:

(2.37) 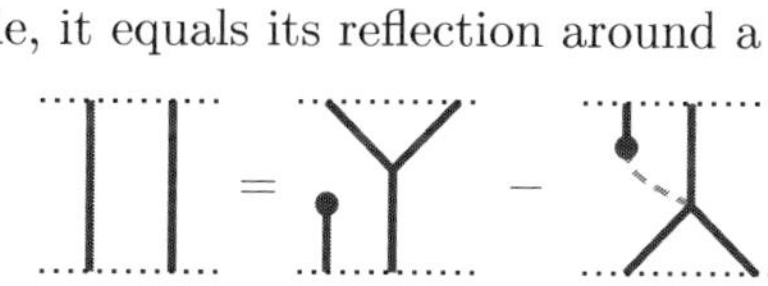

To see that this equation holds, we need the two relations below.

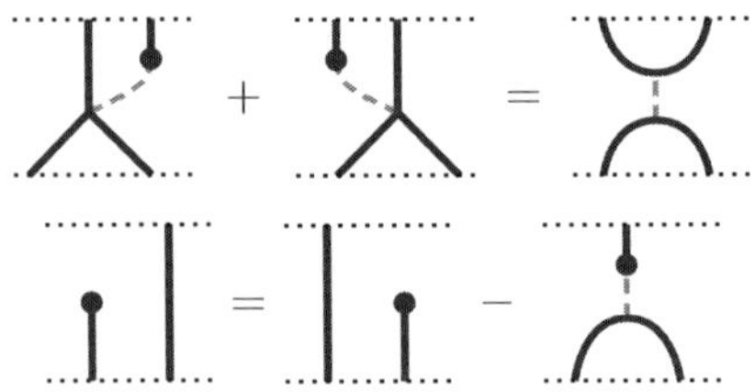

The first one can be proved by direct computation, and the second is obtained from (2.34) by postcomposing all terms with the map m at the appropriate place. Combining these two relations with (2.34) gives (2.37).

The direct sum decompositions in Corollary 2.6 are not canonical. Specializing to $B \otimes B$, there is a canonical short exact sequence below (up to a choice of signs for the maps) with the inclusion given by map (2.31)

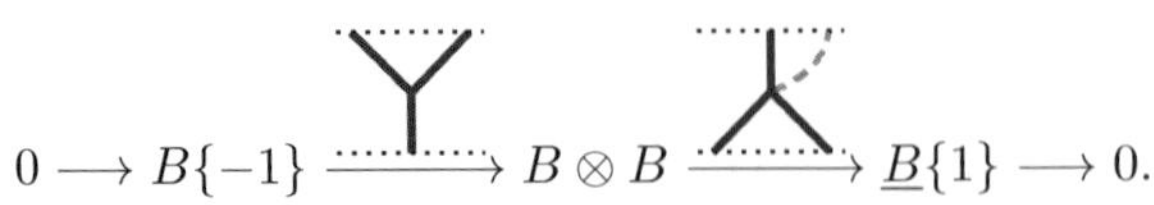

$$0 \longrightarrow B\{-1\} \longrightarrow B \otimes B \longrightarrow \underline{B}\{1\} \longrightarrow 0.$$

This sequence splits, but a splitting is non-unique, due to the existence of a non-trivial degree 2 bimodule map $\underline{B} \longrightarrow B$, see below on the left. Via adjointness, it comes from a degree two homomorphism $R \longrightarrow B \otimes B$, shown below on the right

(the minus sign is added to match our definition of the corresponding adjointness morphism). A particular direct sum decomposition of $B \otimes B$ is given by the following maps, as in the proof of Corollary 2.6.

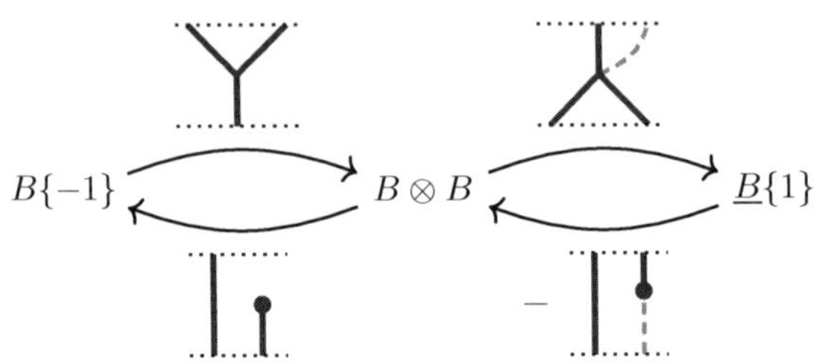

3. Oriented calculus for products of generating bimodules

Our diagrammatics so far explicitly includes bimodules B (blue lines) and $\underline{R}$ (dashed orange lines). Bimodule $\underline{B}$ and the maps that go through it appear implicitly through a combination of diagrammatics for B and for $\underline{R}$. It's natural to extend this diagrammatics, by depicting the identity map of B, respectively $\underline{B}$, by a vertical blue line oriented up, respectively down, see below. Then the biadjointness maps (2.17) can be compactly depicted by oriented cups and caps, with the usual isotopy relations on these cup and caps.

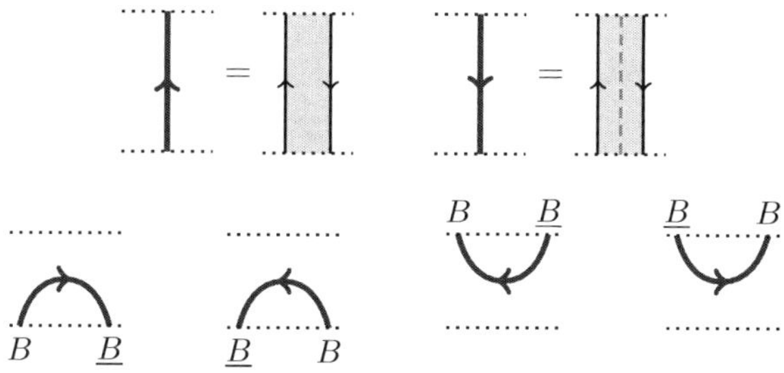

A closed circle, either clockwise or counterclockwise oriented, evaluates to 0, see below.

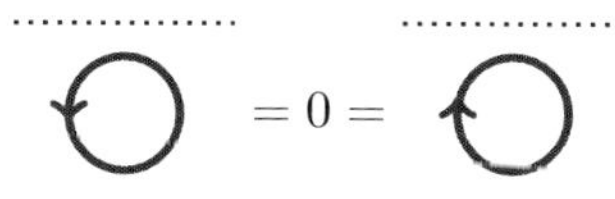

There are additional generating maps and isotopy relations on them (some relations are depicted below, together with the maps).

(3.1)

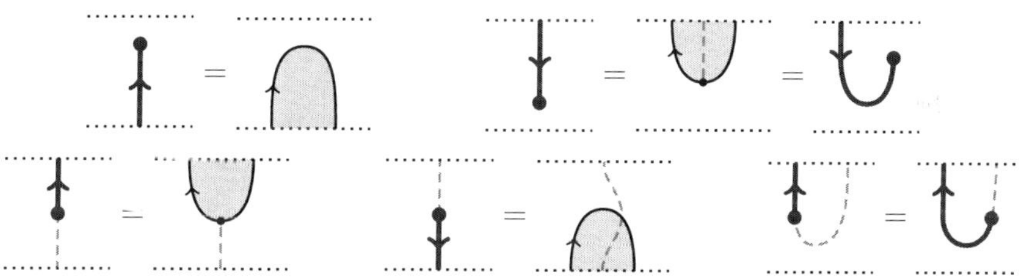

Isomorphisms (2.23) can be depicted by a trivalent vertex where a dashed orange line enters the point of orientation reversal of a blue line, see below, together with the corresponding relations. There are 8 such trivalent vertices, with some relations on them also shown below and other relations are obtained by suitable symmetries (horizontal and vertical reflection and orientation reversal).

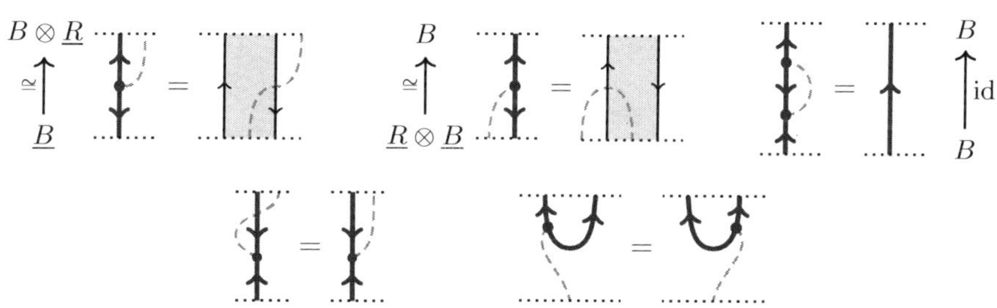

Blue lines in the top row of (2.33) now acquire upward orientation, see below. The rightmost diagram is an exception; dashed orange line is hidden at the cost of orienting the left bottom leg down.

(3.2)

Composing trivalent vertices with cups and caps results in rotated trivalent vertices, see below (where top left diagram is the rightmost diagram in (3.2).

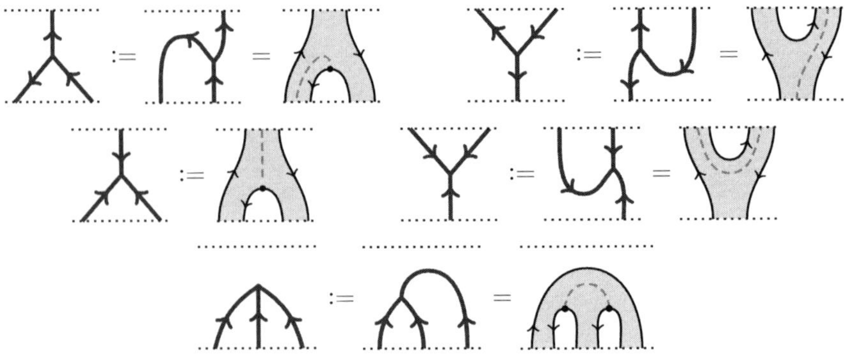

At a trivalent vertex, the three edges either all oriented into the vertex, or one edge is oriented in and two edges out. The number of "out" oriented edges at each vertex is even.

The degrees of various maps are summarized the table below.

map	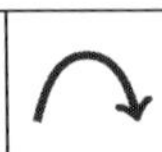	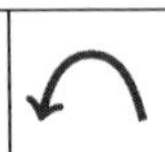				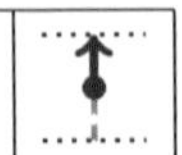
degree	0	0	0	0	1	1

map	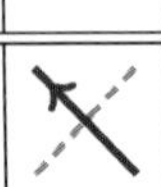	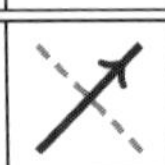	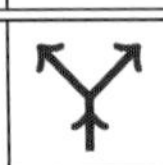			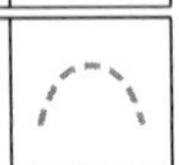
degree	0	0	-1	-1	0	0

It's also convenient to introduce a crossings of a downward-oriented blue line with dashed orange line, a degree 0 map defined as shown below.

Also, the following relations hold.

Similar to the decomposition of $B \otimes B$ and using oriented lines, we obtain the following direct sum decomposition

$$B \otimes \underline{B} \;\cong\; B\{1\} \oplus \underline{B}\{-1\}$$

from Corollary 2.6 diagrammatically (note the minus sign in one of the maps).

(3.3)

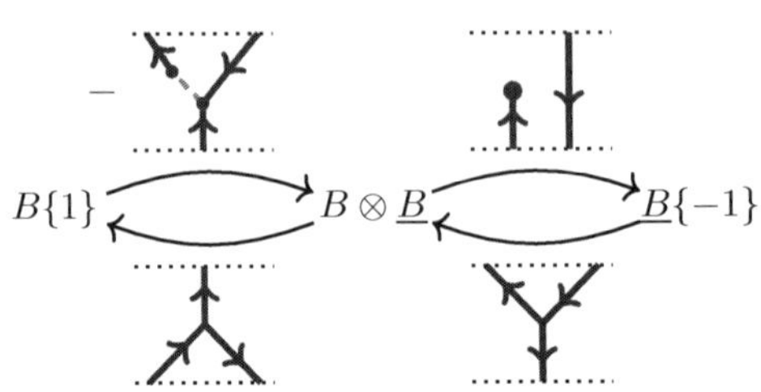

There's flexibility in choosing some arrows in a direct sum decomposition of $B\otimes \underline{B}$, as in the earlier discussion about the equivalent case of decomposing $B\otimes B$.

REMARK 3.1. Let SBim be the monoidal category of 2-variable odd Soergel bimodules generated by bimodules $B, \underline{R}$ and their grading shifts (see more details about SBim in Section 5). Bimodule B has an antiinvolution ϕ given by $\phi(x\otimes y) = y\otimes x$. Bimodule $\underline{R}$ has an antiinvolution ϕ given by $\phi(x\underline{1}y) = y\underline{1}x$. Antiinvolutions ϕ extend to an involutive antiequivalence ϕ : SBim $\longrightarrow$ SBim^{op} of the category SBim that takes B to B and $\underline{B}$ to $\underline{B}$. In our graphical description of SBim some generating maps are invariant under the reflection in a vertical line, such as in (2.25). Our diagrammatical notations for several maps break reflectional symmetry, requiring adding a minus sign to the reflected diagram, including in (2.26). These signs later propagate in formulas: for instance, observe the absence of signs in the direct sum decomposition given by (4.1) and the presence of a single minus sign in the decomposition (3.3).

4. Odd Rouquier complexes and invertibility

Consider the following two-term complexes of graded B-modules, where $\underline{B}$ and B terms are placed in cohomological degree 0. The differential is given by maps (2.25) and the map $\underline{\Delta}$ obtained from (2.26) by tensoring with $\underline{1}$, respectively.

$$\mathcal{R} := 0 \longrightarrow B \xrightarrow{\ m\ } R\{-1\} \longrightarrow 0,$$

$$\mathcal{R}' := 0 \longrightarrow R\{1\} \xrightarrow{\underline{\Delta}} \underline{B} \longrightarrow 0.$$

Complexes $\mathcal{R}, \mathcal{R}'$ can be viewed as odd analogues of the Rouquier complexes.

THEOREM 4.1. *There are homotopy equivalences of complexes of graded B-modules $\mathcal{R}\otimes_R \mathcal{R}' \cong_h R$ and $\mathcal{R}'\otimes_R \mathcal{R} \cong_h R$.*

Here R denotes the identity R-bimodule, viewed as a complex concentrated in homological degree 0.

PROOF. The complex $\mathcal{R}\otimes_R \mathcal{R}'$ is given by forming a commutative square of bimodules below

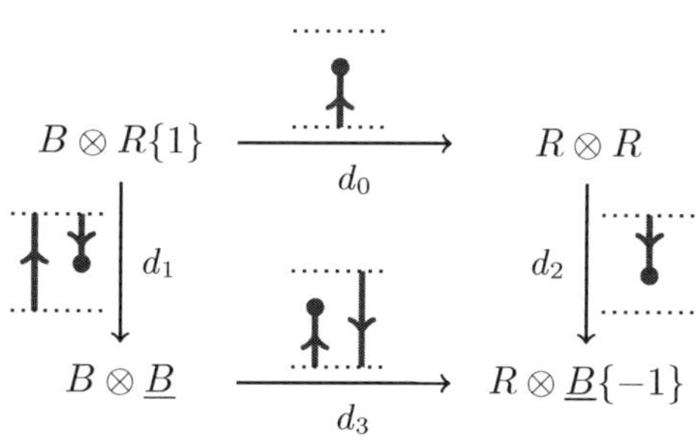

then adding a minus sign to the map d_2 and collapsing the square into the complex $(C,\widehat{d})$ below.

$$C:\quad B\{1\} \xrightarrow{\ \widehat{d}_{-1} = \begin{pmatrix} d_0 \\ d_1 \end{pmatrix}\ } R\oplus(B\otimes \underline{B}) \xrightarrow{\ \widehat{d}_0 = (-d_2, d_3)\ } \underline{B}\{-1\}$$

Introduce maps h_0, h_3 and $\jmath$ between terms in the above commutative square, as shown below.

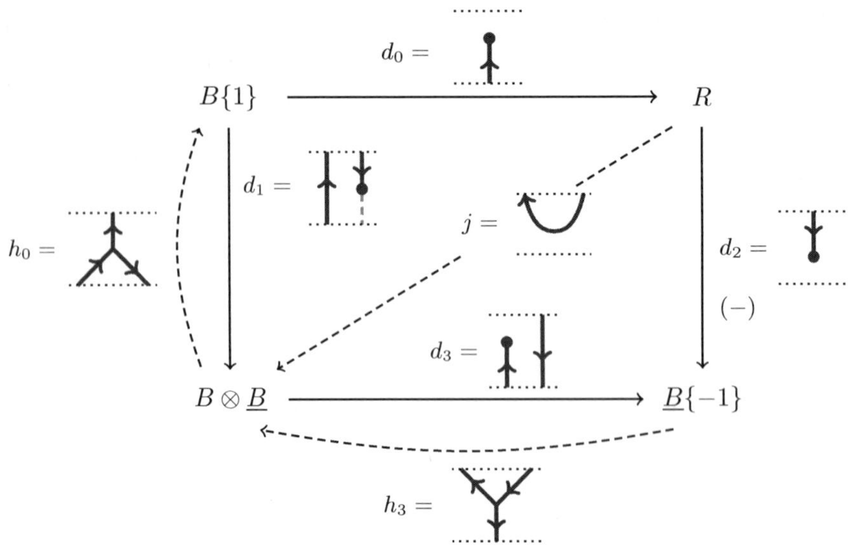

The following relations hold

$$h_0 d_1 = \mathrm{id}_{B\{1\}}, \quad d_2 = d_3 \jmath, \quad d_3 h_3 = \mathrm{id}_{\underline{B}\{-1\}}, \quad h_0 h_3 = 0.$$

Thus, d_1 is split injective, with a section h_0. Likewise, d_3 is split surjective, with h_3 as a section.

We would like to check that $B \otimes \underline{B} = \mathrm{im}(d_1) \oplus \mathrm{im}(h_3)$. Consider the map d'_3 given by

$d'_3 =$ 

Map d'_3 is a rotation of the top left diagram in (3.3). Then

$$d'_3 d_1 = 0, \quad h_0 h_3 = 0, \quad d'_3 h_3 = \mathrm{id}_{\underline{B}\{-1\}}, \quad d_1 h_0 + h_3 d'_3 = \mathrm{id}_{B \otimes \underline{B}}. \tag{4.1}$$

Pairs of maps (d_1, h_0) and (d'_3, h_3) give a direct sum decomposition $B \otimes \underline{B} \cong B\{1\} \oplus \underline{B}\{-1\}$.

Complex C above splits into the direct sum of three subcomplexes:

$$0 \longrightarrow B\{1\} \xrightarrow{d_0 + d_1} B\{1\} \longrightarrow 0,$$
$$0 \longrightarrow \quad R \quad \longrightarrow 0,$$
$$0 \longrightarrow h_3(\underline{B}\{-1\}) \xrightarrow{d_3} \underline{B}\{-1\} \longrightarrow 0,$$

where the middle complex consists of pairs $(a, \jmath(a)), a \in R$. The first and third complexes are contractible, while the middle complex is the identity bimodule R. Consequently, $\mathcal{R} \otimes_R \mathcal{R}' \cong_h R$. A similar computation, changing the order of terms in tensor products and reflecting all map diagrams about horizontal axes, shows that $\mathcal{R}' \otimes_R \mathcal{R} \cong_h R$. □

COROLLARY 4.2. *Functors of tensoring with bimodule complexes $\mathcal{R}$ and $\mathcal{R}'$ are mutually-invertible functors in the homotopy category of complexes of graded R-modules.*

PROPOSITION 4.3. *After removing contractible summands, complex $\mathcal{R}^n$ for $n > 0$ simplifies to the $(n+1)$-term complex, nontrivial in cohomological degrees from 0 to n, with the head*

$$\cdots \longrightarrow B\{5-n\} \longrightarrow \underline{B}\{3-n\} \longrightarrow B\{1-n\} \longrightarrow R\{-n\} \longrightarrow 0$$

and the tail

$$0 \longrightarrow B\{n-1\} \longrightarrow \underline{B}\{n-3\} \longrightarrow B\{n-5\} \longrightarrow \underline{B}\{n-7\} \longrightarrow \cdots$$

for odd n and

$$0 \longrightarrow \underline{B}\{n-1\} \longrightarrow B\{n-3\} \longrightarrow \underline{B}\{n-5\} \longrightarrow B\{n-7\} \longrightarrow \cdots$$

for even n. Under the differential maps, $1 \otimes_s 1$ and $1 \otimes_s \underline{1}$ are sent to $1 \otimes_s \underline{1}x_1 + x_2 \otimes_s \underline{1} \in \underline{B}$ and to $1 \otimes_s x_2 - x_2 \otimes_s 1 \in B$, respectively.

The proposition can be proved by induction on n and a direct computation using Gauss elimination. □

The case of $\mathcal{R}'^n$ is similar:

PROPOSITION 4.4. *After removing contractible summands, complex $\mathcal{R}'^n$ for $n > 0$ reduces to the following $(n+1)$-term complex that lives in cohomological degrees from $-n$ to 0:*

$$0 \longrightarrow R\{n\} \longrightarrow \underline{B}\{n-1\} \longrightarrow B\{n-3\} \longrightarrow \underline{B}\{n-5\} \longrightarrow \cdots.$$

REMARK 4.5. Adding the signed permutation bimodule to $\mathcal{R}, \mathcal{R}'$ gives the 2-strand motion braid group action on the homotopy category of graded R-modules, see A.-L. Thiel [**18**] for the corresponding action of the group of motion braids or virtual braids in the even case for any number of strands.

5. Grothendieck ring

Recall that SBim is the category of 2-variable odd Soergel bimodules generated as the monoidal category by bimodules B and $\underline{R}$ and their grading shifts. Hom spaces in this category are all grading-preserving homomorphisms of bimodules. We can also define the larger spaces

$$\mathrm{HOM}_{\mathrm{SBim}}(M, N) \; := \; \oplus_{n\in\mathbb{Z}} \mathrm{Hom}_{\mathrm{SBim}}(M\{n\}, N).$$

These HOM spaces are naturally modules over the center of R,

$$Z(R) \;\cong\; \mathbb{Z}[x_1^2, x_2^2] \subset R. \tag{5.1}$$

The left and right actions of $Z(R)$ on these hom spaces are not equal, in general.

Indecomposable objects of SBim, up to shifts, are $R, \underline{R}, B, \underline{B}$, with tensor product decompositions

$$\underline{R} \otimes \underline{R} \cong R, \quad \underline{B} \cong B \otimes \underline{R} \cong \underline{R} \otimes B, \quad B \otimes B \cong B\{-1\} \oplus \underline{B}\{1\}.$$

Consider the split Grothendieck ring K_0 of SBim. Grading shift functor induces a $\mathbb{Z}[q, q^{-1}]$-module structure on $K_0(\text{SBim})$. The latter is a free rank four $\mathbb{Z}[q, q^{-1}]$-module with a basis

$$1 = [R], \quad c := [\underline{R}], \quad b := [B], \quad bc = [\underline{B}],$$

generators b, c, and multiplication rules

$$c^2 = 1, \quad cb = bc, \quad b^2 = q^{-1}b + qbc. \tag{5.2}$$

Since the element c is central in $K_0(\text{SBim})$, the latter is a commutative associative $\mathbb{Z}[q, q^{-1}]$-algebra (commutativity fails for analogous algebras for three or more strands).

Define a $\mathbb{Z}[q, q^{-1}]$-semilinear form on $K_0(\text{SBim})$ by

$$([M], [N]) \ := \ \text{gdim}(\text{HOM}(M, N)),$$

where gdim denotes the graded dimension. This form is $\mathbb{Z}[q, q^{-1}]$-linear in the second variable and $\mathbb{Z}[q, q^{-1}]$-antilinear in the first variable. We have $(cm, cn) = (m, n)$, for $m, n \in K_0(\text{SBim})$, and

$$(1, 1) = \frac{1}{(1 - q^4)^2}, \tag{5.3}$$

$$(1, c) = (c, 1) \ = \ 0, \tag{5.4}$$

$$(b, 1) = (1, bc) \ = \ \frac{q}{(1 - q^4)^2}, \tag{5.5}$$

$$(1, b) = (bc, 1) \ = \ \frac{q^3}{(1 - q^4)^2}. \tag{5.6}$$

The inner product $(1, c) = 0$ since $\text{HOM}(R, \underline{R}) = 0$, which follows by a direct computation. The inner product $(b, 1)$ above is computed via adjointness isomorphism

$$\text{HOM}(B, R) = \text{HOM}(R \otimes_{R^s} R\{-1\}, R) \cong \text{Hom}({}_{R^s}R_R, {}_{R^s}R_R)\{1\}.$$

An endomorphism ξ of the (R_s, R)-bimodule R is determined by $\xi(1) \in R$ which we write as $\xi(1) = h_{00} + h_{01}x_1 + h_{10}x_2 + h_{11}x_1x_2$, where $h_{ij} \in \mathbf{k}[x_1^2, x_2^2] = Z(R)$. Commutativity relations $f\xi(1) = \xi(1)f$ for $f \in R^s$ can be reduced to those for generators $x_1 - x_2, x_1x_2$ of R^s, leading to the relations $h_{01} = h_{10} = h_{11} = 0$. Consequently, endomorphisms of this bimodule are in a bijection with central elements of R, via $\xi(1) = h_{00} \in Z(R)$. Passing to the graded dimension results in the above formula $(b, 1) = q(1 - q^4)^{-2}$.

Likewise, the inner product $(bc, 1)$ is the graded dimension of

$$\text{HOM}(\underline{B}, R) = \text{HOM}(R \otimes_{R^s} \underline{R}\{-1\}, R) \cong \text{Hom}({}_{R^s}\underline{R}_R, {}_{R^s}R_R)\{1\}.$$

The generator of the hom space $\text{HOM}(\underline{B}, R)$ is given by the degree 3 map below. There x_1^2 in a box denotes the bimodule map of multiplication by the central element x_1^2 of R. Replacing x_1^2 by x_2^2 in the middle box reverses the sign of the map.

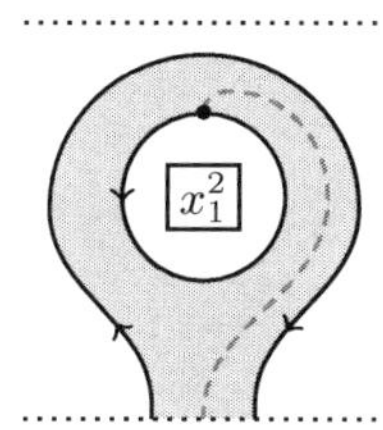

A similar computation to the above shows that an (R, R)-bimodule maps $\underline{B} \longrightarrow R$ are given by $1 \longmapsto f(x_1 - x_2)$, for any $f \in \mathbf{k}[x_1^2, x_2^2]$. The generating map, for $f = 1$, has degree 3, due to shift in the degree of $\underline{B}$ as defined. Consequently, the inner product $(bc, 1)$ is given by (5.6).

Each object of SBim has a biadjoint object (since the generating objects do). Bimodules $B, \underline{B}$ and $\underline{R}, R$ define biadjoint pairs of functors. Denote the corresponding "biadjointness" antiinvolution on $K_0(\mathrm{SBim})$ by τ. It has the properties

$$\tau(1) = 1, \quad \tau(c) = c, \quad \tau(b) = bc,$$

and

$$\tau(xy) = \tau(y)\tau(x), \quad \tau(qx) = q^{-1}\tau(x), \quad x, y \in K_0(\mathrm{SBim}).$$

In general, such "biadjointness" involutions reverse the order in the product, but due to commutativity of $K_0(\mathrm{SBim})$ it does not matter in our case. This involution is $\mathbb{Z}[q, q^{-1}]$-antilinear and compatible with the bilinear form,

$$(xm, n) = (m, \tau(x)n), \quad x, m, n \in K_0(\mathrm{SBim}).$$

Adjointness allows to finish the computation of the inner products (5.5) and (5.6). We can further compute that

$$(b, b) = \frac{1 + q^4}{(1 - q^4)^2}, \quad (b, bc) = \frac{2q^2}{(1 - q^4)^2}.$$

Some of the generating maps for HOM spaces between the four indecomposable bimodules in SBim are shown below, with each map of degree one. Generating maps in the opposite direction, all in degree 3, are not shown.

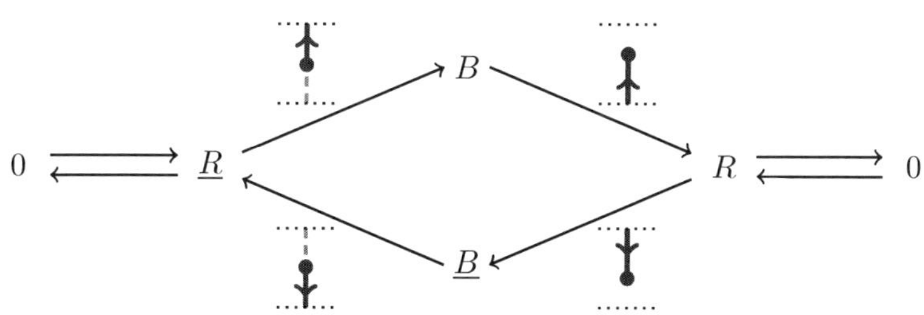

For each of the four arrows between these four bimodules, the HOM space is a one-dimensional $Z(R)$ module with the generator shown. There are no homs between R and $\underline{R}$, and the corresponding compositions are 0. The upper and lower portions of the diagram constitute two short exact sequences, and zero objects are added on the sides to emphasize that.

Note that trivalent vertices appear in this calculus when passing to the tensor products of B's and $\underline{B}$'s, to describe tensor product decompositions into direct sums.

The bilinear form is determined by $\mathbb{Z}[q,q^{-1}]$-linear trace form tr on $K_0(\mathrm{SBim})$, where $\mathsf{tr}(a) = (1,a)$, with

$$\mathsf{tr}(1) = \frac{1}{(1-q^4)^2}, \quad \mathsf{tr}(b) = \frac{q^3}{(1-q^4)^2}, \quad \mathsf{tr}(c) = 0, \quad \mathsf{tr}(bc) = \frac{q}{(1-q^4)^2}.$$

The inner product and trace can be rescaled by $(1-q^4)^2$ to take values in $\mathbb{Z}[q,q^{-1}]$. This corresponds to viewing hom spaces as free graded modules over $Z(R)$ (under either left or right multiplications by central elements) and taking their graded ranks.

6. An obstacle to the Reidemeister III relation

Consider the ring R_3 of supercommuting polynomials in 3 variables,

$$R_3 = \mathbf{k}\langle x_1, x_2, x_3\rangle/(x_i x_j + x_j x_i), \quad 1 \le i < j \le 3.$$

In this section let us denote R_3 by R.

The symmetric group S_3 acts on R, with

$$s_1(x_1) = -x_2, \; s_1(x_2) = -x_1, \; s_1(x_3) = -x_3,$$
$$s_2(x_1) = -x_1, \; s_2(x_2) = -x_3, \; s_2(x_3) = -x_2,$$

and $s_i(fg) = s_i(f)s_i(g)$ for $f, g \in R$.

There are two odd Demazure operators, $\partial_1, \partial_2 : R \longrightarrow R$. Operator $\partial_1 : R \longrightarrow R$ is given by:

- $\partial_1(1) = 0$, $\partial_1(x_1) = \partial_1(x_2) = 1$, $\partial_1(x_3) = 0$,
- the twisted Leibniz rule holds

$$\partial_1(fg) = (\partial_1 f)g + s_1(f)\partial_1 g,$$

and likewise for ∂_2.

The kernels of ∂_1, ∂_2 are subrings $R^1, R^2 \subset R$. For instance, ring R^1 is the subring of R generated by $x_1 - x_2, x_1x_2, x_3$.

Form graded R-bimodules

$$B_i \; := \; R \otimes_{R^i} R\{-1\}, \quad i = 1, 2.$$

Note that $x_3(1\otimes 1) = (1 \otimes 1)x_3$, where $1\otimes 1$ is the generator of B_1.

Diagrammatic calculi of the earlier sections can be repeated separately for B_1 and B_2. In case of B_1 we would need the permutation bimodule, denoted $\underline{R}^1 \cong R\underline{1}_1$, with the generator $\underline{1}_1$ and $x\underline{1}_1 = \underline{1}_1 s_1(x)$ for $x \in R$ and can then form $\underline{B}_1 = B_1 \otimes_R \underline{R}^1 \cong \underline{R}^1 \otimes_R B_1$. We have not tried to develop a diagrammatical calculus of odd 3-stranded Soergel bimodules which would add interactions between products of B_1 and B_2.

Consider graded R-bimodules

$$\underline{B}_i \; := \; R \otimes_{R^i} \underline{R}^i \otimes_{R^i} R\{-1\} \; = \; R \underline{\otimes}_{R^i} R\{-1\}, \quad i = 1, 2.$$

Define the bimodule

$$B_{\widehat{121}} := R \otimes_{R^{[2]}} R\{-3\}.$$

Here $R^{[2]} \subset R$ is the subring of odd symmetric functions in three variables, $R^{[2]} = \ker(\partial_1) \cap \ker(\partial_2)$.

PROPOSITION 6.1. *There exists an exact sequence of graded R-bimodules*

$$0 \longrightarrow B_{\widehat{121}} \longrightarrow B_1 \otimes B_2 \otimes B_1 \longrightarrow \underline{B}_1 \longrightarrow 0 \tag{6.1}$$

This sequence does not split.

The absence of a splitting creates a problem for the Reidemeister III move invariance. When resolving complexes for $\mathcal{R}_1\mathcal{R}_2\mathcal{R}_1$ and $\mathcal{R}_2\mathcal{R}_1\mathcal{R}_2$ there are not enough contractible summands to slim the complexes down to those with the leftmost term $B_{\widehat{121}}$, which is how the isomorphism is proven in the even case. Instead, the leftmost terms are $B_1 \otimes B_2 \otimes B_1$ and $B_2 \otimes B_1 \otimes B_2$, respectively, which are not isomorphic. This prevents the corresponding complexes of bimodules from being isomorphic in the homotopy category. Finding a way around this obstacle is an interesting problem.

7. Comparison with the even case

In this section we use the same notations to denote the corresponding structures in the even case:

- $R = \mathbf{k}[x_1, x_2]$ is the ring of polynomials in two variables. S_2 acts on it by permuting the variables.
- ∂ is the Demazure operator, $\partial(f) = \frac{f - {}^s f}{x_1 - x_2}$.
- $R^s \subset S$ is the ring of symmetric functions.
- $B = R \otimes_{R^s} R\{-1\}$ is the generating Soergel bimodule for two variables.
- $\underline{R}$ is the transposition bimodule, $\underline{R} = R\underline{1} = \underline{1}R$, $x_i\underline{1} = \underline{1}x_{s(i)}$.

For the general theory of Soergel bimodules we refer to [**4, 17**] and for the diagrammatic calculus of Soergel bimodules to [**3, 4**]. There is a natural isomorphism of (R^s, R)-bimodules

$$_{R^s}R \otimes_R \underline{R} \cong {}_{R^s}R$$

due to involution s acting by identity on R^s.

Likewise, there's an isomorphism of (R, R^s)-bimodules

$$\underline{R} \otimes_R R_{R^s} \cong R_{R^s}.$$

Tensoring these equations with the other "halves" of the bimodule B gives bimodule isomorphisms

$$\underline{R} \otimes_R B \cong B \cong B \otimes_R \underline{R}.$$

Odd replacement of these isomorphisms motivates introducing bimodules $\underline{B}$ in that case, see earlier.

Multiplication map $f \otimes g \longmapsto fg$ induces a surjective bimodule map $B \longrightarrow R\{-1\}$ which extends to a short exact sequence of bimodules

$$0 \longrightarrow \underline{R}\{1\} \longrightarrow B \longrightarrow R\{-1\} \longrightarrow 0$$

Tensoring all terms of this sequence with $\underline{R}$ flips the sequence to the opposite

$$0 \longrightarrow R\{1\} \longrightarrow B \longrightarrow \underline{R}\{-1\} \longrightarrow 0 \tag{7.1}$$

These filtrations were emphasized in [**10**]. They allow to think of Rouquier complexes as a sort of *homological perturbation* or *homological quantization* of the permutation bimodule $\underline{R}$. In the homotopy category (and ignoring q-gradings) the complex

$$0 \longrightarrow B \longrightarrow R \longrightarrow 0 \tag{7.2}$$

is not isomorphic to

$$0 \longrightarrow \underline{R} \longrightarrow 0. \tag{7.3}$$

(in both complexes we place the leftmost nontrivial term in degree 0). Homology groups of these complexes are isomorphic, though. Complex (7.2) is giving by thickening complex (7.3) by the contractible complex $0 \longrightarrow R \xrightarrow{\text{id}} R \longrightarrow 0$, which does not change the homology but makes the complex more subtle on the homotopy category level. The transposition relation, which holds for $\underline{R}$ (that $\underline{R} \otimes \underline{R} \cong R$) fails for the complex (7.2). It's substituted by the weaker relation that (7.2) is invertible in the homotopy category, with the quasi-inverse complex $0 \longrightarrow R \longrightarrow B \longrightarrow 0$ given by truncating (7.1). The quasi-inverse is another thickening of $\underline{R}$.

Before "homological perturbation", tensoring with the bimodule $\underline{R}$ is a symmetry of order two, with $\underline{R} \otimes_R \underline{R} \cong R$. Homological perturbation results in an invertible functor of infinite order, while retaining the Reidemeister III relation (the braid relation) in the homotopy category of complexes of bimodules over $\mathbf{k}[x_1, x_2, x_3]$.

Thus, upon this homological perturbation, action of the permutation group S_n on the category of $\mathbf{k}[x_1, \ldots, x_n]$-modules given by tensoring with permutation bimodules $\underline{R}_i$ becomes a much more subtle action of the n-stranded braid group on the homotopy category of $\mathbf{k}[x_1, \ldots, x_n]$-modules by Rouquier complexes.

References

[1] Jonathan Brundan and Alexander Kleshchev, *Odd Grassmannian bimodules and derived equivalences for spin symmetric groups*, **preprint** `arXiv:2203.14149v1` (2022).

[2] Mark Ebert, Aaron D. Lauda, and Laurent Vera, *Derived superequivalences for spin symmetric groups and odd $sl(2)$-categorifications*, **preprint** `arXiv:2203.14153v1` (2022).

[3] Ben Elias and Mikhail Khovanov, *Diagrammatics for Soergel categories*, Int. J. Math. Math. Sci., posted on 2010, Art. ID 978635, 58 pp., DOI 10.1155/2010/978635. MR3095655

[4] Ben Elias, Shotaro Makisumi, Ulrich Thiel, and Geordie Williamson, *Introduction to Soergel bimodules*, RSME Springer Series, vol. 5, Springer, Cham, 2020, DOI 10.1007/978-3-030-48826-0. MR4220642

[5] Alexander P. Ellis, Mikhail Khovanov, and Aaron D. Lauda, *The odd nilHecke algebra and its diagrammatics*, Int. Math. Res. Not. IMRN **4** (2014), 991–1062, DOI 10.1093/imrn/rns240. MR3168401

[6] Alexander P. Ellis and Aaron D. Lauda, *An odd categorification of $U_q(\mathfrak{sl}_2)$*, Quantum Topol. **7** (2016), no. 2, 329–433, DOI 10.4171/QT/78. MR3459963

[7] Seok-Jin Kang, Masaki Kashiwara, and Se-jin Oh, *Supercategorification of quantum Kac-Moody algebras*, Adv. Math. **242** (2013), 116–162, DOI 10.1016/j.aim.2013.04.008. MR3055990

[8] Mikhail Khovanov, *Triply-graded link homology and Hochschild homology of Soergel bimodules*, Internat. J. Math. **18** (2007), no. 8, 869–885, DOI 10.1142/S0129167X07004400. MR2339573

[9] Mikhail Khovanov and Lev Rozansky, *Matrix factorizations and link homology*, Fund. Math. **199** (2008), no. 1, 1–91, DOI 10.4064/fm199-1-1. MR2391017

[10] Mikhail Khovanov and Lev Rozansky, *Matrix factorizations and link homology. II*, Geom. Topol. **12** (2008), no. 3, 1387–1425, DOI 10.2140/gt.2008.12.1387. MR2421131

[11] Mikhail Khovanov and Richard Thomas, *Braid cobordisms, triangulated categories, and flag varieties*, Homology Homotopy Appl. **9** (2007), no. 2, 19–94. MR2366943

[12] Grégoire Naisse and Krzysztof Putyra, *Odd Khovanov homology for tangles*, **preprint** `arXiv:2003.14290v1` (2020).

[13] Grégoire Naisse and Pedro Vaz, *Odd Khovanov's arc algebra*, Fund. Math. **241** (2018), no. 2, 143–178, DOI 10.4064/fm328-6-2017. MR3766566

[14] Peter S. Ozsváth, Jacob Rasmussen, and Zoltán Szabó, *Odd Khovanov homology*, Algebr. Geom. Topol. **13** (2013), no. 3, 1465–1488, DOI 10.2140/agt.2013.13.1465. MR3071132

[15] Krzysztof K. Putyra, *A 2-category of chronological cobordisms and odd Khovanov homology*, Knots in Poland III. Part III, Banach Center Publ., vol. 103, Polish Acad. Sci. Inst. Math., Warsaw, 2014, pp. 291–355, DOI 10.4064/bc103-0-12. MR3363817

[16] ———, *On a triply-graded generalization of Khovanov homology*, 2014, Thesis (Ph.D.)–Columbia University. MR3232379

[17] Wolfgang Soergel, *The combinatorics of Harish-Chandra bimodules*, J. Reine Angew. Math. **429** (1992), 49–74, DOI 10.1515/crll.1992.429.49. MR1173115

[18] Anne-Laure Thiel, *Categorification of the virtual braid groups* (English, with English and French summaries), Ann. Math. Blaise Pascal **18** (2011), no. 2, 231–243, DOI 10.5802/ambp.297. MR2896487

Department of Mathematics, Columbia University, New York, New York 10027
Email address: khovanov@math.columbia.edu

Institut für Mathematik, Universität Zürich, Winterthurerstrasse 190 CH-8057 Zürich, Switzerland
Email address: krzysztof.putyra@math.uzh.ch

Institut de Recherche en Mathématique et Physique, Université catholique de Louvain, Chemin du Cyclotron 2, bte L7.01.02 1348 Louvain-la Neuve, Belgium
Email address: pedro.vaz@uclouvain.be

SELECTED PUBLISHED TITLES IN THIS SERIES

791 **Mee Seong Im, Bach Nguyen, and Arik Wilbert, Editors,** Algebraic and Topological Aspects of Representation Theory, 2024

790 **Peter Crooks and Alexandru I. Suciu, Editors,** Compactifications, Configurations, and Cohomology, 2023

789 **I. S. Krasil′shchik, A. B. Sossinsky, and A. M. Verbovetsky, Editors,** The Diverse World of PDEs, 2023

788 **I. S. Krasil′shchik, A. B. Sossinsky, and A. M. Verbovetsky, Editors,** The Diverse World of PDEs, 2023

787 **Hemen Dutta, Editor,** Mathematical Modelling, 2023

786 **Hemen Dutta, Editor,** Mathematical Modelling, 2023

785 **A. Leroy and S. K. Jain, Editors,** Algebra and Coding Theory, 2023

784 **Dinh-Liem Nguyen, Loc Hoang Nguyen, and Thi-Phong Nguyen, Editors,** Advances in Inverse Problems for Partial Differential Equations, 2023

783 **Alla Detinko, Michael Kapovich, Alex Kontorovich, Peter Sarnak, and Richard Schwartz, Editors,** Computational Aspects of Discrete Subgroups of Lie Groups, 2023

782 **Galina Filipuk, Alberto Lastra, and Sławomir Michalik, Editors,** Recent Trends in Formal and Analytic Solutions of Diff. Equations, 2023

781 **Emanuel Indrei, Diego Marcon, and Levon Nurbekyan, Editors,** Geometric and Functional Inequalities and Recent Topics in Nonlinear PDEs, 2023

780 **Erik Koelink, Stefan Kolb, Nicolai Reshetikhin, and Bart Vlaar, Editors,** Hypergeometry, Integrability and Lie Theory, 2022

779 **Samuele Anni, Valentijn Karemaker, and Elisa Lorenzo García, Editors,** Arithmetic, Geometry, Cryptography, and Coding Theory 2021, 2022

778 **Carlos Galindo, Alejandro Melle Hernández, Julio José Moyano-Fernández, and Wilson A. Zúñiga-Galindo, Editors,** p-Adic Analysis, Arithmetic and Singularities, 2022

777 **Bang-Yen Chen, Nicholas D. Brubaker, Takashi Sakai, Bogdan D. Suceavă, Makiko Sumi Tanaka, Hiroshi Tamaru, and Mihaela B. Vajiac, Editors,** Differential Geometry and Global Analysis, 2022

776 **Aaron Wootton, S. Allen Broughton, and Jennifer Paulhus, Editors,** Automorphisms of Riemann Surfaces, Subgroups of Mapping Class Groups and Related Topics, 2022

775 **Fernando Galaz-García, Cecilia González-Tokman, and Juan Carlos Pardo Millán, Editors,** Mexican Mathematicians in the World, 2021

774 **Randall J. Swift, Alan Krinik, Jennifer M. Switkes, and Jason H. Park, Editors,** Stochastic Processes and Functional Analysis, 2021

773 **Nicholas R. Baeth, Thiago H. Freitas, Graham J. Leuschke, and Victor H. Jorge Pérez, Editors,** Commutative Algebra, 2021

772 **Anatoly M. Vershik, Victor M. Buchstaber, and Andrey V. Malyutin, Editors,** Topology, Geometry, and Dynamics, 2021

771 **Nicolás Andruskiewitsch, Gongxiang Liu, Susan Montgomery, and Yinhuo Zhang, Editors,** Hopf Algebras, Tensor Categories and Related Topics, 2021

770 **Stéphane Ballet, Gaetan Bisson, and Irene Bouw, Editors,** Arithmetic, Geometry, Cryptography and Coding Theory, 2021

769 **Kiyoshi Igusa, Alex Martsinkovsky, and Gordana Todorov, Editors,** Representations of Algebras, Geometry and Physics, 2021

768 **Dražen Adamović, Andrej Dujella, Antun Milas, and Pavle Pandžić, Editors,** Lie Groups, Number Theory, and Vertex Algebras, 2021

For a complete list of titles in this series, visit the
AMS Bookstore at **www.ams.org/bookstore/conmseries/**.